Canada: The State of the Federation 2009

Carbon Pricing and Environmental Federalism

Edited by

Thomas J. Courchene

John R. Allan

Institute of Intergovernmental Relations
School of Policy Studies, Queen's University
McGill-Queen's University Press
Montreal & Kingdston • London • Ithaca

The Institute of Intergovernmental Relations

The Institute is the only academic organization in Canada whose mandate is solely to promote research and communication on the challenges facing the federal system.

Current research interests include fiscal federalism, health policy, the reform of federal political institutions and the machinery of federal-provincial relations, Canadian federalism and the global economy, and comparative federalism.

The Institute pursues these objectives through research conducted by its own staff and other scholars, through its publication program, and through seminars and conferences.

The Institute links academics and practitioners of federalism in federal and provincial governments and the private sector.

The Institute of Intergovernmental Relations receives ongoing financial support from the J.A. Corry Memorial Endowment Fund, the Royal Bank of Canada Endowment Fund, the Government of Canada, and the governments of Manitoba and Ontario. We are grateful for this support, which enables the Institute to sustain its extensive program of research, publication, and related activities.

L'Institut des relations intergouvernementales

L'Institut est le seul organisme universitaire canadien à se consacrer exclusivement à la recherche et aux échanges sur les questions du fédéralisme.

Les priorités de recherche de l'Institut portent présentement sur le fédéralisme fiscal, la santé, la modification éventuelle des institutions politiques fédérales, les mécanismes de relations fédérales-provinciales, le fédéralisme canadien au regard de l'économie mondiale et le fédéralisme comparatif.

L'Institut réalise ses objectifs par le biais de recherches effectuées par son personnel et par des chercheurs de l'Université Queen's et d'ailleurs, de même que par des congrès et des colloques.

L'Institut sert comme lien entre les universitaires, les fonctionnaires fédéraux et provinciaux et le secteur privé.

L'Institut des relations intergouvernementales reçoit l'appui financier du J.A. Corry Memorial Endowment Fund, de la Fondation de la Banque Royale du Canada, du gouvernement du Canada et des gouvernements du Manitoba et de l'Ontario. Nous les remercions de cet appui qui permet à l'Institut de poursuivre son vaste programme de recherche et de publication ainsi que ses activités connexes.

ISSN 0827-0708
ISBN 978-1-55339-197-5 (bound)
ISBN 978-1-55339-196-8 (pbk.)

It is a pleasure to dedicate this volume to Patti Candido, who retired in 2009 after thirty years of invaluable service as the Administrative Assistant to many Directors of the Institute of Intergovernmental Relations.

The Institute of Intergovernmental Relations is pleased to acknowledge our sponsors for the Carbon Pricing and Environmental Federalism Conference that gave rise to this publication.

Co-sponsors:

Queen's Institute for Energy and Environmental Policy

Sustainable Prosperity

Financial Sponsors:

Ontario Ministry of Finance

Ontario Centres of Excellence

TD Bank Financial Group

Kenneth R. McGregor Lectureship Fund

CONTENTS

PREFACE

This 2009 edition of *Canada: The State of the Federation*, entitled *Carbon Pricing and Environmental Federalism*, focuses on the difficulties in formulating effective policies to combat global warming attributable to the emission of greenhouse gases (GHG). Such policies are complicated by the fact that GHG emissions are equally damaging regardless of where they arise; they respect neither provincial nor national boundaries; and free riding – that is, leaving the burden and costs of reduction to others while reaping the same climatic benefits as those who do act – is both feasible and economically attractive. Moreover, with significant differences in the carbon intensity of economic activity in different political units, whether province, state or nation, and, in the case of Canada, with responsibility for environmental protection shared and contested between the federal and provincial governments, it becomes immediately evident why formulating policies to combat climate change is fraught with difficulty and acrimony. Yet, given the overwhelming preponderance of scientific evidence attesting to the anthropogenic sources of global warming, such policies are essential if we are to avoid catastrophic environmental damage.

Such were the considerations that prompted the October, 2008 Conference on Carbon Pricing and Environmental Federalism, held in Kingston by the Institute of Intergovernmental Relations, and co-sponsored by the Queen's Institute for Energy and Environmental Policy, and Sustainable Prosperity. We were fortunate in being able to attract an outstanding group of authors, discussants and session chairs, and we would like to thank Marc-Antoine Adam, Matthew Bramley, Douglas Brown, Stephanie Cairns, Tom Carpenter, Nathalie Chalifour, John Dillon, Stuart Elgie, Andrew Green, Chris Green, Kathryn Harrison, Rick Hyndman, Jeremy Leonard, Peter Leslie, Andrei Marcu, Nancy Olewiler, Bob Page, Bruce Pardy, Bryne Purchase, Barry Rabe, Nic Rivers, and Pierre Sadik for their contributions to what proved to be a stimulating and thought-provoking conference. We would also like to thank the National Roundtable on the Environment and the Economy for permission to include the Executive Summary of their 2009 report *Achieving 2050: A Carbon Pricing Policy for Canada,* and TD Economics for permission to include their *Special Report* summarizing the findings of the study of the costs of combating climate change jointly commissioned by the Pembina Institute and the David Suzuki Foundation. Since both of these reports attracted considerable attention in the period subsequent to the conference, we believe readers will find their inclusion here very useful.

The Ontario Ministry of Finance, Ontario Centres of Excellence, the TD Bank Financial Group, and the Kenneth R. McGregor Lectureship Fund all made generous and much appreciated financial contributions to support the conference or the publication of the proceedings. Without their support neither the conference nor this publication would have occurred.

Conducting a conference and preparing the proceedings for publication is a challenging process that requires the input of many people and causes the organizers to incur many debts. Foremost among the latter are those to the staff of the Institute of Intergovernmental Relations, who managed all aspects of registration and conference administration. In particular, Mary Kennedy was the voice of the Institute to all conference registrants and was the essential administrative link with the conference presenters and discussants. Sharon Sullivan has been a tower of strength in organizing the papers and getting them ready for publication, and we are pleased to acknowledge our debt to the John Deutsch Institute, Queen's University, for making Sharon available to us. We are also indebted to Mark Howes, who added this book to the considerable list of IIGR publications for which he has designed the cover, and to his colleagues at McGill-Queen's University Press. It is a pleasure to thank Nadia Verrelli, one of the IIGR's Research Associates, for her analytical and organizational contributions throughout the planning and publication process. Our research assistants, Margaret McKenzie and Jonathan Aiello both contributed to the smooth running of the conference, and Jonathan has assisted in innumerable ways in getting the papers ready for publication.

Last, but most certainly not least, we would like to acknowledge our debt to Patti Candido, not just for her contribution to this conference and publication, but for her thirty years of service to the Institute of Intergovernmental Relations. Patti has served as Administrative Assistant to every Director of the Institute since the first, and it is difficult to think of the Institute without also thinking of Patti. Indeed, for many, Patti Candido personifies the Institute of Intergovernmental Relations, and her retirement in 2009 leaves a singular void that will be virtually impossible to fill. It is with pleasure that we dedicate this volume of *Canada: The State of the Federation* to Patti Candido.

Thomas J. Courchene,
Director, IIGR

John R. Allan,
Associate Director, IIGR

I

Introduction

1

Introduction and Overview

Thomas J. Courchene and John R. Allan

INTRODUCTION

The nations of the world will gather shortly in Copenhagen to attempt to map out a global strategy to combat the threat of catastrophic global warming. With the Obama administration committed to the adoption of an effective policy to combat climate change and China belatedly signalling at least some recognition that progress on the climate-change front must involve the major emerging economies, there is now some prospect that the unworkable voluntarism of the Kyoto I approach to countering climate change will be superseded, in Kyoto II, by a new and more effective international agreement. Leaving on the sideline, as Kyoto I did, the countries responsible for some three-quarters of the annual additions to greenhouse gas (GHG) emissions, the need for change could not be clearer. What form change may take, however, is anything but clear, as is the policy that Canada will bring to the bargaining table. Particularly complicating the latter is the fact that in Canada and the United States (whose position will strongly influence that of Canada) it is the second tier of government – the provinces and the states – that have taken the initiative and the lead in developing and adopting policies to combat climate change; the federal governments in both countries have been policy laggards.

Further complicating the development of Canadian climate-change policy is the fact that, constitutionally, responsibility for the environment is shared between both levels of government. By itself, this need not be problematic: the federal government, for example, could ensure an adequate pan-Canadian response to the climate challenge and, with strong leadership, co-ordinate this national response with those of participating provinces. Unfortunately, absent strong federal leadership and given increasing provincial impatience with the failure of the federal government to respond in what was seen as a timely and adequate manner – or possibly to forestall such a response – a hodgepodge of largely uncoordinated provincial initiatives has been established. The unfortunate end result has thus been an environmental Balkanization of the Canadian economy.

These and other related issues were the focus of a conference on Carbon Pricing and Environmental Federalism organized by the Institute of Intergovernmental Relations and co-sponsored by the Queen's Institute for Energy and Environmental Policy and by Sustainable Prosperity. The conference was held October 17 and 18, 2008, in Kingston, at Queen's

University. The structure of the conference is reflected in the arrangement of the present volume. Part II, consisting of three chapters, provides a context for the chapters that follow, while Part III focuses on the challenges of carbon pricing in a federal setting complicated by the fact that some countries may not adopt, at least at this time, GHG-reducing policies. Part IV is devoted to an examination of the political challenges in the way of effective policies, while Part V examines the constitutional issues operative in Canada and the possible constraints on policy arising from Canada's membership in the WTO. The political economy of climate change is the subject of the two chapters in Part VI, while in Part VII the final two chapters consider how Canada and the world may move forward in this highly contested area of policy.

PART II: CARBON PRICING: SETTING THE STAGE

The volume opens with three background papers. The first, by Bob Page, is an historical, policy and intergovernmental overview of Canada's experience post-Kyoto climate change. The second, by Nic Rivers is a review of Canadian environmental policy as well as a simulation that generates ballpark estimates of the carbon price needed to bring Canada in line with its Kyoto commitments. (Indeed, it is probably more accurate to say that Rivers' estimates are "out of the ballpark" if the comparison is the range of carbon prices in the various government proposals.) The third paper is more along the lines of an analytical framework in that author Chris Green makes a cogent argument that climate change is inherently a technological challenge rather than a carbon pricing challenge.

The Canadian Policy Struggle with Climate Change: Setting the Context for Carbon Pricing (Bob Page)

Climate change may or may not be the most pressing societal challenge but, as Bob Page's lead paper makes very clear, it is far and away the most complex challenge. Page provides the reader with a *tour d'horizon* of the manifold ways by which the climate-change tentacles impinge on virtually every facet of Canadian public policy, including resource/energy policy, jurisdictional issues (inter-provincial and federal-provincial), global issues (Kyoto and the developed/developing nations), the WTO and the trading regime, revenue sharing from carbon pricing (again inter-provincial and federal-provincial), income distribution issues across citizens, Canada-U.S. relations (including protectionism), as well, of course, the set of issues associated with carbon pricing itself (carbon taxes vs C&T, carbon capture and storage, carbon offsets, conservation, etc.). In addition to weaving the above elements into an historical-cum-public-policy overview of the climate-change dossier, there is a sub-plot to Page's analysis, namely, the conflict between the fossil-energy (and particularly the oil sands) provinces on the one hand and the Kyoto-friendly provinces on the other, a conflict replete with the potential for triggering another NEP-type donnybrook.

Page's reflections on the Kyoto process that left Canada "alone in the Americas with Kyoto obligations!" are especially revealing. Given Canada's position as an emissions-intensive resource exporter and as a country with high population growth, the Kyoto process placed Canada (and the energy-exporting provinces) at a "fundamental disadvantage". An integral part of this disadvantage was that similarly-positioned Australia (a relatively small, open, resource-intensive economy) was assigned a target for 2008–12 that was 8 percent *above* its 1990 emissions, whereas Canada's target was 6 percent *below* its 1990 emissions. Page goes on to note that the Chrétien Liberals finally did respond to this by proposing that Canada be granted special CO_2 export-offset credits that would have raised its permitted emissions by 25 percent, but the European Union and the United States vigorously opposed this, so the matter was quietly dropped.

In the second half of his paper Page addresses a set of thorny issues relating to carbon pricing, the first of which is environmental protectionism:

> While the semantics of the Kyoto negotiations were environmental, the working assumptions were those of trade and competitiveness. Environmental protectionism was evident in the strategy of both the United States and the European Union. The latter wanted to saddle the U.S. producers with additional environmental costs, whereas George Bush cited these costs in rejecting Kyoto.... This environmental protectionism took on a new and more sinister twist in ... January 2008 [when] the European Union announced it would apply a "carbon tariff" on imports from countries with less stringent carbon-emission controls. They argued that they were now forced to establish a "carbon equalization system" to protect E.U. jobs and products from developing countries with no carbon costs or countries like Canada that were not meeting their Kyoto commitment. The World Business Council for Sustainable Development warned its members of the coming "Trade War over Carbon". Carbon taxes can have a variety of forms, including that of protective tariffs, a form that currently seems to be gaining momentum in both the European Union and the United States.

A second highlighted problem area is the oil sands. Again, Page's own words merit quotation:

> The oil sands are a key factor in any carbon pricing debate. Their future expansion complicates Canada's ability to meet its targets for 2020. While technology, such as carbon capture and storage, will help in the long term, it will be some years before the infrastructure is in place and the technology will be commercial. The dilemma is that the public expects emission cuts almost immediately while the technology will take a decade or more to implement. This time gap is at the centre of the Canadian carbon management dilemma. The greatest current threat to the oil sands is potential loss of U.S. markets with new U.S. environmental legislation. The December 2007 U.S. Energy Independence and Security Act (section 526) forbids American government, its agencies, or the armed forces from purchasing high carbon fuel products like oil sands oil. To Alberta this is a much more serious threat than any tax or cap-

and-trade system. Without the U.S. market, the potential for the oil sands does not extend beyond the current levels of production.

A further problem area discussed by Page relates to the challenges on the intergovernmental relations front. Many of these are distributional and, therefore, zero-sum in nature. For example, Quebec wants to be able to sell hydro-offset credits within any national C&T system while Alberta wants all the revenues arising from trading in its permits to stay within Alberta's boundaries. Page's concern with respect to this and similar interprovincial and federal-provincial conflicts is that we lack an effective institutional mechanism like a refurbished Canadian Council of the Ministers of the Environment to handle these controversial, overarching issues that beset climate-change policy.

Current Federal and Provincial Approaches to Climate Change Mitigation: Are We Repeating Past Mistakes? (Nic Rivers)

Whereas Bob Page provides an historical/public-policy overview of the climate change challenge, Nic Rivers offers a comprehensive factual and empirical assessment of Canada's past, present and prospective performance on the carbon-pricing and carbon-abatement fronts. Rivers begins his analysis with a review of the federal government's climate-change history in terms of its commitments, polices and emissions. Over the period from the 1990 *Green Plan* through to the 2008 *Turning the Corner*, Ottawa developed six climate-change packages. Except for a series of voluntary programs and subsidies, the common denominator of all six was/is *non-implementation.* As a result, Canada's emissions have continued to rise to the point where they are now 25 percent above 1990 levels and still rising. Rivers then presents the results of several simulations directed to calculating the carbon price required to have been implemented in 2000 in order to meet the Kyoto commitments by 2010: ignoring the highest and lowest prices, the remaining five range from $99/tonne to $137/tonne, essentially an order of magnitude higher than the $15/tonne proposed in *Turning the Corner.*

Since Ottawa and all provinces have target emissions levels for 2020, Rivers then focuses on a comparison between these targets and projected emissions. Comparing provincial targets for 2020 to actual 1990 emissions, Saskatchewan and Alberta have targeted for an increase (11.7 percent and 34.3 percent respectively), while the other provinces have targeted for decreases between 5 percent and 15 percent. He then simulates 2020 emissions on a "business-as-usual" basis. Given the targets and the business-as-usual emissions levels in 2020, Rivers calculates the marginal price of carbon that would be required if the provinces were to achieve their own targets. These range from roughly $150/tonne for SK and AB (in part because they committed themselves to less aggressive targets), and from $219 to $286 for the remaining provinces, with a $230 carbon price for Canada as a whole.

Rivers' comments on this as follows:

> The European Union Emission Trading System, currently the largest carbon market in the world, has had prices averaging roughly $15-25/tonne over the last three years. Several European countries have imposed carbon taxes of up to $50/tonne on certain activities. The carbon price that is estimated here to be required to meet commitments made by the provincial and federal governments would therefore dwarf the carbon policies already adopted in Europe, the current leader in climate change policy.

Rivers also notes that the existing carbon prices/taxes in Canada fall far short of what would be required to meet targets. For example, despite the fact that the BC carbon tax is ten times that of Quebec, it is not much more than a tenth of what would be required for BC to meet its own commitments by 2020.

In his conclusion Rivers offers the following observations:

- Compared to a decade ago, the provinces are now important climate-change players and will not easily be sidelined by Ottawa;
- There is a recognition that deep emissions cuts will require compulsory policies like carbon taxes and C&T systems;
- Current policies at both levels of government are much less stringent than will be required to meet the commitments made by those governments; and
- Finally, given the above, it is certainly valid to debate whether such dramatic targets should be met. Rivers concludes on this issue with "whatever the benefits of emissions abatement, it is almost certain that they are large enough to warrant application of an emission pricing policy that begins today at a modest level and rises over time to a more substantial level".

These and other issues analyzed by Rivers receive further elaboration in other chapters in this volume, including the paper by Chris Green, to which we now turn, and Appendix 2.

Carbon Pricing and the Technology Imperative (Christopher Green)

Even if carbon prices were to achieve their requisite levels, à la Nic Rivers' modeling, Chris Green argues forcefully that this would not suffice to achieve announced targets since, at base, "climate change is essentially an energy technology problem":

> Pricing carbon, however desirable, is not sufficient to stabilize climate (that is, stabilize the atmospheric concentration of greenhouse gases) without new, scalable, and breakthrough technologies. In *global* terms, pricing carbon without also *directly* addressing the energy-technology challenge is a bankrupt strategy. Yet it is carbon pricing on which most economists and recently converted environmental advocates dwell.

Part of the reason that we underestimate the magnitude of the technology challenge is that the so-called "business-as-usual" estimates already incorporate "huge emissions reductions attributable to technological change", reductions that are assumed to occur "spontaneously", as it were. Hence, the presumed policy role is to close the remaining climate-change gap between these business-as-usual estimates and the emissions target.

Green notes, however, that the policy gap defined in this manner substantially underestimates the true climate-change gap because the amount of technological change already embedded in the business-as-usual estimates overstates what is likely to materialize autonomously:

> Where technologies (i) take many years, even decades, to develop, (ii) are uncertain of success, and (iii) if successful are often characterized by benefits that are far from being fully appropriable, more than the market is needed to convince entrepreneurs to make large upfront investments in R&D many years in advance of any possible return. If current governments cannot commit (distant) future governments to cover anything more than the cost of production of successful technologies, then we have a *time inconsistency* that renders it highly unlikely that the private sector will be willing to make the required *upfront* investments in R&D.

In other words, there is no assurance that the technological changes that occur from year to year will produce anything like the large emission reductions attributed to the technological change already subsumed in these business-as-usual simulation models.

Given the above analysis, it may not be surprising that Green's view is that the combination of carbon pricing and emissions targets is sending us down the wrong track. He is particularly concerned with the problems associated with C&T approaches:

> I would submit that there is something inconsistent about using cap and trade where technological change is crucially important. The arrival of new, scalable technologies is inherently uncertain, so that using cap and trade to meet *date-specific* emission-reduction targets is virtually certain to produce *ad hoc* decisions. For example, pressure to meet emission targets may lead to hasty adoption of inferior technologies ("first generation" biomass in the form of corn for ethanol is an example) or to *temporary* means of reducing emissions simply to meet the target. It is also questionable how far support for cap and trade will go when it dawns on the public that the only sure beneficiaries are financial markets that broker the trades and that are able to capitalize on the inherent price volatility of a quantity-based (fixed supply) approach to carbon pricing. A further nightmare occurs if speculators are able to engage in temporary price manipulation (and take-and-run profits) by buying up a significant share of permits before dumping them. Real world rather than textbook cap and trade assures neither price nor volume certainty, and this fact almost certainly implies important economic inefficiencies. Not much to like here!

Green's preferred way forward is along the following lines:

> If climate change is essentially an energy-technology problem, then I submit that carbon pricing is only part of the story – and in the early stages its role is largely *ancillary*. In a revamped climate policy, there is no place for emission

> targets and no need for cap and trade. What is needed is the widespread adoption of a *low* carbon tax, one that gradually rises over time. The purpose of the carbon tax/fee is to finance an *up-front*, long-term, global effort on the energy technology and infrastructure front. *Commitments* to a gradual increase in the tax/fee send a forward price signal to deploy effective, scalable, competitive, and transferable technologies as they reach "the shelf". Policies that attempt to short circuit this process by setting near- or medium-term emission reduction targets and mandates will be ineffective – or quite likely destructive of long-term efforts to reduce emissions and stabilize climate. Predictably, failure of climate policies will create increasing pressure to consider the adoption of one or another proposal to "geo-engineer" the atmosphere. That brings us face-to-face with still another "inconvenient truth".

These arguments by Green find resonance in the chapter by Rick Hyndman in the following section.

PART III: CARBON PRICING: ANALYTICAL PERSPECTIVES

The paper by Thomas Courchene and John Allan addresses the mechanics of carbon taxation and cap-and-trade systems within the Canadian federation and the international trading environment. Rick Hyndman's contribution elaborates on a creative carbon-pricing system that embodies elements of both carbon-tax and cap-and-trade models.

Carbon Pricing and Federalism (Thomas J. Courchene and John R. Allan)

The role of the chapter by Courchene and Allan (henceforth C&A) is to elaborate on the various approaches to carbon pricing and on the manner in which they interact with Canadian federalism and with multi-level governance generally. After noting that a pure carbon tax will generate carbon price certainty but uncertainty in terms of emissions reductions and that a pure C&T system generates certainty in terms of emissions reduction but at an uncertain carbon price, the authors focus on selected aspects of the two models.

Their preferred carbon-tax model is what they refer to as a carbon-added tax/tariff (CATT), which is a carbon tax analogue of a GST or VAT:

- Under such a system, there will be a tax on the carbon emissions that are added at each stage – hence the "carbon-added" label.
- As the product completes each stage of the production/distribution process, it is taxed on its carbon footprint to that point, and a credit claimed for the carbon taxes on earlier stages. In consequence, only the carbon added in each stage ends up being taxed at that stage.

- Hence, when the product is sold at the final stage, the tax is on the *cumulative* value of carbon emissions, i.e., the sum of the carbon-added taxes at each stage.
- As with the GST, in the case of exports the taxes accumulated to the point of export will be rebated, so that the carbon tax does not diminish our international competitiveness.
- Relatedly, the carbon tax will be levied on the accumulated carbon footprint of each import, including that arising from transporting the product to Canada, thereby safeguarding our competitiveness in Canadian markets.

On the C&T front, the most common version involves setting an overall emissions cap, allocating emissions permits (typically free of charge) to producers up to the overall limit or cap, and requiring firms to buy from other firms any permits required for emissions beyond their allocated limit. This latter feature is the "trade" component of C&T. The genius of the cap-and-trade system is that the overall emissions limit will in fact be attained while the carbon price determined by permit trading will be that which minimizes the cost of emissions reduction and maximizes output.

C&A note that both models can be complemented with carbon offsets (e.g., Kyoto's CDMs (clean development mechanisms) or those for CCS (carbon capture and storage)). And one can even generate hybrids – a C&T model with the government willing to supply carbon offsets a given price is, at this limit, effectively a carbon tax at the specified price. A major difference between the two systems is that border-tax adjustments (BTAs) within a carbon-tax system, and especially within a CATT system, are likely to be more acceptable to the WTO than they would be as part of a C&T system, particularly one in which a major significant proportion of permits are not auctioned. Turning to the federal dimension, both the provinces and Ottawa have engaged in or proposed C&T and carbon-tax systems (e.g., BC and Quebec for carbon taxes and Alberta for C&T at the provincial level, and, at the federal level, Dion's *Green Shift* for carbon taxes and the Conservatives' *Turning the Corner*).

C&A then focus on three areas/issues that are serving to severely complicate Canada's ability to achieve Kyoto-type emissions targets. The first is that since the typical time frames for meeting targets are appropriately very long (e.g., reduce emissions by 60-70 percent by 2050) and since these targets are expressed in absolute levels, countries such as Canada, with higher population growth rates, will face higher *effective* emissions targets. The second design failure is that natural-resource-exporting countries like Canada are enabling the importing countries to appear environmentally benign because Kyoto assigns the carbon footprint arising from the *production* of these resource exports to the originating country, when the footprint should be assigned to the importing countries where the emissions-intensive resources are consumed. The third highlighted problem area is ocean shipping. Since the enormous carbon footprints from ocean shipping are ignored under Kyoto, all countries become, in terms of their carbon footprints, essentially equidistant from the United States, thus negating the advantage that proximity to the United States should confer to Canada under appropriate carbon pricing (e.g., under a CATT).

C&A's conclusion accords with the theme of this volume:

> ...there is no equivalent on the environmental front to the more than 50-year history of federal-provincial fiscal relations dating from the inauguration of the equalization program in 1957. Fiscal federalism includes scores of meetings of federal and provincial bureaucrats each year. The processes of fiscal federalism also include a host of federal-provincial agreements on equalization, on tax-collection harmonization, on a national tax collection agency (the CRA) and even on securing the internal social, economic and fiscal unions. However, over the foreseeable future environmental federalism will likely become every bit as important as fiscal federalism. Indeed, it may embrace key aspects of fiscal federalism. Given this, and the reality that the political and institutional machinery in the area of environmental federalism ranges from weak to non-existent in comparison with the fiscal federalism infrastructure, both Ottawa and the provinces (individually and/or via the Council of the Federation) need to take immediate steps to deepen the intergovernmental infrastructure relating to the substance and the processes of environmental federalism. ...Addressing climate change is a sufficiently daunting challenge in its own right without the complication of tolerating the reality that the structures and processes of environmental federalism are in a state of disarray. Phrased differently, we will have made progress on the climate-change front when "environmental federalism" takes its rightful place in our policy vocabulary.

Carbon Pricing as if GHG Mitigation Matters (Rick Hyndman)

Rick Hyndman's sobering message in terms of relying primarily on carbon pricing to achieve the near-term 2020 targets is that "you can't get there from here". For example, while the 2009 report of the Canada's NRTEE concludes that a carbon price in the range of $100-150/tonne is required, this is not only well above the price that our governments are willing to live with but higher still than what the public will bear. Along similar lines, Hyndman then uses the Kaya identity to show that the GDP contraction that would be required to achieve the U.S. targets under the original Waxman-Markey bill would be in the order of $12-13 trillion over the 2011-2020 period – again well above what the government and the public would tolerate. But if carbon targets are unachievable domestically, might the solution be to take advantage of some version of Kyoto's Clean Development Mechanism (CDM) which would allow developed countries to purchase low-cost carbon reductions from developing nations? Hyndman's answer is no: apart from the political difficulty of having monies for purchasing these (often unverifiable) CO_2 allowances flow to other nations, there simply will not be enough of these CDM offsets to go around.

In the face of public opposition to any significant carbon price and the corresponding unwillingness to submit to the requisite output reductions to achieve announced targets, what then are the alternatives? Hyndman suggests two complementary ways out – i) increased support for low-carbon-emission technology development, and ii) a creative system of carbon pricing for large energy-intensive exporting industries, one that would provide an appropriately high marginal carbon price to guide the technology investment choices but without the high average costs that would flow from more traditional C&T models.[1] As an aside, readers will recognize that Hyndman's view that GHG mitigation is "fundamentally a technology challenge" parallels Chris Green's "technology imperative".

In terms of i) above, Hyndman's way to reconcile a publicly acceptable, low initial price on carbon emissions with the need for large investments in low-carbon-emissions technology is to implement a low-level carbon price on GHG emissions in order *to raise revenues for funding technology R&D.* Hyndman's example uses a $5.00/tonne price which, in the United States, would generate $24 billion annually. His related recommendation is that it is very important to put in place the right governance of these funds. The revenue should go into a technology development trust fund to support the research, development and deployment of transformative technologies and be managed by an independent board, with a public interest mandate, and at arm's length from governments and political interference. (Given the magnitude of the *public* funds that would be raised by even a modest carbon charge, and the possibility of self-serving behaviour by the representatives of emissions-intensive industries who would undoubtedly be represented on the board, some review mechanism to ensure the realization of the public-interest mandate would appear to be clearly essential.) He then adds that "perhaps some revenue should be diverted to provide aid to poor countries to develop cleanly, including national programs to reverse deforestation".

In the penultimate section of his paper, Hyndman offers a proposal for ii) above, i.e., for what he calls the energy-intensive, trade-exposed (EITE) sectors. The essence of the proposal is to generate, at the same time, a high *marginal* price for carbon emissions in order to provide appropriate incentives for investing in low-carbon-emitting technologies and a much lower *average* carbon price in order that the EITE sectors can maintain international competitiveness and minimize production leakage to other countries.

Under his proposed approach an intensity-based performance standard would be set for each major EITE sector, and only firms with emissions intensities in excess of that standard would be subject to taxation, and this – although at a relatively high marginal rate – only on their excess emissions. Should a firm have an emissions intensity less than the performance standard, the resulting emissions "deficiency" could be sold (i.e., traded) at the carbon price established under the system. The (relatively) high *marginal* carbon price would be the price relevant to decisions respecting levels of emissions and investments in emissions-reducing technology – so the economic incentives would be correct – while the affected firms would have to pay the relatively low liability attributable to the taxation of only the excess emissions. The result would thus be a relatively low *average* tax rate or charge that would not impair the international competitiveness of the firm or sector or cause production to move to environmental havens with lax environmental standards. Moreover,

---

[1]It is the case, however, that the same disparity between average and marginal carbon prices may be achieved under C&T if only a relatively small fraction of allowances are provided by means of auction, the remainder being allocated at no charge.

with an intensity-based system, as output increases over time, so too would the untaxed emissions; the system would thus operate more flexibly than one with an absolute (i.e., fixed) cap. If the resulting increase in total emissions were deemed excessive, it could be counteracted by lowering the performance standard – which would have the effect of reducing the "free allocation" and increasing the proportion of emissions subjected to the carbon charge – or by raising the marginal carbon charge.

In summary, Hyndman offers a flexible set of proposals designed to work around the myriad of political and economic constraints (domestic and international) associated with the more traditional proposals for pricing carbon.

PART IV: FEDERALISM, MULTI-LEVEL GOVERNANCE AND CARBON PRICING

Given that it has been the sub-national governments in both Canada and the United States that have been driving climate change, it is clear that the federal or intergovernmental dimension of carbon-pricing policy has to be addressed and assessed. To this end, Kathryn Harrison compares the influence of multi-level governance on climate change in Canada, the United States and the European Union, while Barry Rabe focuses on how the dynamics of federalism shape the making of U.S. climate-change policy.

Multi-Level Governance and Carbon Pricing in Canada, the United States, and the European Union (Kathryn Harrison)

In considering whether federalism or multi-level governance facilitates or deters the adoption of policies to reduce greenhouse gas emissions, Kathryn Harrison's analysis leads her to conclude that, for the European Union, it has facilitated "multi-lateral reinforcement":

> …the impact of multi-level governance in the European Union has been largely positive. Various climate-policy leaders have emerged over time among the member states, and that horizontal dynamic has been matched vertically by activism from the European Council of Ministers, Parliament, and Commission. In response, the European Union has made the greatest progress in adopting policy reforms to price carbon, most notably through its Europe-wide Emissions Trading System.

In the case of the United States, she concludes that federalism has fostered "state action":

> In the United States, federalism also has had a positive impact in facilitating policy innovation and diffusion at the state level, albeit in the face of a policy vacuum at the national level. With respect to carbon pricing, some (though not all) state governments are collaborating to create regional emissions-trading schemes.

In Canada, however, the result is a "joint decision trap":

> In contrast, in Canada the impact of federalism on climate policy has on balance been negative to date. As in the United States, there has been a dearth of action at the national level, but until quite recently Canadian provinces did not respond unilaterally to the same degree as their U.S. counterparts. Federal and provincial governments were deadlocked over how to respond to climate change for almost two decades. Provincial policy innovations have emerged since 2006, led most notably by British Columbia's adoption of a carbon tax and the commitment by BC, Manitoba, Ontario, and Quebec to join with U.S. states in emissions trading. However, those reforms have not diffused to provinces that account for half of Canada's current emissions and the majority of its projected emissions growth.

In her analysis, Harrison distinguishes between interprovincial or horizontal relations on the one hand and federal-provincial or vertical relations on the other. The former can lead to what has come to be called competitive federalism, which can lead to a "pull from the top" or a "race to the bottom". The latter (vertical relations) can lead to the establishment of national standards (which would preclude a race to the bottom). It can also lead to creative and positive-sum intergovernmental cooperation (which Harrison refers to as "horizontal innovation and vertical backup and coordination") or it could lead to the opposite, with the two levels reduced to policy deadlock. She then applies these and other features of her framework to the climate-change history of Canada, the European Union and the United States, with the resulting broad conclusions elaborated in the above quotations.

In somewhat more detail, the actual division of powers in these jurisdictions also played an important role in determining the different outcomes. Confirming the observations by Barry Rabe (see below), Harrison notes that "the long-standing role of state governments in fulfilling federal mandates contributed significantly to the states' administrative capacity to respond to climate change unilaterally, including their familiarity with market-based instruments". In the E.U. case, the fact that regulatory decisions are made by the E.U. Council of Ministers via a "qualified majority vote" facilitated the adoption of the ETS, particularly since the larger states, which carry a greater weight in the qualified majority system, were supportive. And most intriguingly, "the fact that E.U. *taxation* policies, in contrast to regulation, do require unanimity explains the European Commission's greater success with the ETS than in its earlier proposal for a carbon tax". Finally on the Canadian front, given the fact that resources are owned by the provinces, "it is hardly surprising that federal-provincial consensus has been unattainable".

A further important factor relates to the regional distribution of the costs of GHG reductions. In both the European Union and the United States the largest and wealthiest states are "green and keen" (United Kingdom and Germany, and California and New York respectively). In contrast, Alberta accounts for only 10 percent of the Canadian population but roughly one-third of Canada's emissions and over half of its projected emissions growth, while Ontario continues to resist efforts to strengthen emission standards for the transportation sector, which accounts for one-third of Canada's emissions. Harrison observes:

> As a result, the costs of reducing Canada's greenhouse gas emissions will inevitably be borne disproportionately by Alberta, absent a massive compensation program funded by taxpayers in other provinces. Thus, while windfall reductions in the European Union are concentrated in two powerful member states, in Canada the costs of reducing greenhouses gas emissions are disproportionately concentrated in two influential provinces – provinces that to date have exercised an effective veto over measures affecting the industries that are the lifeblood of their economies.

It is hard to disagree with her concluding comment with respect to the Canadian scene: "The regional distribution of costs combined with the division of powers with respect to natural resources suggest that federalism will continue to pose a challenge to Canada's ability to respond to climate change for years to come."

The Intergovernmental Dynamic of American Climate Change Policy (Barry G. Rabe)

Barry Rabe begins his chapter by noting that "perhaps the biggest single surprise as climate policy has continued to evolve is that in the American case and many others it is becoming increasingly evident that climate policy constitutes an issue of federalism or multi-level governance". By way of elaborating on this theme, in particular the bottom-up approach to climate change that characterizes thc United States, Rabe cross-classifies the U.S. states in to high and low emitters on the one hand, and high and low climate-change-policy activists on the other. Clearly the most important of the twelve low-emitting/high-policy states is California, which has "set in motion a carbon cap-and-trade program with wider scope than attempted in any western democracy to date". In addition to taking credit for being "first movers", these states can play a role as policy innovators for others to copy, as was the case when the Obama administration embraced California's vehicle-emissions policy, and may well be the case should Washington follow the examples of California-led WCI and New York-led RGGI and implement a national C&T system. Finally, self interest is never far from the surface, as Rabe notes by pointing out that these states will insist on maintaining 1990 as the policy baseline and on obtaining credit in any national scheme for achieving early reductions.

The 10 states that fall in the high-emissions/high-policy category "tend to view themselves as "mini-Californias", supporting cutting-edge policy experimentation and in the vanguard of national leadership on the climate-change issues". But Rabe reminds us that self interest dictates that they will want to be protected against penalty for any substantial emissions growth (and preferably shift the baseline to 2000) and will want be rewarded for early policy adoption in any future federal climate legislation. More problematical are the 22 states that fall into the category of low emissions and low policy activism, in part because they represent 44 senate seats, which is generally sufficient to block discussion on any legislative proposal. Substantively, Rabe notes that "not only is their emissions growth high and policy adoption minimal, but they may view virtually any federal climate policy as a possible threat to their economic well-being … and they are likely to oppose any policy that would impose significant costs on them and would be particularly mindful of possible redistributive effects that could result from mandates to purchase carbon credits, offsets, or renewable energy credits from outside their state and region.

Finally there are 7 states where there has been virtually no adoption of GHG policies and yet all have emissions rates well below the national average. Rabe points out that in most of these states their low carbon emissions are due to economic stagnation (e.g., Michigan). Therefore:

> …such states will want to make sure that any future policy accords them maximum "credit" for their low rates of emissions growth. Hence, the 1990 baseline will remain sacrosanct and states in this quadrant will welcome any opportunities for credit-trading programs that could deal them a favourable hand, similar to Eastern European nations and Russia which have attempted to maximize the value of their "hot air" credits.

With the above as backdrop, Rabe turns his attention to the disconnect between those policies that are economically desirable and those that are politically feasible. In particular, "those policies that tend to maintain the strongest base of support from policy analysts appear to have the greatest difficulty of being adopted by state legislators and governors", and vice versa. For example, leading economists tend to champion carbon taxes, but no state has opted to make a carbon tax the cornerstone of its climate-change policy. On the other hand, Renewable Portfolio Standards (RPSs) rank low in effectiveness but appear to be the approach of choice for the majority of states.

PART V: CARBON PRICING: CONSTITUTIONAL AND INSTITUTIONAL PERSPECTIVES

The first of the two related goals of this section is to address the constitutional limits to the authority of federal and provincial governments to regulate CO_2 emissions via carbon taxation or C&T regimes. Stewart Elgie deals with the constitutional basis for legislating emissions trading (C&T systems) and Nathalie Chalifour does the same for carbon taxation. Because the constitutional underpinnings of carbon pricing are largely unexplored the authors are forced to break new ground. The result is a set of creative and insightful analyses that, in our view, will serve to inform future court decisions in these areas. The second role of this section is to broaden the analysis of the legitimacy of carbon pricing to embrace potential institutional constraints as they relate to international trade and, in particular, to the operations of the WTO. Here, Andrew Green holds the pen.

Carbon Emissions Trading and the Constitution (Stewart Elgie)

Stewart Elgie begins his assessment of the federal authority to legislate with respect to emissions trading (i.e., C&T systems) by focusing on the two federal powers that would appear most likely to support C&T legislation – the "Peace, Order and good Government" (POGG) or national-interest provision of the preamble to s.91 on the one hand, and the Criminal Law power, s.91(27), which has been used to justify the federal *Canadian Environmental Protection Act* (CEPA) on the other. However, Elgie then suggests two other possibilities: i) because emissions are inherently international, let alone interprovincial, trading in emissions might well fall under the federal Trade and Commerce power, s.91(2); and ii) the Supreme Court might recognize a federal treaty-implementing power (relating to the Kyoto protocol or its successor) in terms of emissions trading, even though this would mean revisiting the 1937 *Labour Conventions* case that gave Ottawa the power to sign treaties that bound the provinces but not the authority to implement those provisions of the treaty that fell under provincial jurisdiction.

Elgie's conclusion in relation to the federal government's authority to legislate with respect of emissions trading runs as follows:

> To sum up, federal legislation to regulate carbon emissions and trading would test the current boundaries of federal constitutional powers. Under any of the four powers reviewed, it would require the courts to answer questions that have not yet been answered – in some cases very significant questions. Up to now, Canada's courts have been able to skirt around the hard questions about the federal government's environmental powers; they have given answers that sufficed for the statute at issue, but which left larger questions unanswered ... Climate change legislation is likely to force these hard questions onto the front burner. Its implications – both ecological and economic – are far reaching. It seems clear that national measures, as part of a larger global effort, are needed to address the problem – and in particular to put a price on carbon. Canada's courts will have to decide if our federal government has such powers. My view is they probably will say yes, provided the federal law is drafted to minimize unnecessary intrusion into provincial powers.

In terms of assessing the constitutional case for provinces to mount C&T systems, Elgie begins by noting that the federal government's apparent authority to legislate over carbon emissions trading does not preclude valid provincial legislation. Beyond this, he recognizes that "the provinces have broad authority to address many aspects of GHG emissions through other provincial powers, including electricity generation, transportation, the construction of buildings and homes (energy efficiency), forestry, agriculture, etc. – all of which are grounded in clear provincial powers". However, this authority may not extend to regulating GHG emissions trading per se since the impacts are largely global, not provincial. Or as Elgie puts it, the issue is "whether a provincial scheme that included *inter*-provincial (or international) emissions trading would be seen as constitutionally valid". His view is that the provincial authority over extra-provincial carbon trading is doubtful, since inter-provincial trade is an area of exclusive federal jurisdictional power. On the other hand, he notes in concluding that the two levels of government could enact coordinated legislation that would integrate federal and provincial GHG trading regimes across the country.

The Constitutional Authority to Levy Carbon Taxes (Nathalie J. Chalifour)

Nathalie Chalifour's contribution is three-fold: "i) to analyze the federal and/or provincial governments' constitutional authority to implement carbon taxes; ii) to draw upon this constitutional analysis to highlight those design features of a carbon tax that might render it *intra vires* of the implementing jurisdiction; and iii) in light of the above, to evaluate the constitutionality of the Quebec and BC carbon taxes".

In terms of Ottawa's authority to levy carbon taxes, Chalifour conclusions can be summarized as follows:

> While the federal taxation power, s.91(3), is clearly a necessary condition for levying a carbon tax, it is not likely to be sufficient since it would be difficult to demonstrate that a carbon tax had revenue raising as its dominant purpose.
>
> While regulation of GHG emissions under the Canadian Environmental Protection Act (CEPA) is justifiable under the federal criminal power, s.91(27), "it seems unlikely that a stand alone carbon tax would fall under the scope of this power given that it is far from a prohibition coupled with a penalty", which are the criteria for relying on s.91(27).
>
> More likely is the federal trade and commerce power, s.91(2), since a federal carbon tax would be of a nature "that the provinces jointly or severally would be constitutionally incapable of enacting and that the failure to include one or more provinces ... in a legislative scheme would jeopardize the successful operation of the scheme".
>
> Along similar lines, "one could argue that carbon taxes are one of the most economically efficient and likely effective means of reducing GHG emissions (and thus addressing climate change), which could argue in favour of a national interest justification", i.e., POGG.

Chalifour's analysis of the provinces' authority to levy carbon taxes leads to the following observations:

> As was the case for the federal government, the greatest hurdle to relying on provincial taxation powers (s.92(2) or the resource taxation power s.92A(4)) would be convincing the courts that the pith and substance of a provincial carbon tax was revenue raising.
>
> A more likely authority is the provincial licensing power s.92(9). And in order to bring a carbon tax within the scope of the licensing power, a province would need to design the charge as apart of a comprehensive code of GHG regulation. In line with the second of the three objectives of her paper, Chalifour argues that this is what BC and Quebec have done in order to enhance the likelihood that their taxes will be viewed by the court as *intra vires.*

Her conclusion merits quotation in full:

> While there are innumerable considerations involved in the selection and design of policy instruments to address climate change, jurisdictional authority is a critical factor in Canada. This paper has shown that both the federal and provincial governments have jurisdiction to implement carbon taxes, as long as they are carefully designed to fit within the appropriate powers. However, it has also shown that the federal and provincial taxation powers – which are often the first to come to mind as possible justifications – are not the optimal sources of authority for a carbon tax. Federally, I have argued that carbon taxes would find their strongest source of authority under the national concern branch of the POGG power, with possible justification under the criminal law and trade and commerce powers depending on design and, of course, court interpretation of those powers. The taxation power is a possible source, but least likely of those analyzed. Provincially, I have argued that the power to charge license fees offers the best source of authority, though there may be room to find authority within the property and civil rights and, possibly, the taxation powers. And indeed, examining the Quebec and BC carbon-tax measures showed that they are best justified under the licensing power (and were probably designed with this in mind).

Carbon Pricing, the WTO and the Canadian Constitution (Andrew Green)

Andrew Green's contribution has a two-fold objective: i) to provide valuable insight into the role and practices of the WTO, and ii) to focus on the principles that the WTO is likely to bring to bear on climate-change policies as they relate to the global trading system. Given that the WTO rules can be viewed as a quasi-constitutional set of constraints on the climate-change policies of domestic governments, Green achieves i) above by exploring the similarities and differences between the Canadian Constitution and WTO agreements and the resulting implications for carbon-pricing policies in Canada. However, our focus here will be on ii), and in particular on border tax adjustments (BTAs).

By way of elaboration, Green notes:

> BTAs may be used to attempt to overcome the political disincentives to putting in place climate policies and to provide an inducement to other countries to take action. They do so by reducing the competitive disadvantage for industries in countries with strict climate policies. BTAs can be placed on either imports or exports. BTAs on imports are taxes on imports from countries with less stringent climate policies. BTAs on exports are rebates of or exemptions from taxes the domestic producers paid under climate policies. In either case, the general principle is that the BTA cannot exceed the level of tax paid if the good were bound for domestic consumption.

He goes on to note that BTAs, as the name implies, may be used to adjust for the competitive impacts of taxes. However, these BTAs would need to be limited to "indirect" taxes, i.e., those that are levied on products rather than on producers. Since most carbon taxes fit this description, or could be made to do so, Green's view is that BTAs related to a carbon taxation regime would likely be permissible under the WTO.

More controversial is whether BTAs could be used for emissions trading systems:

> For BTAs on imports [under C&T systems], the BTA can only offset an "internal tax or other charge". The question then is whether the emissions trading program can be considered an "other charge". There is not much WTO case law on the nature of "other charge". It will depend on the nature of the trading scheme. If the permits are auctioned or firms are required to purchase permits over an allocated level, a panel may view the requirement to purchase a permit as being in the nature of a "charge". If the permits are given away for free, the issue is even more uncertain. A panel could view the provision of permits as a form of subsidy to the recipients as opposed to a charge. Whether panels will find BTAs can be used for emissions trading schemes is therefore uncertain.

In terms of the Waxman-Markey bill – which intends to require importers to purchase emissions units rather than to pay a tax – Green points out that this raises further questions, not the least of which is whether a requirement to purchase allowances constitutes a relevant tax or charge that can be imposed at the border. Green rounds out this discussion of BTAs on C&T systems with the following:

> BTAs relating to emissions trading programs seem even less likely in the case of exports. BTAs on exports can offset a "duty or tax". While an emissions trading program could be seen as a charge, it seems less likely to fit within the apparently narrower terms "duty or tax".

The above analysis referred to products. Green then asks if the same analysis can be carried over to how the products are made, i.e., to "process and production methods" (PPMs). For example, can otherwise indistinguishable steel as a final product be subject to differential BTAs based on carbon emitted in the production process. The WTO is even less clear on this. Green does point out that that the United States used BTAs to impose a charge on imports of ozone-depleting substances and rebated the tax on exports, but the tax was never challenged at the WTO.

Beyond these technical issues, Green introduces the reader to many operational issues with respect to the WTO. For present purposes, one will have to suffice. While the WTO agreements do impose limits on types of BTAs members may put in place, it is not clear that all countries face the same incentives to comply if they are running afoul of the WTO. For example: "if the U.S. public, for example, feels sufficiently strongly about either climate change or the unfairness of the United States taking action on climate change while other countries appear not to be, the U.S. government may not respond to countermeasures by removing non-compliant BTA provisions". Further, it is much more difficult for smaller countries to maintain measures that do not comply with WTO commitments.

Green concludes by offering the suggestion that countries should work toward a multilateral agreement that would take the form of detailed rules about when BTAs can be used and whether they can cover emissions trading and can take account of PPMs in other countries.

PART VI: THE POLITICAL ECONOMY OF CLIMATE CHANGE

This section offers two quite different perspectives on cap-and-trade approaches to carbon pricing. The contribution by Bryne Purchase argues that while carbon taxes are economically superior to C&T regimes, they are politically inferior. Matthew Bramley accepts the inevitability of C&T, and then proceeds to articulate a "pure" version of what such a regime should strive for.

The Political Economy of Carbon Pricing in North America (Bryne Purchase)

The central argument in the contribution by Bryne Purchase is that "politics" cannot be taken out of the policy decisions relating to carbon pricing and climate change. Indeed, he asserts that "more fundamentally, it is the structure of the political market place that determines instrument choice". At the level of the voting public in most or all of the developed world, this is reflected in a political preference for C&T over carbon taxes. In other words, the "technical superiority" of carbon taxes is overwhelmed by the supposed advantages of C&T, and this despite the fact that the latter tends to be characterized by limited coverage, rent seeking, volatile carbon prices, high administrative costs and an inability to extend C&T internationally. Purchase recognizes that this preference is due in part to the very transparency of carbon taxes: voters know that they will bear the incidence of the tax, and even in the presence of revenue recycling they believe they will be net payers. He notes that in spite of the reality that, with the same sector coverage and with the same emissions targets, both carbon taxes and cap and trade imply the same carbon price. The perception or, rather, misperception remains that C&T is all about regulating and taxing large polluting businesses and not ordinary citizens. Purchase then advances a further reason: "The fact that cap and trade requires a new private army of auditors, lawyers, and market experts also creates a powerful professional constituency in its favour."

Purchase then shifts attention to the reality that Canadian federal politics relating to carbon pricing is not conducive to national leadership in spite of the fact that Ottawa has the constitutional authority to implement a national program. As he makes clear by means of various examples, a major part of this has to do with the profoundly divergent interests of the provinces – most particularly those of the energy-producing provinces as against, say, the hydro provinces like Quebec and Manitoba – and with the related reality that the provinces own their natural resources and the revenues derived therefrom. Especially intriguing is that he relates these interests to party politics:

> The three mainstream national political parties also have serious "legacy" constraints on their ability to lead aggressively on this issue. The Conservatives have their power base in the most "at risk" part of the country. The NDP still must appeal to what is left of unions in heavy industry and, of course, to the urban and rural poor. The Liberals have the heritage of the National Energy Policy and Western alienation. And all parties hope to grow in Ontario, a province already undergoing profound economic dislocation.

Purchase concludes this discussion of Ottawa's role in climate change with the following observation:

> Curiously, it was the National Energy Program and the political reaction to that policy that led to the Canada-U.S. Free Trade Agreement, with its Energy Chapter, subsequently confirmed under NAFTA. A North American market in natural gas as well as oil has emerged. As a result, a national energy policy no longer makes any sense compared to a North American energy policy.

The important implication of this is that by attaching ourselves to the U.S. policy framework, as we appear to be doing, Ottawa can rise above the "tortuous and highly risky Canadian political scene" and, as Purchase notes, allow federal politicians to claim that "the devil made me do it"!

Finally, Purchase turns his attention to the politics of climate change south of the border and internationally. One example must suffice. Purchase suggests that, in understanding the politics of U.S. climate change, one might cut to the chase and focus on two high-polluting sectors (electricity production and transportation) and one key region (the Great Lakes Region – Indiana, Ohio, Wisconsin, Michigan, Pennsylvania and Illinois). Within that region, the percentage of coal-fired electricity generation ranges from a low of 48.5 percent in Illinois to 95.8 percent in Indiana. And these same states are the home of the "Detroit three" automakers and their just-in-time parts suppliers. If one adds to this that these 6 states have 12 Senators between them, that they are relatively heavily unionized, and that all went Democratic in the last election, it should not be surprising that a carbon tax is not the instrument of choice, nor that auctioning of permits under the proposed C&T is limited to at most 15 percent of emissions.

The key messages in the Purchase contribution are worth repeating, namely, that "carbon taxes are unambiguously technically superior but perhaps politically inferior", and that when the policy issue at hand is as important and as pervasive in its impacts as is carbon pricing, it may well be the political market place that determines instrument choice.

Key Questions for a Canadian Cap-and-Trade System (Matthew Bramley)

Matthew Bramley's contribution takes the form of a [Pembina Institute] position paper on the desired features of an effective cap-and-trade system. His two organizing principles/questions are: i) Will the carbon price be high enough to transform our energy system? and ii) Will the value of carbon (or the resulting revenues) be distributed rationally and fairly?

In respect of the former, Bramley notes that there is a huge gap in terms of the carbon price recommended by, say, the NRTEE, and what is happening on the ground (e.g., the Waxman-Markey bill). He draws from a recent C. D. Howe Institute study to argue that issues related to competitiveness and leakages are likely to be small and should not stand in the way of a high carbon price. Moreover, the resort to offsets as a way to achieve a lower carbon price is questionable, primarily because only a fraction of CDM projects actually reduce emissions. With respect to oil sands (the "elephant in the room") Bramley notes:

> …the rapid expansion of oil sands production and the high cost of reducing the associated emissions are responsible for driving up Canada's "marginal cost of abatement" of greenhouse gas emissions, which translates into the need for a high carbon price to reduce them. The Pembina Institute believes that it is unfair for the oil sands sector to create a significantly higher carbon price and consequent costs for all other sectors. To prevent this, we believe that the use of carbon capture and storage, or a technology achieving equivalent emissions levels, should be mandatory for all new oil sands operations. New oil sands operations without carbon capture should be viewed as unacceptable in the same way that new coal-fired electricity generation without carbon capture is now widely seen as unacceptable in light of what we know about climate change.

Finally, since Canada is more likely to meet its targets with a cap on 85 percent of our emissions than with a cap on just 50 percent of emission, a C&T system must be as broad as feasible.

Regarding the distribution of carbon value (the value of emission allowances) Bramley offers the following perspective:

> Governments can distribute the carbon value in two forms – by handing out allowances free of charge, or by auctioning off allowances and handing out the proceeds in dollars. People tend to think of these two options quite differently, but they are financially equivalent, because allowances can be converted into dollars – on a carbon exchange or through a broker – at any time. If a firm receives carbon value in the form of free allowances, this is just as much a subsidy as if it receives carbon value in the form of dollars, as a grant or a tax break.

Among the priorities for distributing carbon value should be: prevention of leakage, protection for low-income Canadians, addressing regional balance, investments related to GHG reduction, and technology transfers to developing countries. While Bramley would prefer 100 percent auctioning of allowances, he does recognize that a case can be made for some version of a production subsidy in order to prevent carbon leakage to other jurisdiction. Finally, Bramley is skeptical of producers being able to purchase emissions credits from technology funds (which are allowed under both the federal and Alberta proposals). Apart from the fact that these are "investments in an unknown amount of future reductions occurring at an unknown date", he adds:

> Alberta's greenhouse-gas regulations allow unlimited payments into a technology fund as a compliance option. It was noted above that they distribute most of the carbon value straight back to the industrial emitters through the use of emissions intensity targets set at a level close to business-as-usual emissions. Of the remaining carbon value, most is paid into the technology fund. Since a majority of the fund's board members represent or have recently retired from heavy industry interests, distribution of this value is likely to be dominated by those interests.

In anticipation that the centerpiece of the Copenhagen deal might well be some form of a cap-and-trade system, Bramley views embracing the above proposed system as "a crucial determinant of Canada's credibility at Copenhagen … and a key test of whether the government now recognizes the scale and urgency of the threat of climate change".

PART VII: SUMMING UP AND A LOOK AHEAD

In his role as rapporteur, Peter Leslie offers "Carbon Pricing: Policy and Politics". This is a most comprehensive, carefully-reasoned, and, indeed, insightful and creative contribution. We are most thankful for the major effort that Peter put into this summary paper. However, given our position as both editors and authors, we are leaving to readers the pleasure of perusing Peter Leslie's integrative interpretation of the above contributions, replete with a looking forward perspective.

Post-Copenhagen Addendum

The final chapter in this volume is an Epilogue written by Nancy Olewiler after the conclusion of the Copenhagen Conference. In it she identifies both the successes and, especially, the failures of Copenhagen and explores the implications that may be drawn from the largest conference ever on climate change. In more detail, Olewiler begins her paper by articulating the five main components of the Copenhagen Accord. She then distils from the Copenhagen experience four principal lessons: establishing targets is a very challenging strategy; a debate framed as the economy versus the environment is a false dichotomy; regional inequality and diversity constitute a huge barrier to agreement on climate change; and a combination of policies is needed to tackle climate change.

In the final section of her chapter Olewiler takes these lessons and combines them with various perspectives adopted from the papers in this volume to outline a broad climate-policy strategy for Canada. Without elaboration, the key aspects of the strategy are: dump the 2020 target; implement a low-rate carbon that will rise over time; recycle the revenue by tax reductions to individuals and businesses and by funding low-carbon technology projects; and invest in forest management, starting with the territorial lands of Canada's First Nations. Perhaps most importantly, she suggests that there is more latitude for independent Canadian action, both with respect to timing and design, than is implied by the Government's stated intention of moving in lock-step with the United States.

APPENDICES

Rounding out the volume are two appendices, both drawn from publications already in the public domain. The first of these is the Executive Summary of *Achieving 2050: A Carbon Pricing Policy for Canada,* the 2009 report of the National Round Table on the Environment and the Economy (NRTEE). The second Appendix is adopted from the October 29, 2009 TD Economics Special Report *Answers to Some Key Questions about the Costs of Combating Climate Change: A Summary of the Pembina/David Suzuki Foundation Paper.*

II

Carbon Pricing: Setting the Stage

2

The Canadian Policy Struggle with Climate Change: Setting the Context for Carbon Pricing

Bob Page

INTRODUCTION

This volume deals with issues that are central to any resolution of Canada's long-standing dilemma respecting the formulation of an effective policy on climate change. Despite several decades of consideration, numerous public "consultations" and "framework" documents, Canadians are still in doubt about both the policies to be applied and the level of government that should execute them. Contributing to this public confusion are the evident tensions between the federal and provincial governments and among the provinces, most particularly over some of the controversial provisions of the Kyoto Protocol. Now, in the run-up to the UN Copenhagen Conference on Climate Change, the situation is becoming more urgent as both Canada and the United States are approaching key policy decisions, perhaps in a coordinated way, while their sub-national governments are embarking on initiatives of their own. For example, four of the provinces have signed into climate policy alliances with U.S. states in the Western Climate Initiative, and Quebec has signed a special deal with California. This expanded, cross-border role for the provinces reflects their growing insistence that they have a voice in the formulation of agreements and policies in areas that, constitutionally, fall into provincial jurisdiction. Further complicating the scene is concern over possible adverse interactions between measures aimed at climate change and those directed at countering the most severe economic downturn since the Great Depression. Above all, there is the specter of the potential political costs of "not getting it right" or, *pace* Stéphane Dion, being unable to garner the necessary public support for one's policies. Clearly, whatever the urgency of advancing the climate-change agenda, the way ahead is fraught with difficulties both political and economic, and industry and governments have a stake in "making haste" cautiously.

I have been asked to place climate-change policy issues into context for the remaining papers in the volume. Accordingly, this paper is a combination of my personal observations as a participant in many of these events[1] and my later reflections as an academic. It is not an easy task, given the disparate forces at work – environmental idealism, limits-to-growth economics, federal-provincial rivalries, intergenerational ethics, corporate goals and aspirations, energy supply, etc. One of the first challenges for the scholar is to separate the substance from the rhetoric.

Given the structure of the Canadian economy, it is not surprising that climate change generates huge problems for the federation. There is the inevitable split between the western energy producers and the eastern consumers, including all the controversy over the oil sands and "dirty" oil. As the world price for oil increased the issue of who should be able to alienate resource rents and for what purpose rose to the fore. Many in the east do not trust Alberta to manage environmental regulation given the province's financial stake in the outcome, while many in Alberta see the federal policy on Kyoto as a new Trojan horse to extend Ottawa's jurisdiction in the oil patch. These battles are part of the context for the papers to follow.

Canada's climate-policy voyage has been unique among the western nations, with the possible exception of Australia.[2] Both the provinces and the federal government have had equal difficulty with this file; for example, until recently Alberta wanted no mandatory controls. The policy area has been compared to a "soap opera", with high drama at every stage, but no resolution in the never-ending plot. Hence, in proceeding in the current time frame we need to understand some of the reasons for inaction and to ensure that they can be avoided in the future.

Carbon pricing and federalism involve two separate areas that traditionally have had little interaction. Now, however, we are trying to link energy/resource markets and market forces with the constitutional labyrinth of federalism. Markets and regulation operate on different principles, and there are few precedents to help identify how best to link them – the current federal-provincial tug-of-war over a single national securities regulator may be one example. There will be similar battles over the control and regulation of a national emission trading system. Moreover, the terminology is problematical since concepts such as "carbon tax" and "cap and trade" have acquired political baggage so that they mean different things to different groups of Canadians.

In concluding this opening section, there are several general considerations that should be noted. To begin with, climate change is as much about energy policy as it is about environmental policy, which in turn has created interdepartmental conflict for both the Ottawa and the Alberta governments.[3] To

---

[1]The author has participated in the climate-change negotiations and debates as academic research advisor to government, as a corporate executive (1997–2007) and as Chair, International Emissions Trading Association (Geneva).

[2]As part of my corporate experience I had direct dealings with the Canadian and Australian governments as well as with Alberta and Western Australia, and observed many similarities including the vocabulary of engagement.

[3]The locus for climate change in Alberta was moved from Energy to the Environment and in Ottawa from Natural Resources Canada to Environment Canada. The Ottawa move was particularly filled with acrimony.

promote harmonization and coordination of environmental policies among the territories, the provinces and the federal governments, our intergovernmental apparatus has created the Canadian Council of Environment Ministers. Its effectiveness, however, is greatly limited in controversial areas like climate change where there is no consensus on required action.[4] Similarly, although external to Canada, international environmental institutions like UNEP (United Nations Environmental Programme) and Kyoto are weak instruments for enforcing agreements, which, in turn, tends to promote a lack of respect for their policies (DeSombre 2006, 171).

Second, there were (and to a degree still are) differing levels of acceptance of the science of climate change. In the corridors of Ottawa, the analysis of the IPCC (1995, 3-7) was almost universally accepted, while in Alberta the skeptics found an audience among some oil companies and provincial government officials.[5] It should be noted, however, that most of the work of the critics fell outside the scientific literature.

Third, the very nature of carbon regimes poses some fundamental challenges for policy makers, challenges that have still to be overcome. Carbon is not just a man-made pollutant but part of nearly everything that happens on the planet. It is an essential part of the web of life. Humans exhale CO_2 and plants take it up as essential food. Hydrocarbons power our economic system, and support our consumer-driven life style (Dessler and Parson 2006, 1-45). To manage carbon effectively, we require a complex "systems" approach, one involving the interconnections between energy sources and fuel options. This is well beyond the capability of most provincial governments and is a policy stretch even for Ottawa. While there is general acceptance of the science of CO_2-related global warming, there is still a significant level of uncertainty surrounding the speed and the intensity of the change, and on how these will be affected by efforts to slow the CO_2 increase in the atmosphere over coming decades.

Finally, but not exhaustively, the policy tool kit needed to move Canada toward a carbon-constrained future has a variety of components. The most pressing issue is how to introduce a carbon charge – for example, by a carbon tax or cap-and-trade system – into the economy. This is necessary to force industry and individuals to recognize there is a real cost associated with their carbon emissions and to choose less carbon-intensive alternatives. But we must also understand that there are other tools that complement and enhance carbon management. These include initiatives to promote energy efficiency and energy conservation. They also include offset systems with effective verification to allow credits generation. Technological advance is another key to limiting emissions. Carbon capture and storage for thermal power generation has immediate large-scale potential when it becomes commercially viable. Serious expansion of renewable power sources is yet another option. Clearly, we must

---

[4]However, the Canadian Council of the Ministers of the Environment, located in Winnipeg, has provided useful coordination in other areas (see www.ccme.ca).

[5]Terence Corcoran and *The National Post* provided a continuous platform for the critics. Also see Solomon (2008, 1-8).

allow some flexibility in the implementation of our tool kit, for we are unlikely to get it all right the first time. Moreover, new science may create doubt about some current solutions.

JURISDICTION

Jurisdictional issues have historically played a key role in the evolution of environmental policy. Indeed, regulatory jurisdiction is at the centre of many of the issues in dispute. Constitutional ambiguity lends itself to conflicting claims, especially in areas like energy, where substantial revenues are involved and the provinces assume they have exclusive jurisdiction. This issue is not new; in fact, it goes all the way back to Confederation and the original delegation of powers.

In 1867, the word "environment" was not part of the political vocabulary of the day. As a result, the environment did not find its way into the assignment of powers to the federal or provincial governments in section 91 and section 92 respectively of the *British North America Act*. Hence, the jurisdictional assignment has to be assessed indirectly through the powers that are defined. Even with the Supreme Court as the final arbiter, this is not a neat and tidy solution. With Kyoto, one sees on the one hand the federal government exercising its powers to sign international agreements and on the other the provinces claiming some say in the negotiations and the right to implement those aspects that fall within their jurisdiction. For Alberta and Quebec this has now become an area of great sensitivity. Law Professor Jamie Benidickson (2002, 32) has summarized this issue on the basis of Supreme Court rulings: "It has long been established that the federal government cannot simply by entering into international agreements extend the scope of its domestic jurisdiction." The underlying legal case here is the famous 1937 *Labour Conventions* case (*Canada (A.-G.) vs. Ontario (A.-G.)*, 1937). This was exactly the concern of the Alberta government with the successive federal proposals to implement Kyoto.

To avoid a court challenge over jurisdiction, the Harper government has sought to entrench its climate change regulations by placing them under the criminal code powers of the Constitution, in line with the decision in *Attorney-General for Canada vs. Hydro Quebec* (1997) where the criminal code powers were found able to underpin federal environmental regulation. This case dealt with the toxic substance provisions of the *Canadian Environmental Protection Act*. When the issue came to CO_2, there was one problem: CO_2 is not toxic in any normal sense of the term. Not surprisingly, Alberta was upset when CO_2 was declared toxic, thereby enabling the government to proceed without new legislation having to be passed by the minority House of Commons [Editors' Note: these constitutional issues are elaborated in the contributions by Elgie and by Chalifour in this volume].

On the provincial side there are extensive powers covering environmental issues. Some provinces choose to exercise them while others defer on climate to Ottawa. The powers include property and civil rights, public lands and natural resources. However, in section 92A of the *Constitution Act, 1982* there is the powerful natural resources clause that confers on the provincial legislature the exclusive authority to pass laws for the "development, conservation, and management of non-renewable natural resources and forestry resources in the provinces, including laws in relation to the rate of primary production therefrom". Not surprisingly, Alberta is particularly sensitive to protect its jurisdiction in oil and gas, electricity, forestry, etc.

Under Kyoto, the fossil fuel policies become critical because the largest CO_2 emitters are also major sources for provincial royalties and taxation. Alberta believes this is part of their "sacred trust", while Ottawa wishes to exert more control over the environmental aspects of resource management, including the possibility of accessing revenues for pricing carbon. Given the centrality of these issues it is surprising that there have not been even more fireworks. Part of the reason for the lack of a public explosion is that public opinion outside the province is strongly opposed to the Alberta position, and strong action against Alberta would attract votes in the rest of the country. Even within Alberta, the public overwhelmingly supports strong action on climate change.

Another area where Alberta claims exclusive jurisdiction is the thermal electric power sector. Authority for this also stems from s. 92A: "In each province, the legislature may exclusively make laws in relation to ... (c) development, conservation and management of sites and facilities in the province for the generation and production of electrical energy." The concern here is that the coal-fired power plants are the largest emitters in the province, and these concerns transcend Alberta since in all provinces the electric power sector is regulated by the province, and in most of them the provincially owned crown utility is the major generator. In addition to Alberta, the provinces of Nova Scotia, New Brunswick, Ontario, and Saskatchewan also rely heavily on coal-fired power generation. Although emissions issues are not a factor for nuclear or hydro, federal intrusion would be strongly resisted even by provinces like Quebec that support Kyoto. Only in the case of inter-provincial and international electricity transmission do we find federal regulation by the National Energy Board and, of course, in all matters relating to nuclear power.

HISTORY AND PUBLIC PERCEPTIONS

The disputes over jurisdiction have been compounded by historic events and by the way they have become a continuing part of the political culture in the provinces. When Alberta and Saskatchewan became provinces in 1905, Ottawa retained control over lands and natural resources, thus creating a sense of grievance. Calgary lawyer R.B. Bennett had to battle with the federal Department of the Interior over hydro water rights on the Bow River.[6] Subsequently, as Prime Minister, he fully implemented the agreement – signed before he took office – that conferred in 1930 full provincial status to the two Prairie provinces. There were other irritants in the early decades of the twentieth century, including the advocacy work of the federal Commission on

[6]See the author's forthcoming *Centennial History of The TransAlta Corporation*, due for publication in 2009 or 2010.

Conservation and the national park boundaries.[7] The eastern-based conservation movement clashed with the western frontier-based spirit of development. The Alberta frontier is gone, but the perceptions remain part of the political culture and are rekindled yearly at events like the Calgary Stampede.

The phrase "climate change" has historical significance for Albertans. They believe that they have already experienced it and resent critics who say they have ignored it. In the later 1920s and the early 1930s, severe drought and climate "change" hit the western prairies in ways not experienced in other parts of Canada. Droughts in the last decade brought back memories of the Dust Bowl years and the lack of support the West received. If the federal outreach to the West on climate change had stressed water needs, the Alberta response would have been much more positive. The province is facing declining water supplies and increased water demands. No new water licenses are being granted in the south and new water pricing and water rights trading are under consideration. The basic water source is in the snow pack in the mountains, which has become variable and unreliable for the future. Alberta rivers mainly flow west to east and supply large parts of Saskatchewan and Manitoba. This is a huge ecological and economic issue for the future of the Prairie West.

In the 1960s, a major environmental movement swept across North America and several Canadian provinces responded with new programs. With the creation of the federal Department of the Environment in 1971, Ottawa entered the field and friction with the provinces began to emerge. With the OPEC oil crises, energy supply issues from Alberta began to clash with environmental conservation aspirations within the federal government. Ottawa also reflected the economic nationalism of the day with the creation of Petro Canada as an arm of federal policy within the oil patch. The federal *Canadian Environmental Impact Assessment Act* created tensions over several pipeline and dam projects. Eventually, joint panels with the provinces were established in an effort to improve cooperation.

The next battle was even more bitter, as it combined both jurisdiction and resource revenues. The upward spiral of energy prices, the public controversy about economic rents and windfall profits, and the desire for a larger Canadian ownership stake, all combined to push the Trudeau government into launching the National Energy Program (NEP) in the early 1980s. The federal government wanted to support drilling in the Arctic, to keep low domestic prices for Canadian consumers, to support renewable energy and to increase the federal share of the economic rents. The Province of Alberta saw the NEP as a constitutional revolution in which Ottawa was intervening to erode the province's jurisdiction and to divert economic rents. The resulting bitter oil dispute saw Peter Lougheed confronting Pierre Elliot Trudeau, which provided great theatre for the media but little in the way of unwinding the NEP.

There were two major areas of focus – revenue and jurisdiction. The revenue issue quickly disappeared with the collapse of world oil prices (not because of the Ottawa revenue grab). However, a decade-long deep depression

---

[7]The park boundaries of Banff National Park, until cut back, precluded development in the Canmore area.

followed in Alberta, which was widely attributed to the NEP and left lasting scars. The jurisdictional issues were finally resolved when Mulroney came to power and, to the quiet satisfaction of the Lougheed Tories, abolished the NEP. While Alberta succeeded in defeating the National Energy Program, there were lessons not to be forgotten, and Lougheed became a type of folk hero in the West as the guardian of provincial rights and responsibilities.

While the NEP died in the 1980s, some of the mythology has continued on. Today Peter Lougheed is still active on the Calgary scene as a senior statesman. He has warned of a future battle over climate policy, resource jurisdiction, and the oil sands that may be equal to or greater than that over the NEP. If this were not bad enough, Premier Brad Wall of Saskatchewan has compared liberal Leader Stéphane Dion's carbon tax proposals as the equivalent of the NEP (Wood, 2008). Wall's comments attracted wide media attention across the West. They reflected both the distrust of federalism and the then Liberal leadership. They seemed to confirm the arguments of Preston Manning when he founded and then led the Reform Party. This is now the western wing of the Conservative Party of Canada within Stephen Harper's coalition of western and eastern Tories.[8]

THE KYOTO PROCESS

The Kyoto Protocol (1997) has been *the* most difficult and controversial international environmental agreement that Canada has ever signed. The federal government sees Kyoto as an environmental issue because it is reflecting the consumer interests of most voters, while the western oil-producing provinces are driven by their concerns about energy production and see it as an energy issue. Not surprisingly, Alberta and Quebec see this in different light. Alberta, with its oil-based economy, sees Kyoto and CO_2 regulation as a constraint on its economic growth, while Quebec sees comparative advantage with Kyoto given its hydro-based economy and the potential for selling credits to emission-intensive provinces. Alberta would face heavy costs and probably revenue outflow while Quebec would generate revenue from the credit sales for large hydro if approved. Thermal power faces similar challenges and so Alberta, Saskatchewan, New Brunswick, and Nova Scotia have common cause against Kyoto costs. Caught in the middle between the two camps, Ottawa has tended to delay its final policy determination in the desperate hope that some silver bullet will appear, which is part of their current enthusiasm for carbon capture and storage technology.

The global negotiating began on a positive note with the Rio Conference in 1992. Here the Honourable Jean Charest, then the federal minister of the environment in the Mulroney government, led the delegation with his usual infectious enthusiasm. The modest results did not alarm the provinces. Canada

[8]Attitudes to the environment vary in the two wings of the Conservative party of Canada (Reform vs. Progressive Conservative) and also regionally, East vs. West. While both from Calgary, Stephen Harper and Jim Prentice arguably represent these two wings.

committed to "best efforts" to return emissions to 1990 levels by 2000. This stabilization approach seemed reasonable. The federal policy response called for voluntary measures and public-policy education. When the Liberals resumed power in late 1993, Prime Minister Chrétien came to Calgary and announced that his government would not impose a carbon tax. However, there was growing international concern that Rio was too limited a response to reverse the escalating CO_2 emissions.

In mid-1997, federal-provincial discussions began to heat up in advance of the forthcoming Kyoto summit. The parties met in Regina and the consensus outcome was that Canada should re-affirm its Rio commitment to emissions stabilization at 1990 levels. With Canada's rising emissions this was now a much more serious commitment.

However, Prime Minister Chrétien chose to go beyond this agreement in instructing the negotiating team. At Kyoto, U.S. Vice President Al Gore pressed Canada to go even further. After an exhausting final all-night negotiating session, Canada accepted a 6 percent cut from its 1990 baseline, while the United States agreed to 7 percent and the European Union 8 percent. Australia, with a similar energy exporting economy, was granted an 8 percent *increase* from its 1990 baseline. When the Kyoto terms were announced, some of the provinces were furious. Ralph Goodale, the federal minister of natural resources from Saskatchewan, was reported to have considered resigning (Simpson, Jaccard, and Rivers 2007, 33). The federal government needed the collaboration of the provinces in the implementation of Kyoto, and designed an elaborate sectoral table process to bring the provinces and other stakeholders on board.[9] Meetings were held with 16 issue or sector tables bringing together 450 experts and 225 stakeholders. The author was co-chair of one of the tables. The participants of the 16 tables worked long and hard on a disparate list of recommendations that, subsequently, were largely ignored or rejected.[10] The end result was a sense of discouragement that any national consensus on climate change was possible. Professor Kathy Harrison (2006) summed up the process as a "subsidy program" for Air Canada.

The process revealed many of the challenges faced by federal Canada in coping with climate change. There were evident divisions both within the federal system and between Eastern and Western business (such as between Ontario manufacturers and Western energy producers). There was strong criticism of the federal modelling of Kyoto costs. Specifically, Alberta interests felt the cost estimates were far too low, especially those for greater use of natural gas in place of coal. Also, the complexity of carbon's role in the economy had made the Chrétien government all the more cautious which, in

---

[9]This was the National Climate Change Process co-chaired by David Oulton (federal) and John Donner (Alberta for the provinces) which operated the tables 1999–2000.

[10]I co-chaired the Credit for Early Action Table with Stephen McLellan from Environment Canada. Of my many public efforts, this was the most frustrating – too many layers, too many people, and too many opinions.

turn, encouraged provincial opposition in the hope that Kyoto would just go away. The only consolation was that the United States would face similar costs.

The perceptions changed drastically in 2002, when George W. Bush formally rejected Kyoto. This should have been less of a surprise, since Bill Clinton had refused to send it to the Senate for ratifications because it had little support. However, given Al Gore's role in Kyoto, many Canadians felt a sense of betrayal.[11] Now the competitive issues surrounding Kyoto came into much clearer perspective: Canada was alone in the Americas with Kyoto obligations! The Alberta government and many industry representatives now pressed Ottawa to follow the United States example and not ratify Kyoto. Other provinces, like Manitoba and Quebec, urged ratification. I can remember some very tense confrontations in Ottawa meetings at the time. Environmental NGOs saw it as a moral duty for Canada in terms of our ethical responsibilities to the globe. On the other hand, many in business viewed Kyoto as an economic disaster for Alberta's oil and/or Ontario's car manufacturers. Given the intensity of the rhetoric, the federal government was reluctant to go beyond public education and voluntary measures.

With the ratification issue pending, the provinces attempted to insist that their views be part of the federal decision-making. In Halifax in the fall of 2002, the provincial ministers passed a resolution that the prime minister should consult with the premiers before ratification (Macdonald *et al.* 2004-2005, 183-184). Ottawa ignored the advice but did meet bilaterally with several provinces on various climate issues. While the provinces had no common position on Kyoto, they were trying as always to establish a provincial role in ratifying treaties that involved provincial jurisdiction. Ottawa was equally determined to avoid such a precedent with Kyoto.

The Kyoto protocol was designed to come into force when ratified by countries representing 55 percent of developed country emissions. With the United States total of nearly 40 percent of the required emissions out of the equation, the pressure now mounted on Canada and Russia to ratify. Eddie Goldenberg, the prime minister's most senior advisor, warned that Kyoto targets would not be met (Goldenberg 2006). Some members of cabinet, including Paul Martin, were rumoured to be opposed. John Godfrey, a Toronto pro-Kyoto MP, circulated a petition for ratification with the signatures of 96 of the 172 Liberal MPs. But Chrétien's commitment, unknown to most in Ottawa, was to press ahead. He wished to counter his nemesis George Bush and add to his legacy as PM. At the Sustainable Development summit in South Africa, he suddenly announced Canada would ratify Kyoto even though there was no serious implementation plan in place.

While Quebec was supportive, Alberta denounced not only the decision but the whole process since the Regina meeting, taking the view that the federal government did not have the mandate to proceed because the provinces held the jurisdiction for implementation. For its part, the federal government wished to speed up the climate policy process by accessing directly (without the provinces) some of the key industry players who had already adopted

[11]Personal communication with one of the lead negotiators for Canada at Kyoto.

progressive actions on climate. The author was involved with discussions about "covenants" that would be legal agreements for emissions cuts in return for future guarantees respecting regulation for the company involved. They would become benchmarks for future sectoral policies. These were not received well by the industry associations or by the provinces, which viewed them as federal intrusions to distort provincial regulatory processes.

EMISSIONS AND THE NATURE OF THE CANADIAN ECONOMY

At the heart of the western objections to Kyoto was the nature of the Canadian economy and the methodology for determining the emission cuts. Alberta believed that the system was designed to favour economies that were consumers, not producers, of oil. Canada, like the European Union, was a major consumer, but it was also a major producer, especially for the United States. Thus the Canadian emissions total included both consumption and production, while the production emissions attributable to the E.U. consumption were largely off-loaded into the totals for Russia and OPEC. This put Canada, and especially Alberta, at a fundamental disadvantage. In contrast, if one were to apply a life-cycle analysis – one where the upstream or production emissions attributable to a nation's fossil fuel consumption were included in its total emissions and such emissions excluded from the total of the producing nation – European totals and the Canadian totals would be much closer. Also Alberta has a much faster rate of economic and population growth with the related emissions that this entails. While no system is perfect, the structure of the Kyoto Protocol put Canada at a significant disadvantage given the nature of its economy. And, as noted above, while the Kyoto agreement recognized the energy exports of Western Australia, there was no similar recognition for Alberta production.

In the last two years we have seen growing recognition of the issue of domestic equity in the development of carbon regulations. Tom Courchene and John Allan of Queen's University have reviewed current policies and warned of the consequences (2008). They conclude that this approach (with its failure to properly assign carbon footprints) "will almost assuredly distort rational and efficient decision making", and this to the benefit of consumer nations. "Moreover, if our policy makers are not careful they may lead us into the carbon equivalent of the national energy programs" (Courchene and Allan 2008). Consumers must take some responsibility for their associated production emissions.

The Chrétien government recognized some of this logic and made some efforts to respond. They proposed that Canada be granted special CO_2 export-offset credits to lower the total required emission cuts.[12] The proposal was vigorously opposed by the European Union and the United States, and the matter was quietly dropped in the international negotiations. They also proposed

[12]The original proposal claimed a total of 70 megatonnes or about 25 percent of the Canadian Kyoto target.

large purchases of international offset credits to ease the burden on the Large Final Emitters (thermal electricity, oil and gas, and manufacturing).[13] Alberta opposed these measures even though the province would have been the largest beneficiary.

After Kyoto ratification, the European Union began to move aggressively to the design and implementation of a cap-and-trade emissions trading system. The system included domestic credits from within the European Union and eligibility for credits from developing countries under the Kyoto "Clean Development Mechanism". There was a transition period (2005–2007) followed by the full plan commencing 1 January 2008. The private sector played an important role in the design of the system through the International Emissions Trading Association (IETA), with several Canadian companies (TransAlta and Alcan) playing a leadership role in the organization.

It was only natural that the events in Europe helped to trigger events in Canada. Canadian members of IETA worked closely with the Natural Resources Canada (NRCan), which was preparing the federal climate change plan. They signed a joint formal Memorandum of Understanding on the principles required for a Canadian cap-and-trade system.[14] Ottawa appeared to be headed in the same directions as the European Union in the design of a formal cap-and-trade system.

However, there was a bitter internal struggle within the Government of Canada for the overall control of climate policy. The outcome of this battle saw the transfer of control from the NRCan team to Environment Canada, which then proceeded to repudiate the agreement. Canada was back to square one in terms of emissions trading. By way of alternative approaches, Ontario, Alberta, British Columbia and Quebec began to consider their own emissions trading schemes. The Montreal Stock Exchange began planning a platform for such trading to occur. Ottawa's failure to act tended to promote or at least encourage this public policy fragmentation.

One of the more interesting federal-provincial events was the annual trek of the Canadian representatives to the Conference of the Parties (COP) yearly summit on Kyoto. While Ottawa led the delegation, it included provincial ministers and staff – Alberta, Quebec, and some others. Quebec usually pushed for vigorous implementation of Kyoto and Alberta the reverse. Both, however, agreed on their distrust of the federal government. This Kyoto process continued right up to the most recent COP in December 2008 in Poznan, Poland. Once again, Canada was attacked for its failure to implement Kyoto.

---

[13]The foreign credits would have covered nearly two-thirds of the emission liabilities for LFE companies.

[14]The author, as Chair of the International Emissions Trading Association, negotiated the MOU agreement with George Anderson, the Deputy Minister of Natural Resources Canada.

REFLECTIONS

Protectionism

In observing these international environmental negotiations, I was struck by some of the similarities to trade negotiations.[15] While the semantics of the Kyoto negotiations were environmental, the working assumptions were those of trade and competitiveness. Environmental protectionism was evident in the strategy of both the United States and the European Union. The latter wanted to saddle the U.S. producers with additional environmental costs, whereas George Bush cited these costs in rejecting Kyoto. Ontario and western companies expressed similar concerns to Ottawa.

This environmental protectionism took on a new and more sinister twist in 2008. In January 2008, the European Union announced it would apply a "carbon tariff" on imports from countries with less stringent carbon-emission controls. They argued that they were now forced to establish a "carbon equalization system" to protect E.U. jobs and products from developing countries with no carbon costs or countries like Canada that were not meeting their Kyoto commitment. The World Business Council for Sustainable Development warned its members of the coming "Trade War over Carbon".[16] Carbon taxes can have a variety of forms, including that of protective tariffs, a form that currently seems to be gaining momentum in both the European Union and the United States.

Carbon Taxes

The carbon tax debate in Canada has several manifestations. The basic concept has a lot of political baggage. This emerged most directly in the fall 2008 general election with the Liberal proposal for a carbon tax. One of the issues that emerged was a border tax adjustment or countervail to balance the increased relative costs for Canadian manufacturers or producers as a consequence of the tax. While some believe this is consistent with World Trade Organizations rules, others sharply disagree. Michael Hart and Bill Dymond (2008), among Canada's most experienced trade experts, denounced this proposal claiming it would "wreak havoc" with existing trade agreements. Canada would have to prove that the absence of carbon regulations within the exporting country constituted a "subsidy" under trade laws in order to justify Canadian action. Also, Canada would have to prove the subsidy caused "material injury" to Canadian producers. The option of a border adjustment mechanism is one of the economic complications to a carbon tax.

---

[15]The author was an environmental advisor to the Department of Foreign Affairs and International Trade for the NAFTA negotiations.

[16]Memo, Emissions Trading Revisions, European Union, (Brussels) 23 January 2008 and World Business Council for Sustainable Development (Geneva) Press Release, 10 February 2008.

Another of the complications that has huge federal-provincial implications is the recycling of the revenues from a carbon tax. Dion's proposal was to recycle the revenues back to consumers in the form of personal and corporate income tax cuts. Dion saw this as injecting a carbon price signal into the economy as well as a means of fiscal reform, namely a tax shift from income to consumption taxes. He was strongly supported by many economists across the country. Others believe that any revenues should be recycled back to the provinces where they were raised so that there is no transfer of wealth between provinces. Needless to say, Alberta would fight for the latter structure. The issue of recycling the revenue would probably be as divisive as the issue of the tax itself.

Oil Sands

The oil sands are a key factor in any carbon pricing debate. Their future expansion complicates Canada's ability to meet its targets for 2020. While technology, such as carbon capture and storage, will help in the long term, it will be some years before the infrastructure is in place and the technology will be commercial. The dilemma is that the public expects emission cuts almost immediately while the technology will take a decade or more to implement. This time gap is at the centre of the Canadian carbon management dilemma. The greatest current threat to the oil sands is potential loss of U.S. markets with new U.S. environmental legislation. The December 2007 U.S. Energy Independence and Security Act (section 526) forbids American government, its agencies, or the armed forces from purchasing high carbon fuel products like oil sands oil. To Alberta this is a much more serious threat than any tax or cap-and-trade system. Without the U.S. market, the potential for the oil sands does not extend beyond the current levels of production. The Alberta Premier has demanded a seat at the table in any negotiations with the United States.[17]

Emissions Trading

The Canada emissions trading system was due to begin on 1 January 2010, but recent discussion with officials appear to indicate at least a one-year delay. There are issues of liquidity and fungibility in the design of the Canadian system. There are also constraints on the access to the technology fund and it is unclear where the necessary volume of credits will come from. The final rules for offset credits are still pending, and international credits will be limited to 10 percent of compliance totals. Meanwhile, major provinces are developing their own offsets and trading systems. Four provinces are proposing to join the emissions trading regime of the U.S.-states-initiated Western Climate Initiative.

---

[17]Premier Ed Stelmach, to Hon. Stephen Harper, 15 Dec 2008, courtesy of the Alberta government.

Distributional Issues

Alberta remains very sensitive on the wealth transfers issue, believing that the province is already contributing significantly to the overall Canadian economy and to interprovincial transfers. In return, it is strongly criticized for "dirty oil" by some other provinces that benefit. The Canadian Energy Research Institute released a report in October 2005 that attempted to quantify the national benefits. In 2004 Alberta contributed $30 billion per year to the national coffers and received back value equivalent to $17 billion – a net contribution of $13 billion. In terms of the allocation of Alberta's resource revenues, 41 percent went to federal government, 36 percent to Alberta and the rest to the other provinces. On an overall per capita basis, each Alberta resident's net contribution to "financing Canada", as it were, is $4,000 with Ontario the next highest net contributor at $1,700 per capita.[18] With the increase in world oil prices in the past three years, the relative share for Alberta has increased while that for Ontario has decreased. The Alberta government will resist further revenue or wealth transfers except where they may involve new technology or other partnerships.

Many officials in Ottawa and some other provinces still feel that Alberta can do much more given the corporate profits and provincial royalties. However, in the final weeks of 2008, the collapse in the world price of oil from $147/barrel in June to about $35/barrel in December has production declining and provincial revenues evaporating.[19] Alberta has put up $4 billion for carbon capture and storage and transportation projects. It was expecting federal support, which so far has not materialized. With the double hit of world oil prices and the collapse of capital markets, there will be further delays in the implementation of the new climate technology and less willingness to accept new carbon taxes or levies until prices rise significantly.

Alberta is assuming today it can strike a deal with the federal government to qualify for "equivalency" under CEPA so the province would protect its jurisdiction. The province has had extensive discussions with Ottawa, and senior officials told me they were very close to a "political" deal on equivalence. Alberta is prepared to modify its GHG regulations to close some of the gap between Alberta and the federal government. However, will this be enough to make it politically acceptable in Ottawa for Ontario and Quebec voters? There are a variety of issues that could derail the negotiations. They include the federal effort to require carbon capture and storage technology in the 2012 to 2018 period; the absolute target of 20 percent emission cuts by 2020; the Alberta desire for no emissions trading beyond its border; and the federal proposal to exclude raw bitumen exports to Asia. In its search for equivalency, how far is Alberta prepared to go? There are no quick or easy solutions. One last complication is the Harper desire to go to Washington to negotiate a North

---

[18]Canadian Energy Research Institute, Calgary 2005, and *The National Post*, 2 September 2008.

[19]World oil prices have rebounded in 2009 reaching $70.00/barrel in early September.

American cap-and-trade system as part of a wider energy security and environmental protection deal. Whether this will materialize or not is uncertain. My own discussion with U.S. officials elicited little interest as yet in such a deal. However, they did leave open the option of links between separate Canadian and American cap-and-trade systems.

CONCLUSION

This paper documents the complicated web of events and assumptions that provide context to this volume. Any new carbon tax or cap-and-trade system will have a difficult political stage to operate on. New policy initiatives will require creativity, flexibility, and balance between eastern consumers and western producers. The history of the climate process does not give us much confidence in going forward, except the urgent necessity of addressing this issue for domestic and international reasons. Nonetheless, climate change is the most pressing environmental challenge facing our generation. Certain lessons do emerge from the above to guide policy architects.

The physical complexity of carbon regimes makes policy design difficult, especially in avoiding unintended consequences. The current fragmented policy regime – six provinces plus Ottawa – will lead to further federal-provincial tensions and disputes. We do not seem to have any effective institutional mechanism like a refurbished Canadian Council of the Ministers of the Environment to handle controversial overarching issues like climate change. The provinces have participated in a variety of processes without any clear policy outcomes. Unlike other environmental issues, climate policy cannot be downloaded on the provinces without significant loss of face for the federal authorities. Moreover, the decision of the federal government to use its criminal code powers under CEPA would appear to eliminate the potential court challenges to federal climate policy.

Wealth transfer is an upsetting prospect for Alberta, but a major gain for Quebec should it be able to sell hydro offset credits. This could be an annual $2 billion windfall for Quebec. The devil is in the detail of any tax or cap-and-trade system, and disputes over recycling of the revenue from a carbon tax are almost inevitable. There are significant differences between the provinces, which preclude a common front, and differences between federal departments, which serve to weaken the federal policy process.

With the collapse of world oil prices and the decline in government revenue, further delays in the fundamental retooling of our fossil-fuel technology seem very likely. Also, the decision of the Harper Government to pursue a North American accord will add complications for the Canadian provinces. Assuming that the minority Harper government survives, environment minister Jim Prentice will need all of his considerable negotiating skills to deliver effective climate policy.

REFERENCES

Benedickson, J. 2002. *Environmental Law*. Toronto: Irwin Law Inc.

Canada A.-G. v. Hydro Quebec, [1997] 3 SCR 213.

Canada A.-G. v. Ontario A.-G., [1937] A.C. 326 P.C.

Courchene, T. and J. Allan. 2008. "The carbon footprint", *Globe and Mail*, 17 July.

DeSombre, E.R. 2006. *Global Environment Institutions*. London: Routledge.

Dessler, A.E. and E.A. Parson. 2006. *The Science and Politics of Global Climate Change*. New York: Cambridge University Press.

Goldenberg, E. 2006. *The Way It Works*. Toronto: McClelland and Stewart.

Harrison, K. 2006. *Proceedings of the Annual Meeting of the International Studies Associations*. San Diego, California (March).

Hart, M. and B. Dymond. 2008. "Canadians should be wary of green protectionism", *The National Post*, 2 September.

IPCC. 1995. *Climate Change 1995: The Science of Climate Change*. New York: Cambridge University Press.

Macdonald, D. *et al.* 2004–2005. "Implementing Kyoto: When Spending is Not Enough", in G.B. Doern (ed.), *How Ottawa Spends*. Montreal and Kingston: McGill-Queen's University Press.

Simpson, J., M. Jaccard, and N. Rivers. 2007. *Hot Air: Meeting Canada's Climate Change Challenge*. Toronto: McClelland and Stewart.

Solomon, L. 2008. *The Deniers*. Minneapolis, MN: Richard Vigilante Books.

Wood, J. 2008. "Green shift a throwback to NEP: Wall", *The Regina Leader-Post*, 18 September.

3

Current Federal and Provincial Approaches to Climate Change Mitigation: Are We Repeating Past Mistakes?

Nic Rivers

INTRODUCTION

With the recent prominence of climate change in federal and provincial policy discussions, it is tempting to conclude that the problem is new to both scientists and to policy makers. This is not the case. In fact, the theoretical basis for climate science was being developed more than 100 years ago, and the science had firmly entered the public domain over 40 years ago (Arrhennius 1896; Presidents Scientific Advisory Panel 1965). Among policy makers in developed countries, climate change has been a focal point for 20 years.

Like most other rich countries, Canada has a two-decade history of multilateral engagement on climate change mitigation. In 1988, Toronto hosted the World Conference on the Changing Atmosphere, the first international scientific conference on climate change. During these talks, Canada developed a target of reducing its output of greenhouse gases by 20 percent from 1988 levels by the year 2005. Later that year Canada pushed to have climate change included on the agenda for G7 talks. At these discussions, Canada made another international commitment: this time to stabilize national emissions at 1990 levels by the year 2000. These commitments were echoed in national policy documents in 1990, and reiterated internationally at the 1992 Earth Summit in Rio de Janeiro, when Canada ratified the United Nations Framework Convention on Climate Change. This latter agreement was superseded by the Kyoto Protocol, under which Canada committed to a 6 percent reduction from 1990 emissions levels during the period 2008-2012. Canada signed the Kyoto Protocol in 1997 and ratified it in 2002. More recently, in 2005, Canada hosted the first meeting of the parties to the Kyoto Protocol, the largest international climate change conference since the Kyoto Protocol was first adopted in 1997. In sum, there has been no lack of Canadian international engagement and target-setting on climate change.

Canada's experience on domestic climate policy extends nearly as long as its international engagement. The first domestic policies aimed at climate change mitigation were implemented by the federal government as part of its omnibus Green Plan of 1990. The plan, which involved over 200 environmental policy initiatives and a budget of $3 billion over five years, included $175 million for 24 emission reduction policies, mostly focused on energy-efficiency and alternative energy programs. Independent researchers noted that virtually all policies in the Green Plan emphasized the provision of information to motivate Canadian businesses and consumers to take voluntary actions for environmental improvement (Hoberg and Harrison 1994; Gale 1997). In 1990, the government also initiated the National Action Strategy on Global Warming, with the aim of information sharing among municipalities, provinces, and industry to foster emission reductions.

In 1995, the federal government launched the National Action Program on Climate Change, which included information programs and some modest subsidies; government estimated that it would reduce greenhouse gas emissions by 66 million tonnes by 2010 (Environment Canada 1995).[1] The primary program was the Voluntary Challenge and Registry, under which companies would submit an action plan for greenhouse gas reduction and provide regular progress reports, all on a voluntary basis. The Federal Buildings Initiative was also announced, which proposed retrofitting federal government buildings to higher energy efficiency standards, as was the National Communication Program, which sought to educate Canadians about climate change.

After signing the Kyoto Protocol, the Canadian government launched its *Action Plan 2000 on Climate Change*, a set of initiatives designed to reduce annual domestic emissions of greenhouse gases by 49 million tonnes by 2010 (Government of Canada 2000). This program included limited subsidies for renewable energy supplies as well as a host of energy information programs for consumers and businesses, such as free energy efficiency audits in small business establishments.

Prior to ratifying the Kyoto Protocol in 2002, the Canadian government released the *Climate Change Plan for Canada*, a composite of policies projected to reduce total emissions by 100 million tonnes by 2010 (Government of Canada 2002). A key component, covering about half of the projected emissions reductions, involved a system of negotiated covenants with large point-source emitters to set emission intensity caps for each sector and then allocate tradable permits on this basis. This planned policy, however, was never implemented. Other components in the plan were a mix of information and modest government subsidies to motivate voluntary actions by firms and households. These programs included financial support for public transit coupled with voluntary targets for increased transit use, high efficiency insulation by commercial building developers, and for achieving the target of 10 percent renewables for new electricity generation, and improved vehicle efficiency. Through these programs as well as education programs, the government estimated that each Canadian would reduce their average annual emissions by one tonne.

---

[1]Throughout this document, greenhouse gas emissions are measured in units of carbon dioxide equivalent.

More recently, in 2005, the Canadian government launched *Project Green*, yet another policy initiative to reduce greenhouse gases (Government of Canada 2005). Project Green's key policy proposal – the development of an intensity-based cap-and-trade system for large final emitters – reflected a continuation of the policy development initiated in the 2002 plan. Again, however, this policy was never implemented. Other key policies included in Project Green included the development of a voluntary memorandum of understanding with vehicle manufacturers, subsidies for clean energy and energy efficiency, the development of a "Climate Fund" for purchases of international and domestic emission offsets, and further subsidies for encouraging provincial participation in reducing emissions.

The change in government in 2006 precipitated the most recent federal climate plan – called *Turning the Corner* – which was announced in 2006-2007 (Government of Canada 2008). While the plan has been repeatedly criticized for being weak and ineffective, in practice it closely resembles the previous two federal government plans described above. In particular, the policy aimed at reducing emissions from large industry is very similar to previous plans (as with previous plans, however, this policy has not yet been implemented), and policies aimed at the rest of the economy largely continue the focus on voluntarism and modest subsidies. For example, subsidies on home appliances and home retrofits, as well as subsidies for renewable energy are mostly unchanged from previous plans, and subsidies for public transit are consistent with the approach of previous governments.

In total, Canada's federal government has developed six packages of policies aimed at reducing emissions over the past two decades. While specific policy details within each of the packages have differed, the approach as a whole has not. The policies that have been implemented have almost exclusively been voluntary programs or modest subsidy programs (Jaccard *et al.* 2006). These types of policies have been widely criticized by economists and policy experts as being ineffective for significant greenhouse gas mitigation (Jaccard *et al.* 2006; Takahashi *et al.* 2001; OECD 2003; Khanna 2001). Additionally, policies implemented over the past two decades have covered only small fragments of the economy, leaving emissions from some major sources completely uncovered by any policy, and therefore free to increase.

It should therefore come as no surprise that despite repeated implementation of voluntary and subsidy policies over the past two decades, Canada's emissions have continued to increase. As Figure 1 shows, the most recent estimates of Canada's greenhouse gas emissions show that total emissions are now 25 percent higher than they were in 1990. As a result, Canada has missed by large margins all of the international and domestic commitments it made to reduce emissions over the past two decades. And as indicated on the figure, government forecasts suggest that emissions are expected to continue increasing until at least 2010, in defiance not only of international commitments, but also of increasingly urgent calls from scientists for large and immediate emission reductions, especially from developed countries like Canada.

Figure 1: Canada's Climate Change Commitments, Policies, and Emissions

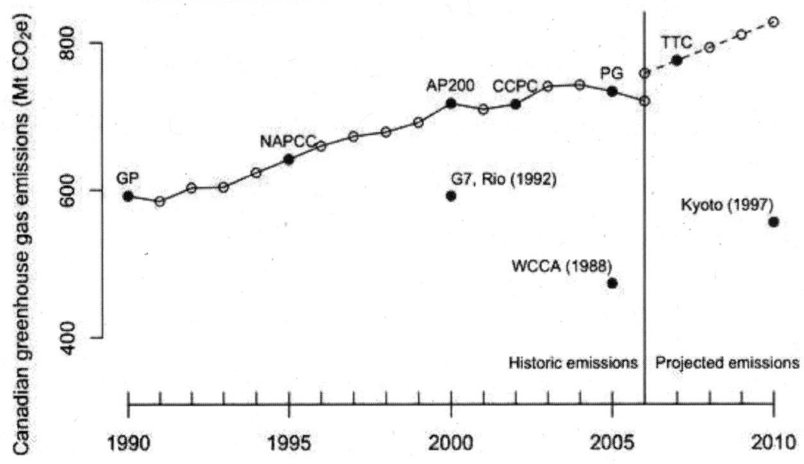

Notes: Historic emissions data from Environment Canada (2008a). Projected future emissions data from Natural Resources Canada (2006). Actual emissions are hollow points connected by solid line, forecast emissions are hollow points connected by broken line, commitments are solitary solid points (with the year commitment was made in brackets), and policies are connected solid points. Dates beside commitments indicate the year in which the commitment was originally made. GP = Green Plan, NAPCC = National Action Program on Climate Change, AP2000 = Action Plan 2000 on Climate Change, CCPC = Climate Change Plan for Canada, PG = Project Green, TTC = Turning the Corner, WCCA = World Conference on the Changing Atmosphere. Note that the Kyoto Protocol, which is shown as a 2010 commitment, actually requires average emissions from 2008 to 2012 to fall to 6 percent below 1990 levels.

Canada's failure to develop effective and efficient policies to reduce emissions over the past two decades is disappointing on its own, but it is made all the more so because it was predictable. The emergence of the Kyoto Protocol led to dozens of studies worldwide, conducted in the late 1990s and early 2000s, of the policies that would be required to ensure compliance with the treaty. Most of these studies express policies in units of dollars per tonne, which represents the stringency of taxation (or other market-based policies) that would be required to achieve a given level of emissions reduction. As shown in Table 1, most of the studies that were conducted suggested that Canada would have required a carbon price of at least \$100/t CO_2e, implemented in the year 2000 and maintained until after 2010, to have a chance at meeting its commitments under the Kyoto Protocol.

Canada, of course, never implemented such a carbon price. And even when carbon pricing was proposed – through the much discussed but never implemented cap-and-trade system for large final emitters – it was at levels far below where the studies in Table 1 suggested that it would need to be. For example, both the *Climate Change Plan for Canada* and *Project Green* (the

Table 1: Estimated Policy Stringency for Kyoto Achievement

Model	*Required Carbon Price for Kyoto Achievement (CDN_{2002}\$/t CO_2e)*
CIMS	\$137/t
MARKAL	\$57/t
BMRT	\$112/t
Rutherford	\$99/t
SGM	\$129/t
GTEM	\$309/t
DRI	\$113/t

Notes: Study results reported in Wigle (2001) and Natural Resources Canada (2000). All figures have been converted to 2002 Canadian dollars using market exchange rates for currency conversions and the Canadian consumer price index to convert to 2002 dollars.

2002 and 2005 climate plans described above, respectively) proposed a cap-and-trade system for large final emitters, which would effectively set a price on carbon emissions. However, both proposed capping the carbon price at \$15/t CO_2e, around ten times below the levels suggested as necessary in the studies outlined in Table 1. Additionally, the proposed cap-and-trade systems only covered industrial emissions, leaving the other half of the economy uncovered by the carbon price. By implementing emission reduction policies so far below what was required for reducing emissions, let alone for meeting the series of international and domestic commitments made by successive governments, Canada assured its failure years ago.

Over the last two years, as climate change has risen in priority in government policy agendas, both the federal and provincial governments have developed a new series of commitments to reduce emissions. As in previous years, they have accompanied those promises with a new suite of policy proposals aimed at meeting the commitments. The aim of this paper is to assess the proposed policies in light of the commitments that have been made.

EVALUATION OF CURRENT PROVINCIAL AND FEDERAL CLIMATE CHANGE PLANS

Ten years ago, the federal government was the primary government level engaged in climate change policy. Since then, rising public pressure and provincial self-interest have induced all levels of government to participate in the development of climate change policy, so that it is no longer the exclusive domain of the federal government. As a result, both provincial governments and the federal government (as well as some municipal governments) have made commitments to reduce emissions over both the medium term (10 to 15 years) and the long term (25 to 50 years). This paper focuses on the medium term – the year 2020 – since policies aimed at significant emission reductions in the year 2020 need to be implemented in the very near future to ensure enough time to impact a significant fraction of the stock of capital that produces emissions. Preliminary judgement of the likely success in meeting mid-term commitments can therefore be made by examining current and proposed climate policies.

Emission Reduction Commitments

Figure 2 shows the medium-term emission reduction commitments that have been made by the provincial and federal governments. Interestingly, all provincial governments and the federal government have made commitments in the last two years to meet a given target level of emissions by the year 2020. The stringency of the targets can be judged in different ways: by comparing the targets to emissions in some historic reference year (the year 1990 is usually chosen), or by comparing the targets to a projection of emissions in the year 2020. While each of these yields the same formal outcome, the comparisons are illuminating.

Figure 2: Provincial and Federal Government Medium-Term Emission Reduction Targets

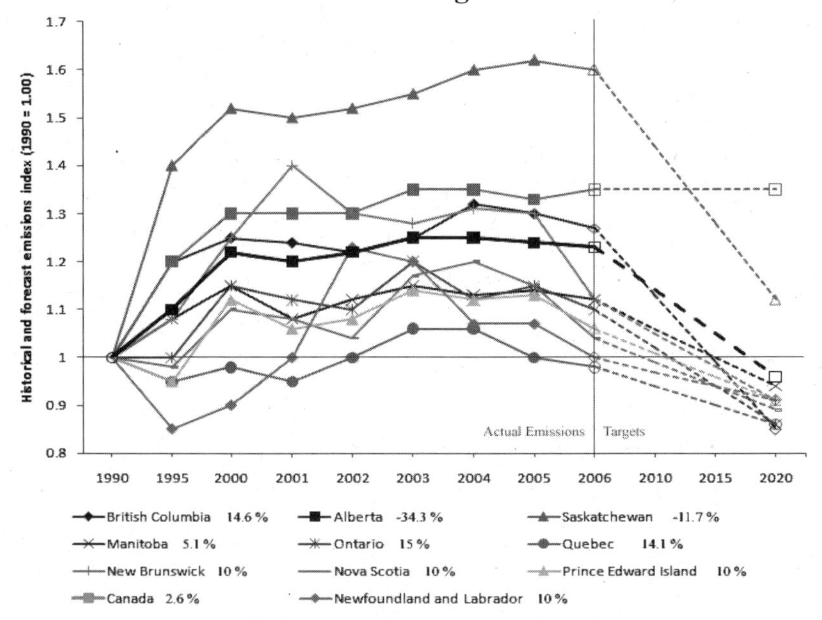

Notes: Historic emissions data from Environment Canada (2008a). Emission reduction targets from Bollinger and Roberts (2008), various provincial climate change plans, and documentation from the *Western Climate Initiative*. Values in text represent 2020 emission reduction targets relative to 1990 levels; negative values indicate that the government has promised to increase emissions relative to 1990 levels.

The year 1990 is the first year in which governments committed to the systematic development of greenhouse gas inventories, and is consequently used as a benchmark year for most international climate change negotiations. Under the Kyoto Protocol, for example, Canada committed to reduce emissions to 6 percent below 1990 levels between 2008 and 2012. As shown in Figure 2, Canada's most recent commitment is to reduce emissions to 2.6 percent below 1990 levels by the year 2020 – somewhat less than mandated under the Kyoto Protocol, and over a significantly longer time frame. This fact has led to widespread criticism of the federal government target as unambitious by environmental groups. Alberta and Saskatchewan have made commitments that entail increases in emissions from 1990 levels, while most other provinces have promised reductions of 5 to 15 percent from 1990 levels by 2020.

While evaluating emission targets based on 1990 emissions is informative, and allows straightforward comparison with commitments made in other regions, it does not convey the magnitude of the challenge imposed by the targets, because it does not factor in secular growth in emissions caused by population growth, economic growth, or structural change. For example, a commitment to reduce emissions by 10 percent from 1990 levels would be more difficult for a fast-growing economy, whose emissions were on track to increase significantly, than for a shrinking economy, whose emissions were shrinking even without policies to constrain them. For this reason, the stringency of targets is best evaluated by comparing targets with a forecast of how emissions would evolve in the future in the absence of policy – a so-called business-as-usual forecast.

Figure 3 shows the most recent federal government business-as-usual forecast for each province and for the country as a whole. The forecast suggests that in the absence of policies specifically aimed at reducing emissions, emissions are likely to maintain their past trajectory of rapid growth in most provinces. Emissions in Alberta and Newfoundland, where oil extraction (as well as the rest of the economy) is poised for fast growth, are expected to grow especially rapidly. For Canada as a whole, emissions are projected to grow by 1.8 percent per year until 2020 (compared to a 1.6 percent annual growth rate from 1990 to 2005). As a result, overall emissions are expected to be almost 60 percent above 1990 levels by 2020.

In this context, the emission reduction commitments shown in Figure 2 appear much more challenging. Figure 4 shows the implied reduction in emissions relative to the business-as-usual forecasts in Figure 3. Alberta, Saskatchewan, Manitoba, and Nova Scotia have effectively promised to reduce emissions by about 30 percent over the coming decade; Ontario, Quebec, New Brunswick, Prince Edward Island, and Canada as a whole have all promised to reduce emissions by around 40 percent; British Columbia has promised to reduce emissions by about 50 percent; and Newfoundland and Labrador has promised to reduce emissions by about 60 percent relative to where they would otherwise be.

These commitments are made even more daunting when one understands the nature of the business-as-usual forecasts in Figure 3. Business-as-usual

Figure 3: Provincial and Federal Government Business-as-Usual Emission Forecasts

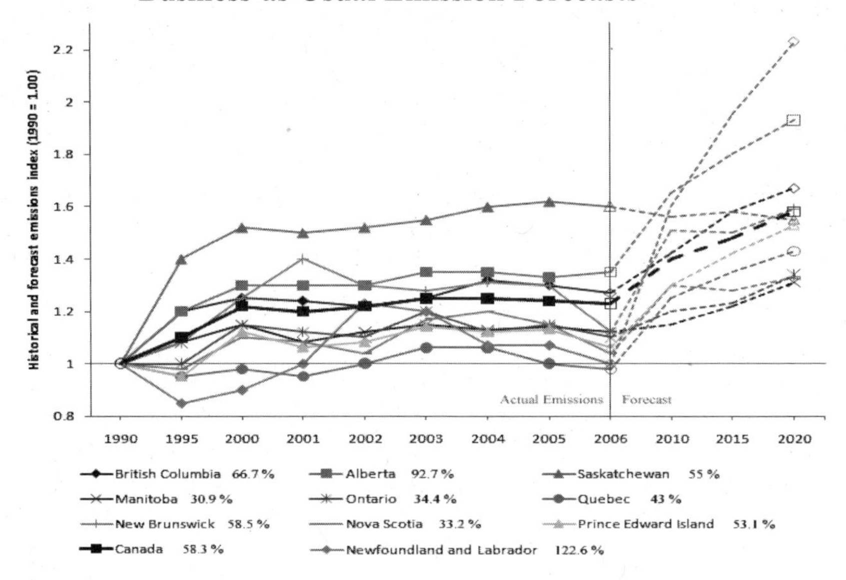

Notes: Historic emissions data from Environment Canada (2008a). Projected future emissions data from Natural Resources Canada (2006). Historic emissions are displayed using solid lines, while forecast emissions are displayed using dashed lines. Values in text indicate projected increase in emissions by 2020 relative to 1990 emissions.

Figure 4: Required Emissions Reductions from Business-as-Usual Scenario to Reach Medium-Term Commitments

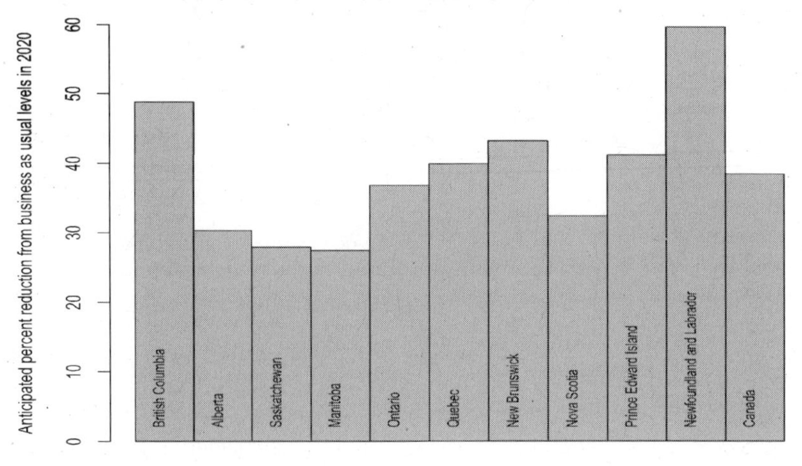

Notes: Author calculations.

forecasts are an estimation of how the economy would evolve in the absence of new policies aimed specifically at reducing emissions. However, they do include measures to reduce emissions that occur naturally as the economy evolves over time. For example, the business-as-usual forecasts already include significant improvements in energy efficiency, which occur naturally over time (Natural Resources Canada (2008) estimates that the average energy efficiency of the Canadian economy has improved by about one percent per year over the past two decades). They also include natural changes in technologies. For example, the forecasts in Figure 3 include the proposed phase-out of coal-fired power generation in Ontario, and the increased integration of Manitoba's and Ontario's electricity sectors. As a result, the required reductions from the business-as-usual scenario shown in Figure 4 have to be on top of these natural changes in emissions, which have already been included in the business-as-usual forecast. When this factor is considered, the emission reduction commitments shown in Figure 4 are especially challenging.

Emission Reduction Policies Required for Meeting Commitment

Table 1 showed estimates made a decade ago of the policy stringency that might be required to meet Canada's Kyoto Protocol obligations, which were thought at the time to represent roughly a 20 to 25 percent reduction from business-as-usual emissions. The failure to meet the Kyoto Protocol targets is directly due to the failure of government to implement policies similar in stringency to those in Table 1.

Similar calculations can be made today, regarding the policies that would be required to meet the commitments that federal and provincial governments have made for emission reductions through 2020. For these calculations, I use the CIMS energy-economy model, which has been used extensively for policy analysis in Canada (Rivers and Jaccard 2006; Bataille *et al.* 2006).

The solid lines in Figure 5 show estimates made using CIMS of the marginal cost of greenhouse gas abatement for Canada as a whole and for the provinces. As with most empirically derived marginal abatement cost curves (and consistent with economic theory), the cost of reducing emissions is low for the first units of emission reductions, but rises as more and more emissions are reduced, such that it becomes very steep for large emission reductions. For emission reductions of 40 percent from business-as-usual levels, the model suggests that a carbon price of roughly \$250/t CO_2e would be required.

The dashed lines in the figure show the commitments made by each of the provinces and by the federal government, taken from Figure 4. The point at which the dashed line representing the commitment crosses the marginal abatement curve is the estimate of policy stringency required for meeting the commitment. These estimates are presented for each region in Figure 6. As described earlier, the marginal cost figures presented here can be thought of as the strength of carbon tax, or the trading value of emission permits in a cap-and-trade system, that would be required to meet the commitments described above.

Figure 5: Marginal Abatement Cost Curves for 2020

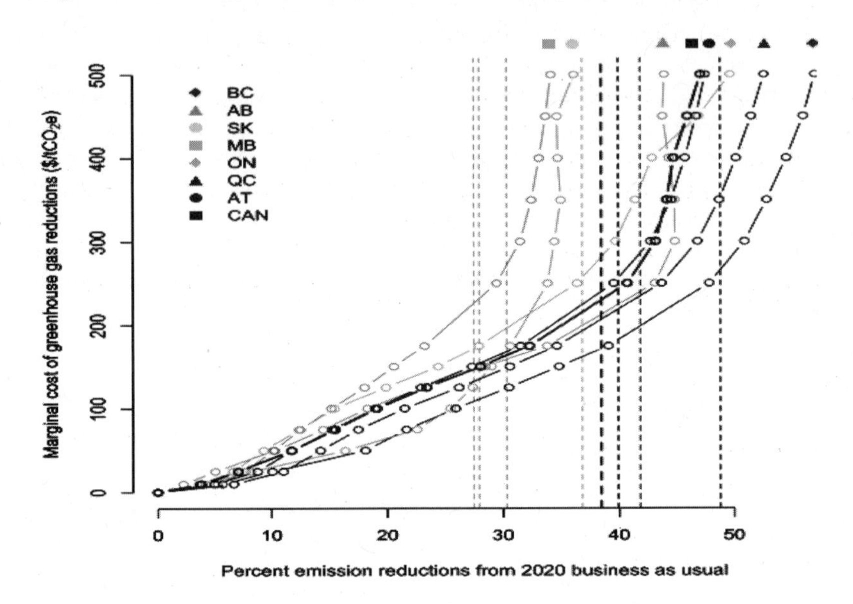

Notes: Author calculations using CIMS model and data described in previous figures. Atlantic provinces (AT) are treated as an aggregate in CIMS. Marginal costs are in 2005 Canadian dollars.

Figure 6: Estimated Marginal Cost of Meeting 2020 Emission Reduction Target

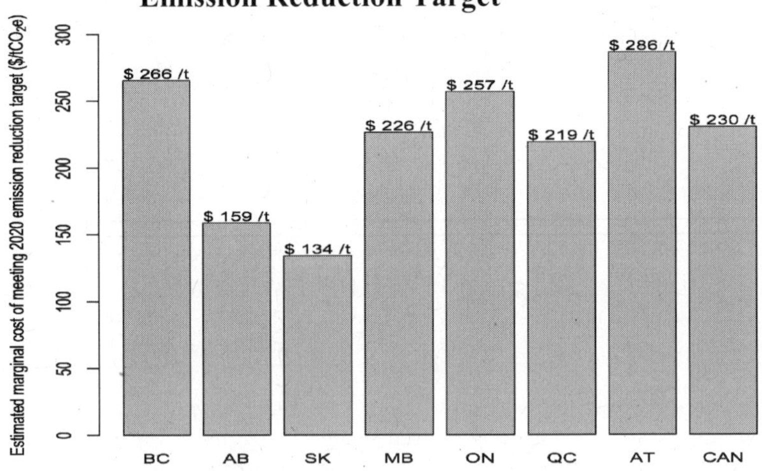

Notes: Author calculations using CIMS model and data described in previous figures. Atlantic provinces (AT) are treated as an aggregate in CIMS. Marginal costs are in 2005 Canadian dollars.

The policy estimates presented in Figure 6 are striking. For Alberta and Saskatchewan, reducing emissions to the level promised would require a carbon price of around \$150/t CO_2e, implemented by 2010 and maintained until 2020. These provinces have made less aggressive promises than the others, and because their emissions are already very high, have many opportunities for reducing emissions at relatively low cost. Still, the modelling suggests that stringent policies would be required to meet targets in these provinces. The modelling conducted here suggests that the other provinces would require even more stringent policies to meet their targets: from \$220 to \$290/t CO_2e. For Canada as a whole, the modelling suggests that a carbon price of \$230/t CO_2e would be sufficient to reduce emissions to the level promised by the current government. Again, to meet the target, the model suggests that such a price would need to be in place by 2010, and maintained through 2020.

It is helpful to provide some context around these prices. A \$100/t CO_2e carbon price would increase the price of gasoline by about 24 ¢/L (30 percent), and the price of natural gas by about \$5.00/GJ (50 percent). The policies estimated here as being necessary to meet provincial and federal commitments would therefore cause significant increases in residential and industrial fuel prices.

The European Union Emission Trading System, currently the largest carbon market in the world, has had prices averaging roughly US\$15-25/t CO_2e over the last three years. Several European countries have imposed carbon taxes of up to US\$50/t CO_2e on certain activities. The carbon price that is estimated here to be required to meet commitments made by the provincial and federal governments would therefore dwarf the carbon pricing policies already adopted in Europe, the current leader in climate change policy.

Emission Reduction Policies Proposed for Meeting Commitment

A decade ago, modelling was available that showed that stringent carbon pricing policies would be required to meet the government's Kyoto commitment in 2010. None of those policies were developed, and the targets were missed. Today, this study and others show that more stringent carbon pricing policies would be required to meet medium-term commitments made by provincial and federal governments. Once again, failure to implement policies consistent with the commitments will result in failure to comply with the commitments. This section describes current and proposed policies in light of the evaluation conducted in the previous section.

Table 2 shows the carbon pricing policies that have been proposed and implemented by the federal and provincial governments over the last two years. Only two jurisdictions – Quebec and British Columbia – have implemented a carbon tax. Quebec's tax of \$3/t CO_2e is applied on fossil fuel distributors, so covers most emissions in the economy. However, the level of the tax is almost 100 times lower than the level estimated in Figure 6 as necessary to meet Quebec's 2020 commitment. British Columbia's recently introduced carbon tax is significantly higher, starting at \$10/t CO_2e in 2008 and increasing incrementally to \$30/t CO_2e in 2012. The British Columbia tax, like its Quebec

Table 2: Current and Proposed Emission Pricing Policies

Region	*Type*	*When*	*Flexibility Mechanisms*	*Coverage*
British Columbia	Carbon tax ($10 → $30 / tCO_2) Absolute cap and trade (WCI)	2008 2012/2015	None Limited offsets	~70% ~50/80%
Alberta	Intensity cap and trade (12% intensity reduction)	2007	Unlimited domestic offsets; unlimited technology fund at $15/t CO_2	~55%
Saskatchewan	None	NA	NA	NA
Manitoba	Absolute cap and trade (WCI)	2012/2015	Limited offsets	~50/80%
Ontario	Absolute cap and trade (WCI)	2012/2015	Limited offsets	~50/80%
Quebec	Carbon tax ($3 / tCO_2) Absolute cap and trade (WCI)	2007 2012/2015	None Limited offsets	~70% ~50/80%
New Brunswick	None	NA	NA	NA
Nova Scotia	None	NA	NA	NA
Prince Edward Isl.	None	NA	NA	NA
Newfoundland	None	NA	NA	NA
Canada	Intensity cap and trade (18% intensity reduction by 2010 + 2%/yr)	2010	Limited technology fund at ~ $20/t; unlimited domestic offsets	~50%

Notes: Various provincial and federal climate change plans. Pembina Institute (2007).

counterpart, is applied on fossil fuel distributors, so will cover about three-quarters of the province's total emissions. However, despite the fact that the tax is much higher than the Quebec tax, even by 2012 it will still be ten times below the level suggested by the modelling in this study as necessary to meet the province's medium-term commitment.

Other provinces, as well as the federal government, have proposed or implemented cap-and-trade systems. Like a carbon tax, a cap-and-trade system puts a price on greenhouse gas emissions (albeit indirectly). Alberta is the only province to have actually implemented a cap-and-trade system (others are still under development). Its system, which came into force in 2007, affects all firms with emissions that exceed a given threshold – roughly 55 percent of total emissions in the province. While in theory the emission price under a cap-and-trade system is determined by the market for emission credits, Alberta's cap and trade system imposes a maximum carbon price of \$15/t CO_2e on emission permits. From Figure 6, this is roughly ten times below the level estimated as necessary for achieving Alberta's targets.

Canada's proposed cap-and-trade system – due to enter into force in 2010 – is similar in many ways to Alberta's. It requires that large firms, accounting for about half of the total emissions in the country, acquire tradable permits for each unit of emissions. Like Alberta's system, the federal system initially caps the maximum price of emissions at roughly \$20/t CO_2e, again ten times below the price suggested in Figure 6 as consistent with government targets. However, price caps are slowly increased over time, and gradually phased out, so that the policy should become more stringent close to 2020. The government suggests that the price of permits might rise to \$65/t CO_2e by then, much higher but still far below what the modelling suggests is necessary for meeting mid-term goals (Environment Canada 2008b). Independent modelling suggests that other flexibility mechanisms included in the policy, notably the domestic offset provision, could keep the price much lower (Jaccard, Rivers, and Peters 2008). Either way, the policy will only cover one-half of all emissions in the economy, and will impose a carbon price much lower than required to meet government targets.

Of course, carbon pricing policies – taxes and cap-and-trade systems – are not the only type of greenhouse gas mitigation policy available. Governments can also use traditional regulatory instruments, as well as subsidies, voluntary measures, or other evolving types of policy instruments like feebates or product standards.[2] A full assessment of government climate change plans should include an evaluation of these instruments. However, it is unlikely that inclusion of these instruments would significantly alter the conclusions reached here – namely that, at present, government plans are wholly inadequate for meeting current government emission reduction commitments.

---

[2]Feebates combine a fee on products with an undesirable quality (e.g., high emissions, high vehicle weight, low energy efficiency) with a rebate on products with a desirable quality (e.g., low emissions, low vehicle weight, high energy efficiency). Product standards mandate specific qualities in products produced by a given sector (e.g., electricity without emissions).

As stated in the introduction and in other studies, voluntary measures and subsidies have proven to be ineffective in the past for significant emission reductions. So while current government plans do contain these elements, it is unlikely that they will produce significant emission reductions. Current government plans also feature traditional regulatory instruments, like energy efficiency standards for appliances, building codes, and process regulations for industrial facilities. Many of these, however, do not require technology choice or behaviour change significantly different than current or projected best practice, suggesting that they are unlikely to spur the innovation of new technologies, and that they are unlikely to significantly reduce emissions. Finally, evolving policy instruments, like feebates and product standards, while promising, have not been embraced by governments in current climate plans, and so are unlikely to play a major role in reducing emissions by 2020. In sum, while the focus on carbon pricing policies is certainly not a complete evaluation of current plans, it is indicative of the stringency of the overall policy approach.

CONCLUSION

The development of climate policy in Canada has significantly changed over the past decade. While the federal government was the main actor then, now all provinces are engaged in climate change mitigation, both in terms of making commitments to reduce emissions and in terms of developing climate policies to comply with those commitments. It is now unclear which level of government will assume the greatest responsibility for climate change mitigation in the future, but it is clear that provincial governments are unlikely to be sidelined as in the past.

A second key change is the renewed recognition by governments and the public that deep cuts in emissions will require implementation of compulsory policies, and especially emission pricing policies like taxes or cap-and-trade systems. Three provincial governments have already implemented emission pricing policies, and several of the rest, as well as the federal government, have proposed the implementation of carbon pricing policies in the coming years. This new understanding of the types of policies that are likely to be effective for mitigation of emissions should make current policies much more successful than those of a decade ago, which focused overwhelmingly on ineffective voluntary initiatives and modest subsidies.

Some things, however, have not changed. As in previous years, the evaluation in this paper suggests that current government policies at both the federal and provincial levels are much less stringent than will be required to meet the commitments made by those same governments. The analysis here suggests that medium-term commitments made by governments in Canada are likely to require dramatic departures from projected trends: 30 to 60 percent reduction in emissions from projected levels. With only about a decade for this change to take place, very stringent policies are required nearly immediately to ensure compliance. Current government policies are likely to fall well short of the level required to ensure compliance, such that the commitments for the year

2020 are extremely unlikely to be met, in most if not all provinces, and for the country as a whole.

It is certainly valid to debate whether such dramatic targets should be met. If the analysis conducted here is accurate, meeting the targets will require especially stringent policies because of the short time available, and could impose relatively high costs on the Canadian economy. Environmental policies should only be pursued to the point where the cost of the policy matches the benefit (in terms of improvement in environmental outcome) due to policy implementation. While this study does not consider the benefits of emissions abatement, it is possible that they would not be as large as the costs of such policies, and that delay of a few years would allow the same emission reduction to be reached at a lower price, consistent with the benefits of reducing emissions. Further study on the benefits of emission reductions would help to make this clear, since current evaluations are inconclusive (Tol 2005; Weitzman 2008). However, whatever the benefits of emissions abatement, it is almost certain that they are large enough to warrant application of an emission pricing policy that begins today at a modest level, and rises over time to a more substantial level.

What is particularly unfortunate about the current phase of government plans (as well as the last phase of a decade ago) is that they do not facilitate such a debate on whether such targets are appropriate. By maintaining a focus on emission reduction targets, they divert attention from policies, which are what ultimately generate reduction in emissions. And without focus on policies, it is likely – as shown here – that the policies will be inadequate to meet the targets proposed by government.

References

Arrhennius, S. 1896. "On the Influence of Carbonic Acid in the Air on the Temperature of the Ground", *Philosophical Magazine* 41: 237-276.

Bataille, C., M. Jaccard, J. Nyboer, and N. Rivers. 2006. "Towards General Equilibrium in a Technology-Rich Model with Empirically Estimated Behavioral Parameters", *Energy Journal* 27: 93-112.

Bollinger, J. and K. Roberts. 2008. *Building on our Strengths*. Calgary: Canada West Foundation.

Environment Canada. 1995. *National Action Program on Climate Change*. Ottawa: Government of Canada.

— 2008a. *Canada's Greenhouse Gas Inventory: 1990-2006*. Ottawa.

— 2008b. *Turning the Corner: Detailed Emissions and Economic Modelling* [Homepage of Government of Canada], 25 November. At http://www.ec.gc.ca/doc/virage-corner/ 2008-03/571/tdm_toc_eng.htm.

Gale, R. 1997. *Canada's Green Plan*. Berlin: Springer-Verlag.

Government of Canada. 2000. *Action Plan 2000 on Climate Change*. Ottawa: Government of Canada.

— 2002. *Climate Change Plan for Canada*. Ottawa: Government of Canada.

— 2005. *Project Green: Moving Forward on Climate Change – A Plan for Honouring our Kyoto Commitment*. Ottawa: Government of Canada.

— 2008. *Turning the Corner: Taking Action to Fight Climate Change*. Ottawa: Government of Canada.

Hoberg, G. and K. Harrison. 1994. "It's Not Easy Being Green: The Politics of Canada's Green Plan", *Canadian Public Policy* 20(2): 119-137.

Jaccard, M., N. Rivers, C. Bataille, R. Murphy, J. Nyboer, and B. Sadownik. 2006. *Burning Our Money to Warm the Planet: Canada's Ineffective Efforts to Reduce Greenhouse Gas Emissions*. Toronto: C.D. Howe Institute.

Jaccard, M., N. Rivers, and J. Peters. 2008. *Assessing Canada's 2008 Climate Policy* [Homepage of Simon Fraser University]. Available at http://www.sfu.ca/pamr/files/fall2008/PDF/AssessmentofCanadasClimatePolicySep26-08.pdf.

Khanna, M. 2001. "Non-Mandatory Approaches to Environmental Protection", *Journal of Economic Surveys* 15(3).

Natural Resources Canada. 2000. *An Assessment of the Economic and Environmental Implications for Canada of the Kyoto Protocol*. Ottawa: Government of Canada.

— 2006. *Canada's Energy Outlook: The Reference Case 2006*. Ottawa: Government of Canada.

— 2008. *Energy Efficiency Trends in Canada*. Ottawa: Government of Canada.

Organisation for Economic Co-operation and Development (OECD). 2003. *Voluntary Approaches for Environmental Policy: Effectiveness, Efficiency, and Usage in the Policy Mix*. Paris: OECD.

Pembina Institute. 2007. *Highlights of Provincial Greenhouse Gas Reduction Plans*. Pembina Institute. Drayton Valley, Alberta.

Presidents Scientific Advisory Panel. 1965. *Restoring the Quality of Our Environment*. White House. Washington, D.C.

Rivers, N. and M. Jaccard. 2006. "Useful Models for Simulating Policies to Induce Technological Change", *Energy Policy* 34(15): 2038-2047.

Takahashi, T., M. Nakamura, G.C.v. Kooten, and I. Vertinsky. 2001. "Rising to the Kyoto Challenge: Is the Response of Canadian Industry Adequate?" *Journal of Environmental Management* 63(2): 149-161.

Tol, R.S.J. 2005. "The Marginal Damage Costs of Carbon Dioxide Emissions: An Assessment of the Uncertainties", *Energy Policy* 33(16): 2064-2074.

Weitzman, M.L. 2008. *On Modeling and Interpreting the Economics of Catastrophic Climate Change*.

Wigle, R. 2001. *Sectoral Impacts of Kyoto Compliance*. Ottawa: Industry Canada.

4

Carbon Pricing and the Technology Imperative

Christopher Green

CLIMATE CHANGE IS AN ENERGY TECHNOLOGY PROBLEM

My approach to climate policy is informed by my understanding that climate change is essentially an energy-technology problem. That is not the "conventional wisdom", but reflects more than 15 years of research on the issue. It is also a problem that will not be easy to solve. Pricing carbon, however desirable, is not sufficient to stabilize climate (that is, stabilize the atmospheric concentration of greenhouse gases) without new, scalable, and breakthrough technologies. In *global* terms, pricing carbon without also *directly* addressing the energy-technology challenge is a bankrupt strategy. Yet it is carbon pricing on which most economists and recently converted environmental advocates dwell.

In assessing the value of carbon pricing, we need to consider the size of the energy technology challenge to climate stabilization. Unfortunately, economists are prone to assuming that getting carbon prices right will induce the requisite technology change. But this is a reliable assumption only where the requisite technologies are already "on-the shelf". Where the technology challenge is large, with many of the required technologies neither ready nor scalable and/or requiring basic research and development, the assumption is not reliable. Why?

Where technologies (i) take many years, even decades, to develop, (ii) are uncertain of success, and (iii) if successful are often characterized by benefits that are far from being fully appropriable, more than the market is needed to convince entrepreneurs to make large upfront investments in R&D many years in advance of any possible return. If current governments cannot commit (distant) future governments to cover anything more than the cost of production of successful technologies, then we have a *time inconsistency* that renders it highly unlikely that the private sector will be willing to make the required *upfront* investments in R&D (Montgomery and Smith 2007).

Unfortunately, the task of assessing the energy-technology challenge has been badly fumbled by the Intergovernmental Panel on Climate Change (IPCC).

IPCC Working Group (WG) III has grossly *understated* the magnitude of the technology challenge. It has done so by (i) *overstating* the potential for global energy intensity decline (Baksi and Green 2007); (ii) *disregarding* the many limits to renewable energy supply scalability (Hoffert *et al.* 1998, 2002; Caldeira, Jain, and Hoffert 2003); and (iii) *overlooking* the need for "enabling" technologies, that is technologies such as utility-level storage for intermittent wind and solar energies, that are required to make possible the scale-up of carbon-emission-free energy-supply technologies (Green, Baksi, and Dilmaghani 2007). Worst of all the IPCC has all but ignored the huge amount of energy technology improvement built into the baselines it uses to assess the climate stabilization challenge (Pielke, Wigley, and Green 2008). Among other things, these missteps have led to potentially large *understatements* of the *cost* of stabilizing climate. In the absence of an energy-technology revolution – a revolution that cannot be assumed but must be addressed directly – the cost will be many times higher than the 1-3 percent of global GDP that has been widely reported. Unfortunately, these issues are far from the minds of most economists, or any discussion that I have seen of Canadian climate policy.

Of course, just as climate change is a global problem that may invite free riding, the same is true of R&D expenditures. In practice, Canada does not need to contribute to the research and development of new energy technologies as long as countries such as the United States, Japan, China, India, and Germany do the work. Just as Canada has resided comfortably under the U.S. defense umbrella, it can live under the technology-development "umbrella" of the United States and other countries that do serious energy-technology research. But whether or not Canada participates in the development of new technologies (I believe it should), without their development and transferability, climate cannot be stabilized. In this case, Canadian carbon pricing would make a meaningless contribution so far as climate change is concerned.

Why are new energy technologies crucial and why is carbon pricing alone insufficient to induce their development? These are big questions. Before attempting to answer them, I should note that what I have to say is not what the world has been told. The IPCC (Metz *et al.* 2001, 2007) has repeatedly claimed that the required technologies "exist", and are either ready or awaiting commercialization. In 2001, the IPCC claimed that the barriers to stabilization are socio-economic and institutional, not technological (Metz *et al.* 2001: 8). This view has been slightly modified in the IPCC's Fourth Assessment Report where heavy emphasis is placed on induced technological change to assure the development as well as adoption of the required carbon-emission-free technologies. The claims of the IPCC are widely accepted. This is as true in Canada as it is in Europe, and in the economics profession as it is among born-again environmentalists. They are in error.

THE MAGNITUDE OF THE CHALLENGE

Hoffert *et al.* (1998) was the first serious attempt to measure the magnitude of the technology challenge to stabilizing the atmospheric concentration of carbon.

They estimated that stabilizing the atmospheric concentration of carbon dioxide at twice its pre-industrial level would require 10-30 TW of carbon-free energy by 2050, and 30-40 TW by 2100. (A TW is 31.56 Exajoules of energy per year.) To put these numbers in perspective, the world currently consumes 16+ TW, only 2.0+ are carbon-free, and over 90 percent of that is from hydro and conventional, open cycle, nuclear generated electricity. Hydro sites are limited, while "conventional" nuclear faces rising costs, limited supplies of U-235 and problems of waste disposal. The latter two problems could be substantially reduced by closing the fuel cycle through reprocessing and/or breeding of nuclear fuels. To do so requires technological breakthroughs, including a "spiking" of the large amounts of plutonium associated with a closed-fuel cycle. Other currently available carbon-free energies are not yet scalable: making them scalable requires technological breakthroughs.

Yet, the IPCC views the technology challenge very differently. One reason is the emission scenario *baselines* the IPCC uses in its analyses. Built into the IPCC emission scenarios are huge emission reductions attributable to technological change. These emission reductions are usually ignored in analyses that use emission scenarios as baselines for calculating or estimating what it will take in technology terms and cost to stabilize climate. The IPCC's own emission scenarios (IPCC 2000) illustrate the point.

Figure 1, which is drawn from the most recent report of IPCC WG III (Metz *et al.* 2007), portrays four of the IPCC emission scenarios (A2, A1F1, B2 and A1B. For present purposes, only the scenarios B2 and A1B are discussed.). Consider each of the sets of lines in the lower panels of Figure 1. The emission scenarios are indicated by the B2 and A1B lines. The climate stabilization paths are the B2 550 and A1B 550 lines. The upper lines are emissions that would occur *if there were no technological change* or *improvement in the efficiency of existing energy technologies*. (These upper lines are constructed *as if* technology remained the same or were "frozen" over the course of the 21st century.)

Now consider the "gaps" between the lines. As is readily observable, the *gap* between the B2 and A1B lines on the one hand and the two 550 climate-stabilization lines (paths) on the other, is relatively small (although substantial in technology-change terms). This is the stabilization *gap* on which climate policy has focused. In contrast, the *gap* between the upper ("frozen" technology) lines and the B2 and A1B emission scenarios is much larger (shown as the gray area). Most of this gap represents emission reductions attributable to technology change that is assumed to occur "spontaneously" – that is without policy intervention. Clearly, a lot of *spontaneous* technology change is embedded in the emission scenarios, B2 and A1B – and the other IPCC emission scenarios too.

Of course technology will not remain unchanged ("frozen"). It is nonetheless crucial to take account of the technology already built-into the emission scenarios. Why? The answer is because using emission scenarios as baselines *understates the full technology challenge and, by extension, the economic cost of achieving stabilization.* The IPCC, by constructing emission scenarios with large amounts of built-in technological change, made stabilization appear much easier and less expensive than it will be (Green and

Figure 1: Impact of Technology on Global Carbon Emissions in Reference and Climate Mitigation Scenarios

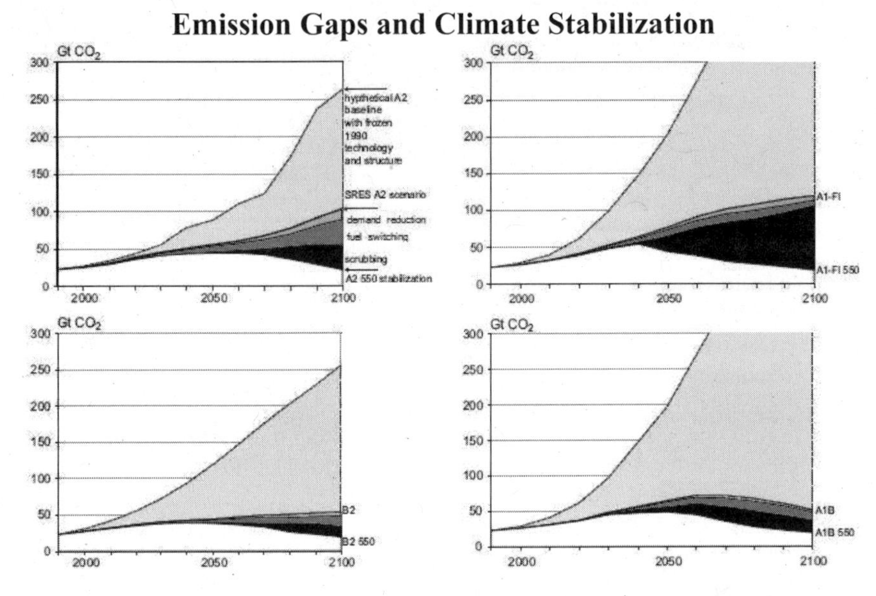

Note: Global carbon emissions (GtC) in four scenarios developed within the IPCC SRES and TAR (A2, B2 top and bottom of left panel; A1Fl and A1B top and bottom of right panel). The light-gray shaded area indicates the difference in emissions between the original no-climate policy reference scenario compared with a hypothetical scenario assuming frozen 1990 energy efficiency and technology, illustrating the impact of technological change incorporated already into the reference scenario. Darker-gray shaded areas show the impact of various additional technology option deployed in imposing a 550 ppmv CO_2 stabilization constraint on the respective reference scenario, including energy conservation (darker-gray), substitution of high-carbon by low- or zero-carbon technologies (darkest-gray), as well as carbon capture and sequestration (black). Of particular interest are the two A1 scenarios shown on the right-hand side of the panel that share identical (low) population and (high) economic growth assumptions, thus making differences in technology assumptions more directly comparable.

Source: Adapted from Nakicenovic *et al.* (2000), Riahi and Roehrl (2001), and Edmonds (2004).

Lightfoot 2002). Further, failing to account explicitly for spontaneous technology changes increases the likelihood of "double counting" the contribution of specific technology changes, once between the frozen technology curve and the emission scenario *baseline*, indicated by the gray area, and again between the *baseline* and the stabilization path. The potential for "double-counting" technologies is overlooked by most analysts.

We can view the issue another way. Recently, Pielke, Wigley, and Green (2008) calculated the magnitude of the "spontaneous" carbon-emission reductions built into the IPCC SRES scenarios, and in other scenarios used in the most recent report of IPCC WG III. These calculations are shown in Figure 2. The total vertical bar for each scenario (the sum of the medium, dark and light portions) represents the cumulative (2000-2100) emissions under each of the scenarios. (As an aside, while it is clear that these various scenarios can differ markedly in their assumed level of total emissions, this is not the issue at stake in the context of Figure 2.) The medium gray (lower) portion of each bar represents the emissions that are consistent with the stabilization of the atmospheric concentrations at 500 ppm. (The pre-industrial concentration of CO2 was 275 ppm. The current atmospheric concentration of CO2 is about 387 ppm.) The light gray (top) portion of each scenario represents the *reduction in CO_2 emissions* that the scenario creators have built into their simulations. These "spontaneous" reductions come primarily from the technological assumptions built into the scenarios. This leaves the dark gray (middle) portion of each bar as the remaining emissions that must be reduced by climate-change policy in order to bring total emissions down to levels consistent with climate stabilization (i.e., 500 ppm).

Figure 2: Cumulative Emission, Emissions Reductions and Technological Change in Emission Reduction Scenarios

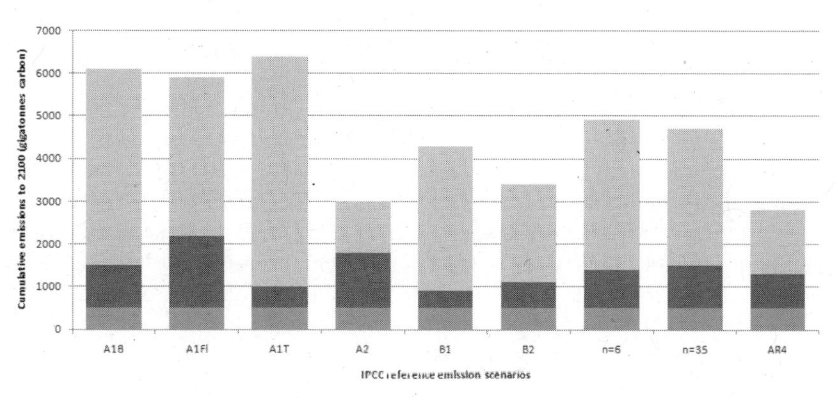

Note : A range of "built-in" emissions reductions (light gray) in the scenarios used by the Intergovernmental Panel on Climate Change (IPCC). Total cumulative emissions to 2100 associated with a frozen-technology baseline are shown for: six individual scenarios, the means of these scenarios (n=6), and for all 35 IPCC scenarios, and the median of the scenario set (AR4). Additional reductions will have to be achieved by climate policy (dark gray), assuming carbon-dioxide stabilization at about 500 part per million (p.p.m.), leaving allowed emissions for this stabilization target (medium gray).
Source: Pielke, Wigley, and Green 2008, Figure 1.

What is very clear from these scenarios is that the policy-generated reductions in CO_2 (the dark gray component) is only a part of the overall required reduction, and typically a much smaller part than the light gray component linked mainly to built-in or spontaneous technological change. Most analysts who address the economic costs of and technological requirements for stabilizing climate seem unaware of the distinction. Their focus is on the dark gray part of the bars. But Pielke, Wigley, and Green (2008) argue that for at least three reasons the upper or light gray portions cannot be ignored. First, there is no assurance that the technological changes that occur from year to year will produce anything like the large emission reductions built-into the Figure 2 scenarios. Second, the technologies that reduce emissions represented by the dark gray may include some of the technological changes (usually unspecified) that are assumed to produce the spontaneous emission reductions. If so, there would then be "double counting". Third, to overlook what is built into the scenarios is likely to *understate* substantially the economic cost of stabilizing climate.

That the issues raised by Pielke, Wigley, and Green (2008) are not simply "theoretical" is clear from developments in the first half of the present decade. Not only have emission rates soared from an annual average rate of 1.2 percent in the 1990s to a 3.1 percent rate 2000-2006, the assumed reductions in the carbon intensity of energy and the energy intensity of output that are common to virtually all emission scenarios did not materialize in the 2000-2005 period (Pielke, Wigley, and Green 2008). In fact, the *global* carbon intensity of energy rose, and so did the energy intensity of output when world GDP is calculated using market exchange rates. The result is that the global carbon intensity of GDP rose too.

The main explanation for the reversals resides in momentous changes now taking place in the developing world. The rise in the carbon intensity of energy and in energy intensity are connected to, and probably account for, much of the tripling in the annual rate of change in *global* emissions from 1.2 percent per year in the 1990s to 3.1 percent per year in 2001-2007. At the heart of these emission-growth-rate changes is the development success story coming out of Asia. That story is increasingly associated with a huge shift in the location and relative importance of energy-intensive industries, ones which rely heavily on power generated from combusting coal.

The best example is China which, in 2006, accounted for 48 percent of the *world's* production of cement, 49 percent of the world's production of flat glass, 35 percent of its steel, and 28 percent of its aluminum (Rosen and Houser 2007). The list could go on, but it suffices to say that these are among the world's most energy-intensive materials and the industries that produce them, with energy-to-output ratios ("energy intensities") about 10 times higher than those of most other manufacturing industries. Much of the energy comes from coal-fired power. The important point is that as development proceeds, rural populations move to cities, and to an increasing extent into high-rise buildings on broad streets traversed by overpasses and underpasses, and all supported by modern city (and intercity) infrastructure. All of these consume very energy-intensive materials such as steel, cement, flat glass and copper. This is a process that is likely to continue for decades, not only in China, but all over populous

Southeast and South Asia, the Middle East, and eventually Africa, until well beyond the middle of the century.

As a result, we have only begun to see the surge in global energy use that the transformational development process now implies. And with that development process and energy surge will come a GHG emissions surge that will only terminate with a transformation of the world's energy systems. Not only will that transformation be a slow process, but the required energy technologies, for the most part, are neither ready nor *scalable*. And when they are ready and scalable, it will likely require a huge technology transfer to the developing world before there will be a substantial payoff in CO_2 emissions reductions. If the climate policies of the developed world (including Canada) ignore the underlying causes of emission growth in the developing world, and the implications these have for energy technology development and transfer, then most of any progress made by developed nations to reduce their emissions will go for naught when measured in global terms.

THE ROLE OF CARBON PRICING

In my view, Canadian climate policy went off the tracks when we adopted emission-reduction targets. That, of course, was largely dictated by our support for, and eventual ratification of, the Kyoto Protocol (KP). Even though the Canadian government has indicated that it will no longer be bound by Kyoto, it, like the other political parties, still feel bound to adopt targets that it says its policies will achieve. But commitments to targets are rarely "credible" (Schelling 1992); they are even less credible when calculated from current or past benchmarks rather than from what future emissions would be in the absence of emission-reduction policies. Moreover, basing emission-reduction mandates on domestically-produced emissions places a heavier burden on countries such as China and Canada whose exports are much more energy intensive than the energy intensity of their imports. Not only that, emission-reduction mandates have failed altogether to account for emissions from international air and ocean travel and shipping.

Commitments to emission-reduction targets have created a bias in favour of policies that quantitatively limit emissions ("cap and trade") as opposed to price-based policies (carbon taxes). But "cap and trade" is especially problematic. Even an upstream cap-and-trade system will be an administrative nightmare. Once cognizance is taken of the need for "escape valves" (if abatement costs and carbon prices rise too sharply), "cap and trade" becomes even more complicated. "Escape valves" are likely to take the form of "offsets" and "credits" that are almost impossible to monitor properly. Because there is no end to hucksters, or to the resources needed to police the system, there is an important incentive for fraud.

Further, I would submit that there is something inconsistent about using cap and trade where technological change is crucially important. The arrival of new, scalable technologies is inherently uncertain, so that using cap and trade to meet *date-specific* emission-reduction targets is virtually certain to produce *ad hoc* decisions. For example, pressure to meet emission targets may lead to hasty adoption of inferior technologies ("first generation" biomass in the form of corn for ethanol is an example) or to *temporary* means of reducing emissions simply to meet the target. It is also questionable how far support for cap and trade will go when it dawns on the public that the only sure beneficiaries are financial markets that broker the trades and are able to capitalize on the inherent price volatility of a quantity-based (fixed supply) approach to carbon pricing. A further nightmare occurs if speculators are able to engage in temporary price manipulation (and take-and-run profits) by buying up a significant share of permits before dumping them. Real world rather than textbook cap and trade neither assures price nor volume certainty, and this fact almost certainly implies important economic inefficiencies. Not much to like here!

Carbon taxes, while administratively far superior to "cap and trade", nevertheless have severe deficiencies if treated as the main means of inducing large reductions in carbon emissions. Carbon taxes that start *high* and/or rapidly rise are likely to be both politically and economically toxic, particularly where energy-intensive industries are concerned. Experience with carbon taxes in Europe suggests that it is necessary to introduce much lower rates of tax, or exemptions for, energy-intensive industries (Metcalf 2009). A major concern of either carbon-pricing instrument (taxes or tradable permits) is the fear, and probability, of "carbon leakage" if implemented unilaterally. Substantial "leakage" would contribute to rising emissions elsewhere in the world even though emissions are reduced at home.

Economists who extol carbon pricing usually add a *fine-print* caveat that the carbon price should be universal or "harmonized". It is not difficult to finger a basic reason why high and/or rapidly rising carbon prices (whether the product of carbon taxes or "cap and trade") will not be harmonized. Many leading emitters are heavily dependent on coal. A tax of just $10 per tonne CO_2 implies a $28.60 tax per tonne of coal. (There are 2.86 tonnes of CO_2 in a tonne of coal.) Coal sells at anywhere from $15 to $110 per tonne. Proposing, as Canada's National Roundtable on the Environment and Economy (NRTEE) has recently done (NRTEE 2009), an emission-reduction plan in which carbon prices soar to $115/t$CO_2$ ($329 per tonne of coal) by 2020 is, in my view, a non-starter. And Canada is much less dependent on coal than the United States, China, India, Russia, Australia, and Poland. Any illusions the NRTEE may have entertained of "harmonization" with the United States died with the gutting of the Waxman-Markey climate change bill by coal state Democrats. (Incidentally, back-of-the envelope calculations suggest the 2020 emission-reduction target in the *original* Waxman-Markey bill was unachievable and trying to achieve it could have cost the United States several trillion dollars in lost GDP over the course of the decade 2010-2020.) Needless to say, a high and/or rapidly rising carbon price invites repeal or reversal at the next election.

The economic, administrative, and political limitations of climate policies reliant on carbon pricing are perhaps a silver lining. The effectiveness of "cap and trade" and carbon taxes as stand-alone policies has been vastly oversold. Given the magnitude of the technology challenge to climate stabilization, a realistic climate policy for Canada and the world is one that focuses on technology and energy-infrastructure development. In the initial stages the role of carbon pricing should be *ancillary*, providing for secure and on-going funding of *basic* R&D, the testing of new technologies, and contributing to the construction of the required or supporting infrastructure. A low carbon price, say $5, or at most $10, per ton CO2, is all that is needed. However, it is important that *commitments* are made to gradually raise the carbon tax/fee (say, at a rate that implies a doubling of carbon prices every 10 to 15 years). As the carbon tax/charge/fee slowly but inexorably rises it would send a *"forward price signal"* to deploy effective, scalable technologies as they "reach the shelf". (A variant of the carbon tax/fee on all carbon emissions is the plan adopted by Alberta which sets a tax (or fee) on marginal emissions only. In principle, the Alberta approach would allow a higher and more rapidly rising tax/fee. Rick Hyndman's paper speaks to these issues.)

An initially low, gradually rising tax (fee) built around an R&D/technology policy has a far better chance of being widely adopted – and effective – than the emission-reduction-mandate approach. Large developing-country emitters (such as China and India) understandably refuse to take on emission-reduction commitments because of an energy-intensive development trajectory (see above). However, as long as there are no emission caps, these countries may find it in their interest to enter into a technology race financed by a low, very slowly rising charge on carbon emissions. Some developed countries may willingly, or under pressure, commit to emission-reduction targets. However, especially in countries where energy-intensive industries are important, the commitments are not credible – unless the requisite carbon-emission-free technologies are ready and rapidly deployable. That is currently not the case!

While economists tend to fixate on (carbon) prices, there is also, in my view, an important role for standards. These may take the role of efficiency standards for appliances, houses and, yes, even motor vehicles. New houses could be required to be built with in-ground piping so that their space conditioning can be at least partly via ground-based (geo) heat pumps. Much can be done to gradually improve energy efficiency and provide services from naturally produced sources if we would simply focus on a variety of actions, some market-based, some not, rather than on the drum beat of targets and mandates.

CONCLUSION

Carbon pricing is not the be all and end all of the economics of climate change. Carbon pricing cannot be a stand-alone climate policy. Canada should not be a "price-taker" so far as global climate policy is concerned. In framing a sensible policy we need to think of what it will take to reduce *global* emissions – not just ours. In this respect, a rethinking of *global* climate policy is long overdue. A rethinking should include: (a) recognition that R&D and adaptation, as well as mitigation, are components of a climate policy, and (b) a willingness to reconsider the time-related mix of these policies.

In my view, the most important climate policy decision facing Canada is our stance (and, of course, that of the United States too) at Copenhagen 2009. There is likely to be great pressure, especially from the European Union, to gain broader acceptance of renewed, Kyoto-type emission-reduction mandates, and ones with a shorter time horizon (say 2012-2020) than Kyoto (1997-2012). If this happens, another decade will be wasted with futile exhortations, the blame game, and reneging on non-credible commitments. Canada, and the United States too, needs to avoid this trap, and instead offer a different vision and plan of attack.

If climate change is essentially an energy-technology problem, then I submit that carbon pricing is only part of the story – and in the early stages its role is largely *ancillary*. In a revamped climate policy, there is no place for emission targets and no need for cap and trade. What is needed is the widespread adoption of a *low* carbon tax, one that gradually rises over time. The purpose of the carbon tax/fee is to finance an *up-front*, long-term, global effort on the energy technology and infrastructure front. *Commitments* to a gradual increase in the tax/fee send a forward price signal to deploy effective, scalable, competitive, and transferable technologies as they reach "the shelf". Policies that attempt to short circuit this process by setting near- or medium-term emission-reduction targets and mandates will be ineffective – or quite likely destructive of long-term efforts to reduce emissions and stabilize climate. Predictably, failure of climate policies will create increasing pressure to consider the adoption of one or another proposal to "geo-engineer" the atmosphere. That brings us face-to-face with still another "inconvenient truth".

REFERENCES

Baksi, S. and C. Green. 2007. "Calculating Economy-Wide Energy Intensity Decline Rate: The Role of Sectoral Output and Energy Shares", *Energy Policy* 35: 6457-6466.

Caldeira, K., A.K. Jain, and M.I. Hoffert. 2003. "Climate Sensitivity Uncertainty and the Need for Energy without CO_2 Emission", *Science* 299: 2052-2054.

Edmonds, J.A. 2004. "Technology Options for a Long-Term Mitigation Response to Climate Change". 2004 Earth Technologies Forum, 13 April, Pacific Northwest National Laboratory, Washington, D.C.

Green, C. and H.D. Lightfoot. 2002. "Making Stabilization Easier than it will be: The Report of IPCC WG III", *C^2GCR Quarterly*, Number 2002-1.

Green, C., S. Baksi, and M. Dilmaghani. 2007. "Challenges to a Climate Stabilizing Energy Future", *Energy Policy* 35: 616-626.

Hoffert, M.I., K. Caldeira, A.K. Jain, E.F. Haites, L.D.H. Harvey, S.D. Potter, M.E. Schlesinger, S.H. Schneider, R.G. Watts, T.M.L. Wigley, and D.J. Weubbles. 1998. "Energy Implications of Future Stabilization of Atmospheric CO_2 Content", *Nature* 395: 881-884.

Hoffert, M.I., K. Caldeira, G. Benford, D.R. Criswell, C. Green, H. Herzog, A.K. Jain, H.S. Kheshgi, K.S. Lackner, J.S. Lewis, H.D. Lightfoot, W. Manheimer, J.C. Mankins, M.E. Mauel, J. Perkins, M.E. Schlesinger, T. Volk, and T.M.L. Wigley. 2002. "Advanced Technology Paths to Climate Stability: Energy for a Greenhouse Planet", *Science* 298: 981-987.

Intergovernmental Panel on Climate Change. 2000. *Special Report on Emissions Scenarios*. Cambridge, U.K.: Cambridge University Press.

Metcalf, G.E. 2009. "Market-Based Policy Options to Control U.S. Greenhouse Gas Emissions", *Journal of Economic Perspectives* 23 (Spring): 5-27.

Metz, B., P. Bosch, R. Dave, O. Davidson, and L. Meyer. 2007. *Climate Change 2007: Mitigation,* Contribution of WG III to the Fourth Assessment Report, Intergovernmental Panel on Climate Change. Cambridge, U.K.: Cambridge University Press.

Metz, B., O. Davidson, R. Swart, and J. Pan. 2001. *Climate Change 2001*: *Mitigation,* Intergovernmental Panel on Climate Change, Third Assessment Report, Cambridge, U.K.: Cambridge University Press.

Montgomery, W.D. and A.E. Smith. 2007. "Price, Quantity, and Technology Strategies for Climate Change Policy", *Human-Induced Climate Change: An Interdisciplinary Assessment.* Cambridge, U.K.: Cambridge University Press.

Nakicenovic, N., *et al.* 2000. *Special Report on Emissions Scenarios. Working Group III, Intergovernmental Panel on Climate Change (IPCC)*. Cambridge, U.K.: Cambridge University Press.

National Roundtable on Environment and Economy (NRTEE). 2009. *Achieving 2050: A Carbon Pricing Policy for Canada.* Ottawa.

Pielke, R. Jr., T.M.L. Wigley, and C. Green. 2008. "Dangerous Assumptions", *Nature* 452: 531-532.

Riahi, K. and R.A. Roehrl. 2001. "Energy Technology Strategies for Carbon Dioxide Mitigation and Sustainable Development", *Environmental Economics and Policy Studies* 3(2): 89-123.

Rosen, D.H. and T. Houser. 2007. *China Energy: A Guide to the Perplexed.* Centre for Strategic and International Studies and the Peterson Institute for International Economics.

Schelling, T.C. 1992. "Some Economics of Global Warming", *American Economic Review* 82(1): 1-14.

III

Carbon Pricing: Analytical Perspectives

5

Carbon Pricing and Federalism

Thomas J. Courchene and John R. Allan

INTRODUCTION

This paper focuses on alternative approaches to carbon pricing and the manner in which they interact with Canadian federalism and, more generally, with multi-level governance. As used here, carbon pricing will encompass both carbon-tax and cap-and-trade (C&T) regimes as well as approaches that embody features of both systems. Also included in the paper will be an elaboration of our CATT (carbon-added tax/tariff) model, which is a carbon-tax analogue of the GST or VAT. Attention will then be directed to the variety of ways in which carbon pricing interacts with federalism. This is especially relevant for Canada and the United States because for both countries it is the sub-national governments that have taken the leading role on the climate policy and carbon-pricing fronts. Among the issues addressed in this context are which level of government can constitutionally legislate and/or implement carbon pricing and which level ought to be able to appropriate the proceeds from carbon taxes or auctioned permits. Beyond North America, the multi-level governance issue embraces not only Canada's relationship to the Kyoto Protocol and other international carbon-pricing regimes, but also the manner in which carbon-pricing systems are able to level the international carbon playing field. Because all of the above issues will be addressed in other papers in this volume, our analysis will at times be more indicative than comprehensive.

The remainder of the chapter then addresses a series of carbon-pricing challenges, such as whether emissions limits ought to be expressed in absolute or per capita terms, whether the exporting or importing country ought to bear the carbon footprint of emissions generated prior to exportation, whether border adjustments are warranted and, if so, how they might be made WTO-compatible, and whether and how carbon footprints from shipping should come under the Kyoto umbrella.

To round out this introduction it is convenient to highlight a key difference between "pure" carbon-tax and C&T models: while carbon taxes provide carbon price certainty but uncertain levels of CO_2 reductions, C&T systems provide certainty with respect to CO_2 reductions but at uncertain carbon prices. In both models, government can control with precision the relevant policy parameter,

---

This paper draws freely from our joint papers and those by Courchene listed in the references.

namely, the carbon price in the tax model or the amount of emission permits in the C&T model. The response to the chosen parameter value, however, is determined by markets, and both models are characterized by considerable uncertainty concerning the availability and cost of the necessary emission-abatement technology. In the case of the carbon-tax model, this uncertainty translates into imprecision in the amount of abatement a given carbon price will effect, while in the C&T model the imprecision is in respect of the resulting carbon price associated with the chosen level of abatement. Expressed differently, rather than having a clearly defined supply (i.e., cost) curve for emission abatement, we have an ill-defined, positively sloped band of price-quantity relationships, and the resulting uncertainty complicates the formulation of climate-change policy.

This difference between the two models is captured in Figure 1, the Rick Hyndman version (2009, and this volume) of the seminal Weitzman (1974) model. The "uncertain" emissions reduction curve is upward sloping, with the uncertainty captured by the flared cone shape of the curve. Thus, the higher are carbon prices, the more will be the reduction in CO_2 emissions and the more will be the associated uncertainty.

In more detail, assume that there is a carbon tax of P_0 dollars per tonne. The resulting emission reductions will fall between C_1 and C_2. In other words, a carbon tax leads to a certain price for CO_2 but an uncertain response in terms of emissions reductions. To be sure, if the actual CO_2 reductions are less than the desired reductions, government can always increase the carbon tax. Under a pure C&T model the authorities can decree a given CO_2 reduction. Let this be C_0

Figure 1: Polar Versions of CT&T and Carbon Tax Systems

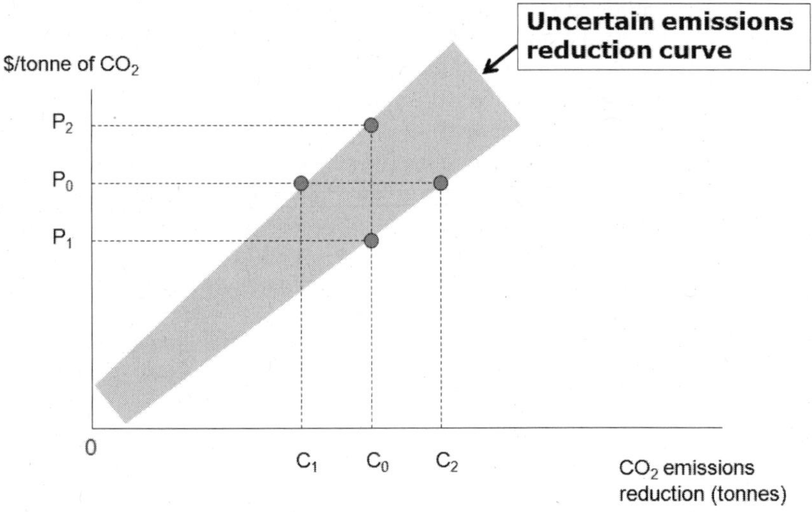

Source: Adapted from Hyndman (2009).

in Figure 1. However, the *ex ante* carbon price can vary from P_1 to P_2: C&T guarantees emissions certainty but only at an uncertain price for carbon. Hyndman (this volume) notes that this analysis can be extended to introduce an "uncertain demand curve for CO_2 reductions" which would imply uncertainty with respect to both price and quantity, but Figure 1 will suffice for present purposes.

With this as backdrop, the analysis now focuses in more detail on carbon-pricing models.

CARBON PRICING: TAXATION

A carbon tax is a levy imposed on CO_2 emissions and, more generally, on the emissions of greenhouse gases (GHGs) with the intent not only of reducing emissions but also providing incentive for encouraging low-carbon alternatives/technologies. The resulting tax revenues can be utilized in various ways – to cushion the fossil-fuel price increases for low-income Canadians, to provide funding for developing low-carbon technologies including those designed to capture and store/sequester carbon (henceforth CCS), to engage in a "tax shift" to reduce other taxes in order to offset the economic impact of the carbon tax, and so on. This ability to both reduce emissions and offset selected negative impacts of the carbon tax is typically referred to as the "double-dividend" associated with a carbon tax.

The Mintz-Olewiler Carbon Tax

Among the earliest analytical proposals for a comprehensive carbon-tax was that by Mintz and Olewiler (2008). The starting point was to recognize that the existing ten cent per gallon federal tax on gasoline (at the pump) is equivalent to a carbon tax of $42 per tonne of CO_2. Their proposal was to extend this tax of $42 per tonne to all other fossil fuels – oil, heating oil, coal, natural gas, kerosene, etc. In effect, this would level the carbon-tax playing field across alternative fuels. While their tax relates to CO_2 emissions arising from the use of fossil fuels, the principle would apply as well to any emissions related to the initial production of these fuels. Mintz and Olewiler argue that the playing field also ought to be levelled internationally via border tax adjustments (BTAs) – carbon taxes would be imposed on imports and rebated on exports. Finally, they note (under the heading "revenue neutrality") that "revenues received by the government should be used to reduce the most distortionary aspects of the tax structure and to provide relief for distributive purposes" (2008, 24).

The Carbon-Added Tax/Tariff (CATT)

Our variant (Courchene and Allan 2008a) of a carbon tax is modelled along VAT/GST lines. Under such a system, there will be a tax on the carbon emissions that are added at each stage – hence the "carbon-added" label. As the product completes each stage of the production/distribution process, it is taxed on its cumulative carbon footprint to that point, and a credit claimed for the carbon tax on earlier stages. In consequence, only the carbon added in each stage ends up being taxed at that stage. Hence, when the product reaches the final stage, the tax is on the *cumulative* value of carbon emissions, i.e., the sum of the carbon-added taxes at each stage. As with the GST, in the case of exports the tax accumulated to the point of export will be rebated, so that the carbon tax does not impact our international competitiveness. Relatedly, a carbon tax will be levied on the accumulated carbon footprint of each import, including that arising from transporting the product to Canada.

Thus, *the CATT is export-import neutral* (it does not affect the international competitiveness of Canadian production in either domestic or external markets). However, while the carbon tax *rate* on imports will be identical for all imports irrespective of the country of origin, similar imports from different countries may carry different *amounts* of carbon taxes depending on their carbon emissions prior to landing in Canada. Importers thus have an incentive to source their products from suppliers with smaller carbon footprints. Given that value-added taxes are fully acceptable under WTO rules (indeed, they are not even considered as "border tax adjustments" since they are effectively regulated/imposed in the domestic production/consumption process), the presumption is that a carbon-added tax should, in principle, fall well within WTO guidelines. This also appears to be the presumption one would draw from the Andrew Green paper (this volume) on WTO-compatible border-tax adjustments. In this context it is necessary to note that, should the implementation of effective carbon-pricing regimes require BTAs that are problematic under the WTO, the appropriate remedy may be a reconsideration of the policies and practices adopted by the WTO. That organization has evolved to deal with issues relating to trade supervision and liberalization, and this in an era when climate change and global warming were essentially absent from domestic and international policy agendas. If the policies and practices that best served these objectives get in the way of necessary responses to the climate-change crisis, it is surely the former that may have to be reconsidered and, if necessary, altered.

To be sure, the measurement problems associated with the implementation of a CATT will likely be severe (depending on how detailed the classification is), so that conforming to WTO standards may be problematical. However, the increasing concern about carbon has already launched a "carbon auditors" industry that may well make the measurement issues less severe than they might at first appear. Moreover, some recent papers provide a less-information-intensive approach to assigning carbon-added taxes to imports (e.g., Ismer and Neuhoff 2007).

There is another way of viewing the carbon-added tax. Since most exports come from global multinationals, the international component of the carbon tax is, in the first instance, a tax on the carbon footprint created by the exports of these global multinationals rather than on the exporting countries themselves. That is, Wal-Mart's imports from its suppliers in China would be subject to the carbon tax whether or not China itself is a "signatory" to the carbon-tax regime. One assumes that, other things equal, these multinationals will select production locations in order to minimize carbon taxes and hence carbon footprints (including shipping). Similarly, one also assumes that all nations will recognize that introducing low-carbon-emitting fuels and production processes will make them more attractive places in which to produce and from which to export. Finally, while the carbon tax is offered as a global approach to climate change, its advantages (e.g., export-import neutrality) would also apply at a regional (E.U. or NAFTA level), or even at a national level.

By way of a final comment on CATT at this juncture, one might note that the developing countries at the 2009 G8 meetings in Italy have rejected the notion that they be assigned a target for emissions cuts. However, under a CATT, say a joint E.U. and U.S./Canada CATT, the developing countries would be included *de facto* since the carbon footprints of their exports to the European Union and Canada/United States would be subject to the carbon tax at the point of importation. They would thus have obvious incentives to control their own emissions in order to avoid penalizing their domestic exporters and to be a more attractive location for global multinationals intent on exporting.

CARBON TAXATION AND FEDERALISM

With its 2007 levy of 0.8 cents on every litre of gas and 0.9 cents on each litre of diesel fuel, Quebec became the first jurisdiction in Canada (indeed, in North America) to impose a carbon tax. This is an origin-based carbon tax and it is levied on roughly 50 large emitters and distributors. However, Canada's carbon-pricing star is surely British Columbia. The province introduced in 2008 a consumption-based carbon tax of $10/tonne to apply to all fuels (e.g., the tax on gasoline will be 2.4 cents per litre and 2.76 cents for diesel fuel). This levy will rise to $30/tonne in 2012. A further innovation in the BC carbon tax is that it was made *revenue neutral*, by using the revenue generated to reduce personal and corporate taxes and to provide low-income tax credits to help offset the rise in fuel prices.

At the federal level, Stéphane Dion's *Green Shift* – a revenue-neutral carbon tax designed to achieve over four years a $40/tonne CO_2 tax on all fossil-fuel emissions – was the focus of intense media attention and contributed to the Liberals' election defeat and the demise of Dion as Liberal leader. Since the existing federal tax on gasoline is already $42/tonne, it would have been unaffected; in consequence, the *Green Shift* would have served to effectively level the playing field by increasing the carbon tax on all other fossil fuels. While this follows the Mintz-Olewiler model, two other features of the tax do not. First, *Green Shift* effectively ignored international competitiveness issues: there was no provision for a tax on imports or rebates on exports. Second, while the proposal was revenue neutral in the sense that the overall budget would be unaffected, some of the proceeds were to be devoted to enlarging the *expenditure* side of the budget, i.e., to enhancing aspects of Canada's social envelope. Thus the claim that *Green Shift* was "revenue" neutral came under considerable attack.

Turning to the interaction between carbon taxation and federalism and, in particular, to the issue of who can levy carbon taxes, the answer according to

Chalifour (this volume) is that both the federal and provincial governments have the jurisdictional authority to implement carbon taxes, as long as the taxes are carefully designed to fit within their appropriate powers. However, several caveats are in order.

The first of these concerns a possible CATT. Since a CATT would be an indirect tax, it could not be implemented by the provinces unless Ottawa were to delegate administrative responsibility as it has in the case of the GST in Quebec.[1] A second concern is that even if Ottawa were to mount a CATT, it still might be (ought to be?) the case that the resulting carbon-tax revenue could be shared with the provinces on some appropriate basis. (As a relevant aside, Stéphane Dion was amenable to allowing BC to receive credit for its carbon tax under *Green Shift*, although how this was to be done was not spelled out.) By involving both levels of government, joint administration or revenue sharing could complicate any attempt to make carbon taxes revenue neutral, most particularly if provinces chose different offsets to the carbon tax.

A third caveat is that a Mintz-Olewiler carbon tax implemented by a province could not be made export-import neutral either domestically or internationally. In terms of the latter, this is a consequence of the provinces being unable, constitutionally, to apply border taxes to imports or provide rebates on exports. These same restrictions also apply to interprovincial trade: British Columbia cannot assess its carbon tax on products arriving from out-of-province, nor can it prevent multi-province firms operating in BC from transferring production to their own plants in, say, Alberta, for "export" to BC. Finally, not only does a provincial carbon tax fall short in terms of levelling the playing field but, as well, it provides an incentive for "emissions leakages" (i.e., transferring production to non-carbon-tax jurisdictions for reshipment to, say, BC), so that a provincial carbon tax loses some of its effectiveness in terms of curtailing emissions.

CARBON PRICING: CAP-AND-TRADE

The Cap-and-Trade Model: Analytics

Tableau 1 presents an overview of variations on the C&T theme. In its most common version, C&T involves setting an overall emissions cap, allocating emissions permits (free of charge) to producers up to the overall limit or cap, and requiring firms to buy from other firms any permits required for emissions beyond their allocated limit. This latter feature is the "trade" component of C&T. The genius of the cap-and-trade system is that permits will be bought and sold in a manner that will maximize output for any given overall emissions cap. In this stripped-down version of a C&T system, the permit price that will be

[1]Although Chalifour (this volume) notes that under s. 92A a province has the right to levy an indirect tax on natural resources, this would have to be tied to "primary production there-from", which would be a much narrower base than would characterize a full-blown CATT.

Tableau 1: Cap-and-Trade Basics

Under a cap-and-trade system the authorities will determine the limit (cap) on the pollution allowed, and will issue "pollution permits" for this amount. If a company wants to produce more than it has permits for, it has to acquire additional permits from other companies, from approved "carbon offset" sources, or from the government itself in the form of, say, "price caps". This is the "trade" component of cap and trade and it will set the price of the permit.

Selected Features:

1. Cap-and-trade systems are usually limited to large emitters.
2. The cap will presumably be set in line with overall climate-change goals. The cap can be an "absolute" cap or an "intensity" cap (i.e., a cap on the emissions per unit of output). While an intensity cap may not reduce absolute emissions, it does facilitate new entrants, especially new entrants with lower-carbon technologies.
3. The allocated permits are almost always distributed free of charge. However, most environmental NGOs, among others, would recommend that the permits be auctioned. This could be introduced at the outset, or the price of the permits distributed to the existing firms could be escalated gradually over time. This will yield revenues that can be used for other purposes, e.g., to provide low-income relief from rising energy prices, to reduce other taxes, to invest in low-carbon research. In this case, a cap-and-trade system takes on some of the characteristics of a carbon tax.
4. In order to increase efficiency as well as to avoid price "spikes", the geographical trading area should be as large as is feasible.
5. Some systems include a "price cap" (i.e., a limit on how high the permit price can rise). This can undermine the integrity of the system. So can a guarantee that carbon offsets are available at a fixed price, since this is in effect a price cap. For example, in Alberta's cap-and-trade system, there is a widely available offset set at $15/tonne if the payment goes to a low-carbon-technology fund. If this price cap is effective, then the cap-and-trade system effectively becomes a variant of a carbon tax system.
6. Cap-and-trade systems can be complex to implement and manage since, at a minimum, the trading component needs to include brokers, bankers, insurers, lawyers, carbon auditors, etc., as well as a host of regulations (replete with regulators).
7. Border tax adjustments to level the competitive playing field are likely to run into WTO problems more than is the case with border tax adjustments under a carbon tax (Andrew Green, this volume).

Note: This table draws in part from Horne (2008).

determined on the market will be such that the overall emissions limit is in fact attained; i.e., C&T generates emissions certainty but at an *ex ante* uncertain permit price.

Since the government would receive no revenues from this version of a C&T, any spending to shelter low-income individuals from the effects of the policy would have to come from the consolidated revenue fund. However, as Tableau 1 indicates, there are various ways to modify C&T systems in order to provide governments with revenues and/or to contain the amplitude of the price of carbon permits. One of these is for the government to auction some or all of the emissions permits, which, in turn, would allow the C&T to acquire the "double dividend" feature of a carbon tax (see note 3 of the tableau). Another is to impose a maximum $/tonne "price cap", at which price the government would sell additional permits. When this price cap is binding, the C&T reverts to a version of a (typically low-rate) carbon tax. Yet another approach to the price cap would take the form of a "carbon offset", an option to buy the right to emit one tonne of carbon from a designated firm or agency committed either to designing or delivering low-carbon technologies or to engaging in carbon capture and sequestration (CCS), as provided, for example, in both the Alberta and federal C&T proposals. Finally, there is a Kyoto-sanctioned version of a carbon offset, namely the CDM (Clean Development Mechanism), which would allow domestic firms in signatory nations, in lieu of purchasing permits from other domestic firms, to substitute commitments from developing countries to reduce emissions. These CDM offsets feature prominently in the European ETS, the E.U. version of a C&T system.

A further hurdle facing C&T systems is how to level the competitive playing field, i.e., how to ensure that a C&T can be made export-import neutral in ways that would be WTO-consistent. Unlike the CATT system, which would assess and apply a uniform border tax adjustment (BTA) on the basis of the carbon footprint of the import, under C&T the BTA would vary by country depending on the nature of the country's carbon pricing regime. While a recent 2009 WTO report suggested that BTAs for carbon-pricing policies should be acceptable, complications are bound to arise and lead to WTO challenges (Tamiotti *et al.* 2009). For example, how does one assess international equivalence in the face of fluctuating permit prices? Or, how, without clearly benefitting domestic producers, does one assess equivalence when most or some substantial part of the emissions permits are distributed free of charge? And neither of these examples takes account of the variety of practices one will find across the exporting countries.

In spite of all the above complications, it is nonetheless the case that, as alluded to earlier, C&T appears to be the preferred form of carbon pricing in industrial, financial and political circles. The reasons for this are hardly surprising. Industry is likely to favour C&T since the typical version embodies free distribution of all, or at least most, of the permits. For example, even the Waxman-Markey bill has 80 percent of permits distributed free of charge. These permits constitute valuable property rights, analogous to dairy quotas, and, apart from their monetary value, they serve both to protect insiders from new entrants and, in the presence of a BTA, domestic producers from foreign competitors. The financial-market players will also be big fans of C&T because, as noted in point 6 of Tableau 1, the requisite regulatory and trading infrastructure will involve a spate of bankers, brokers, insurers and auditors.

The political rationale for preferring C&T over carbon taxation is based on the prevalent illusion – that Harper's attacks on Dion's *Green Shift* did much to encourage – one that carbon taxes represent levies on consumers whereas the costs of C&T are assumed to be borne by the big emitters. Yet the reality is that achieving a given emissions reduction will require a comparable increase in the price of CO_2 emissions under both regimes, with comparable implications for consumers. What is also underrated in terms of a C&T regime is its susceptibility to rent-seeking and "green protectionism", as we are likely to discover if Waxman-Markey ever sees the light of legislated day.

The Cap-and-Trade Model: Federalism

In 2007, Alberta became the first jurisdiction in North America to implement a C&T system for reducing CO_2 emissions. (The earlier C&T system in the eastern United States focussed on acid rain.) Specifically, the Alberta model has the following features: an intensity cap (i.e., a cap on CO_2 emissions per unit of output) rather than an absolute cap; most if not all the permits will be distributed free of charge; the permit-trading area is limited to Alberta (i.e., funds cannot flow out of Alberta); and the price of permits is effectively capped at $15/tonne since emitters can purchase carbon-offset credits at this price from an accredited (Alberta-based) R&D agency for the development of low-carbon technologies. For reasons already addressed, none of these provisions auger well for the effectiveness of this C&T regime. For example, the requirement that all revenue flows from permit and offset trading must remain in Alberta will render the market very small and subject to volatility (although the availability of offsets at $15 will limit this as well as the effectiveness of the overall program). This is problematical in any event but even more so in the Alberta model where the government will not derive revenue from the C&T regime and therefore will have to draw on its consolidated revenue fund to provide any assistance to those rendered vulnerable by this volatility. Furthermore, and in line with the earlier analysis of carbon taxation, Alberta will not be able to level the competitive playing field either domestically or internationally, with the result that, to the extent that resources are mobile, there will be "emissions leakage" to other jurisdictions.

With its 2007 *Turning the Corner* regulatory framework, the federal government committed itself to a 20 percent cut in 2006 GHG emissions by 2020, in part by promising to introduce a carbon-intensity (i.e., emissions-per-unit) cap and a permit-trading system in 2010. As in the case of Alberta, permits would be distributed free of charge and there would be an inappropriately low $15/tonne price cap on permit trading. Ottawa's 2008 policy update promised carbon capture and storage requirements for the oil sands (with the 2009 federal budget devoting considerable funding to CCS) as well as several command-and-control measures (no dirty-coal electricity plants after 2012, vehicle emission standards, etc.). All of this appears to have fallen by the wayside since the Harper government now seems intent on linking its climate-change policy with whatever C&T version emerges in the United States.

By far the most comprehensive (90 percent coverage of GHG emissions) and "federal" C&T system is that proposed by the Western Climate Initiative (WCI). The WCI is a collaboration of cross-border jurisdictions – seven U.S. states, arguably headed by California, and four Canadian provinces (BC, MB, QB and ON) – committed to implementing a cross-border C&T system with a goal of a 15 percent reduction in 2005 emission levels by 2020. A key federal feature of this model is that each jurisdiction will be assigned emissions limits and the corresponding permits. In 2012, at least 10 percent of these permits will be auctioned, rising to 25 percent by 2020. The WCI will also allow a "rigorous offsets system" related to CCS, reforestation, waste management and other activities related to carbon sinks and low-carbon technologies. Accredited WCI offsets can be purchased by firms operating in any state/province and the intention is to allow partners to agree to recognize offsets located anywhere in NAFTA space. The partner jurisdictions may also accredit offsets from developing countries issued under the Kyoto CDMs. The one constraint with respect to offsets is that they cannot exceed 49 percent of the total emissions reductions from 2012-2020. A further important federal feature of this model is that "the WCI Partner jurisdictions have designed a program that can stand alone, provide a model for, be integrated into, or be implemented in conjunction with programs that might ultimately emerge from the federal governments of the United States and Canada" (http://www.westernclimateinitiative.org/designing-the-program). Indeed, the flexibility goes even further: the WCI is committed to integrating the BC carbon tax with (or into) the proposed C&T, although how this is to be done has still to be identified.

This WCI proposal is a quite remarkable document in terms of Canadian climate-change policy in that four provinces with over 75 percent of our population have signed on to a comprehensive, rigorous and effective emissions-reduction system that essentially ensures that the watered-down Alberta and federal C&T proposals cannot be resuscitated. One can take this further: the above reality plus the fact that the WCI embraced a joint Canada-U.S. carbon-pricing regime made it much easier (and perhaps inevitable) that Canada would attempt to buy into the emerging U.S. system. And as if to ensure this, the Ontario-Quebec joint cabinet meeting in the summer of 2008 (prior to the formal announcement of the WCI C&T) committed these provinces to an effective C&T system based on Kyoto's 1990 baseline, thereby effectively relegating the weak Alberta and Ottawa C&T systems to the sidelines.

We now turn to several carbon-pricing challenges that are important in their own right but that, as well, have implications for Canadian environmental federalism or for Canada's role in the global carbon-pricing context.

EMERGING OR NEGLECTED ISSUES IN CARBON PRICING

Kyoto Targets and Population Growth

At the 2009 G8 summit in L'Aquila, Italy, environment minister Jim Prentice stated that Canada would not commit to the agreed-upon 80 percent reduction in emissions by 2050. Rather, it would stick with its earlier commitment to reduce emissions by 60-70 percent by 2050: "This is a realistic target [given] the climate we have, the industrial base we have, and our population growth" (O'Neil 2009). Earlier, Simpson, Jaccard, and Rivers (2007, 249-50) expressed similar concerns about Canada's Kyoto commitments:

> Canada made a commitment at Kyoto that was the most difficult and expensive in the world. It took no account of our economic growth, population growth, cold temperature, vast distances and fossil fuel production; it contravened our federal system, because Ottawa broke a fragile federal-provincial consensus on the eve of the negotiations. So the first lesson must be to remember these factors – and therefore undertake commitments and implement serious domestic policies that take these distinctly Canadian factors into account, because no other country has all of these characteristics.

It may be argued that the utilization of a base year in the Kyoto Protocol already accounts for such climatic and other structural disadvantages that a signatory nation may experience: while these may inflate its emissions relative to those of other countries, the inflation is included in both the base and target years, so no further adjustment is necessary. To the extent, however, that there are significant structural changes *between* the base and target years, the choice of a particular base year may be especially disadvantageous to a country. Thus, where there has been rapid growth in, say, population, since the base year, failure to allow for this may be very punitive. Similarly, a rapid growth in energy-intensive exports may render the achievement of a common target (e.g., "reductions equal to 10 percent of the 1990 benchmark") disproportionately challenging. The population issue is the focus of this section, while the implications of our industrial base and, in particular, energy exports are addressed in the following section.

By way of elaborating on the population issue, and again drawing from Simpson, Jaccard, and Rivers, over the 1990-2005 period Canada's average annual population growth rate was just over one percent, or about two-and-a-half times that of France, four times that of Germany and five times the average Italian growth rate, with commensurate impacts upon total emissions. Given that the Kyoto targets are *absolute aggregate targets*, and not per capita targets, this effectively sets the bar significantly higher for Canada than for the slower growing European countries; i.e., Canada will have to make appreciably greater per capita cuts in emissions than slower growing countries to meet these absolute targets. Moreover, the selection of 1990 as the Kyoto benchmark year was very beneficial to Germany. When the two Germanys were reunited, the newly merged country inherited the East-German-era plants that generated pollution levels that were extraordinary by Western standards. By simply closing these plants or by upgrading them to Western norms, Germany achieved its 2008-12 Kyoto targets by 1992![2]

Harrison and Sundstrom (2007) elaborate further, noting that Germany's "windfall reduction" had its counterparts in the United Kingdom (because of the ongoing conversion of its electricity generation plants from coal to offshore gas) and in Russia (because of economic collapse and the unwinding of heavy-polluting industries). In order to present a more realist assessment of what has transpired under Kyoto, Harrison and Sundstrom (14) rework GHG comparisons on a per capita basis (see Table 1), on which they comment as follows (2007, 14):

---

[2]http:www.numberwatch.co.uk/german_kyoto_protocol_hoax.htm, viewed 13-07-09.

> While there is tremendous variation in performance, from a 33 percent decline in emissions in Russia to a 27 percent increase in Canada [column 1], the variation in population growth evident in the next column suggests that emissions trends reflect more than just policy efficacy. Canada, the US, and Australia have experienced much greater increases in emissions in large part because they have experienced much greater population growth than other jurisdictions. Indeed, when one compares trends in *per capita* emissions [column 3], it is striking that the only country to see a decline other than the three that experienced "windfall" reductions (Germany, the UK, and Russia) is the US, which has been vilified for its decision not to ratify the Kyoto Protocol. In fact, with the exception of Germany and the UK, the rest of the EU has experienced per capita emissions comparable to Canada and Australia [compare the last row of column 3 with the first two rows].

By way of further elaboration, over the 50-year period 1956-2006, Canada's population increased by 120 percent. While it is highly unlikely that our growth over the 40-year period 2010-2050 will be anywhere near this rate, existing forecasts suggest that 20 percent may be in the probable range. This rate of growth is surely to be much higher than that of continental Europe and would rationalize, if not justify, environment minister Prentice's commitment to a 60-70 percent reduction in emissions rather than the 80 percent reduction agreed upon by the G8 at L'Aquila.

Table 1: Comparison of Greenhouse Gas Emission Trends

Country	*Emissions Growth (without LULUCF) 1990 to 2004* %	*Population Growth 1990 to 2004* %	*Increase in Emissions (without LULUCF) per capita, 1990 to 2004* %
Australia	+24.3	+17.0	+6.3
Canada	+26.6	+17.0	+8.2
Japan	+6.5	+3.1	+3.4
Russia	-33.1	-3.0	-3.1
United States	+15.8	+17.1	-1.2
E.U. 15	-1.0	+4.5	-5.3
Germany	-17.4	+3.8	-20.3
United Kingdom	-14.3	+4.8	-18.2
Rest of E.U.	+12.8	+4.7	+7.8

Sources: Emissions data from UN FCCC emissions profiles (http://unfccc.int/ghg_emissions_data/items/38954.php). Population data from US Census Bureau (http://www.census.gov/ipc/www/idbrank.html).
Note: LULUCF is *L*and-*U*se, *L*and-*U*se *C*hange and *F*orestry.
Reproduced from Harrison and Sundstrom (2007, Table 4).

Unfortunately, this is not the only area where Canada's Kyoto negotiators settled for arrangements that clearly disadvantaged us relative to other nations. Assigning carbon footprints on exports is another problem area.

Canada's Energy Exports: Whose Carbon Footprint?

In a recent *Globe and Mail* article (Courchene and Allan 2008b), we argued that the Kyoto Protocol is predicated on an incorrect assignment of carbon footprints, one that attributes to producer nations the entire footprint generated by the extraction and processing of resources some or all of which are subsequently exported to other countries. It was our contention that the total footprint associated with a given quantum of a particular resource should be allocated among countries in proportion to their consumption of the resource. Thus each consuming country would be charged with the emissions generated by its consumption of the resource (which is the Kyoto practice), plus those attributable to the extraction, processing and transportation of the resources that it consumes (all of which, under Kyoto, are assigned to the exporting/producing nation). Our argument was that this practice is particularly disadvantageous to resource-exporting nations such as Canada, Australia and Russia, while it makes countries such as Japan and many members of the European Union that import the vast quantities of resources they consume each year appear much better environmental citizens than they truly are.

Two observations are in order here. The first is that a case might be made that the use of a reference year (e.g., 1990 for Kyoto) solves the problem; i.e., with resource exports in both the base and terminal years, the issue is already addressed. However, the problem with 1990 as the reference year is the dramatic recent (i.e., post-Kyoto) growth of India, China *et al.* and their demands for Canada's resources, and the fact that the associated carbon footprint as the resources leave Canada remains with us. Over the last decade, for example, there has been a four-fold increase in our exports – which are resource intensive – to China and India, with the entire carbon footprint attributable to their production being assigned to Canada. Such rapid growth in energy-intensive exports represents a structural change that is not accounted for or recognized by the use of a base year. Clearly, what is necessary is a change in the emission-attribution rules, one that would charge the importing nations with the carbon footprint due to the *production* of their imports.

The second observation is both analytical and self-serving: under our CATT model, carbon footprints are taxed as if they were properly assigned. For example, under a destination-based CATT, the cumulated carbon-added taxes would be rebated at the point of export, effectively transferring the entire tax base – i.e., the entire carbon footprint – for use by the importing country. Similarly, the entire carbon footprint embedded in our imports would, on a destination basis, be subject to the domestic CATT. Tax revenues would thus be allocated internationally as they would be were the carbon footprints properly assigned. It must be acknowledged, however, that achieving the correct assignment of taxes is not the same thing as achieving the correct assignment of carbon footprints. Unless the relevant international agencies can be persuaded to change their carbon-attribution practices, Canada would continue to be penalized by an unfair carbon burden, namely, the emissions attributable to the production or extraction of resources that are exported and consumed elsewhere. At least, under a CATT, the associated tax revenues would be properly assigned.[3] In this context, we would note that the California policy of determining whether an energy import meets its announced standards by assessing the carbon footprint of the entire production chain is tantamount to internalizing the carbon footprint: where the footprint attributable to production is judged to be too high, the importation is not permitted. California thus behaves as it would if the emissions attributable to the production of its energy imports were in fact attributable to it, declining those imports where the deemed attribution is considered excessive. Such an approach may not sit well with the oil-sands producers or Alberta, particularly if California were to include in its calculations, as it should, the carbon released or not absorbed as a result of the massive deforestation that accompanies oil-sands development. They may draw some comfort, however, from the following discussion of Kyoto's complete ignoring of the carbon footprint of shipping, a practice that contributes to a comparative disadvantage in their primary export market, the United States.

Shipping, Copenhagen and Kyoto

The carbon footprint of marine shipping is not included in Kyoto, nor is it included in any existing carbon pricing or C&T regime of which we are aware.[4] This omission is perhaps attributable to the fact that the true scale of climate-change emissions from shipping is almost three times higher than previously believed (Vidal 2008; Mittelstaedt 2009). Rather than being of the same order as aviation emissions, marine emissions are almost twice as large, accounting for about 5 percent of total CO_2 emissions, some 30 percent of nitrous oxides, and almost 10 percent of sulphur oxides. Moreover, without control measures, these percentages will all increase greatly by 2020, a hardly surprising result given that ships are permitted to burn fuel more than 3,000 times dirtier than the fuel required to be burned in U.S. and European diesel cars and trucks. Were shipping emissions to be integrated into Kyoto, one could expect some creative solutions – shipping hybrids (fuel and sails, and fuel and sun, e.g., solar panels on the top deck). Of relevance to Canada – and especially to Alberta – is that any comparison of GHG emissions from oil-sands oil and that from the Gulf States would then have to include in their GHG footprints those from marine shipping. This would go some way to reduce the comparative disadvantage of non-conventional oil.

---

[3]Even in the absence of a correct assignment of carbon footprints, a correct assignment of tax bases and revenues – as would be the case under a generalized destination-based CATT – will ensure an incentive system for importers to seek their imports from low-carbon-footprint sources and thus for exporters to reduce their footprints.

[4]Aviation is also excluded, but it appears that this will be addressed in Copenhagen.

Such an inclusion in Kyoto II is essential: Why generate a measurable and taxable GHG footprint by sending a container by rail from Halifax to Vancouver when you can avoid the carbon tax by sending the container via GHG-exempt marine transport through the Panama Canal? The bottom line here is that there is a strong case for Canada to take these issues to Copenhagen (en route to Kyoto II) this December and, perhaps more importantly, to press the case for them to be considered in any joint Canada-U.S. carbon pricing system.

Once again, addressing the shipping footprint is rather automatic under a CATT since the tax on imports and domestic products and processes would apply to the CO_2 emissions all along the production/processing chain. Although somewhat more problematic, this is also the case for carbon taxation generally. The problem with Kyoto on this score is that it essentially deals directly *with countries and not with emissions per se.* Since shipping is "stateless", as it were, and in any event typically flies flags of convenience, it has largely escaped attention under the country focus of Kyoto. As noted earlier, if the key developed nations were to adopt a CATT (including generous provisions for transferring state-of-the-art low-carbon technologies to developing nations as well as rigorous CDMs), the non-participants in the Kyoto process would be effectively co-opted into emission reductions as they sought to protect their exports.

Finally, attention now turns to two "federalism" challenges related to carbon pricing. The first deals with alternative ways to mount joint federal-provincial carbon-pricing regimes, whereas the focus of the second is on fiscal federalism and carbon pricing, most particularly on revenue sharing.

ENVIRONMENTAL FEDERALISM

With the above *tour d'horizon* of theory, practice and challenges as backdrop, we now direct attention to the emerging intergovernmental interplay, or to what the title of this volume labels "environmental federalism". A convenient launch point is the forewarning from former Alberta Premier Peter Lougheed of the distinct possibility that the proliferation of uncoordinated and even conflicting federal-provincial and interprovincial policies has the potential, if mismanaged, to precipitate another NEP-type donnybrook. The number and heterogeneity of initiatives must be staggering for companies that must deal with them. Alberta has adopted a C&T system that effectively imposes an origin-based emissions charge (essentially a production tax) while insisting that all revenues from environmental levies remain within the province. BC and Quebec have imposed destination-based carbon taxes (essentially taxes on consumption). Quebec has also joined with Ontario in challenging Ottawa and their sister provinces to adopt stricter GHG controls via a nation-wide C&T system, and, along with BC and Manitoba, both Quebec and Ontario have joined the WCI C&T system. Meanwhile, although initially intent on launching its own national, intensity-based C&T and CCS systems, Ottawa has more recently suggested that Canada link up with the U.S. system (presumably in part to avoid the softwood-lumber-type contentiousness and litigiousness being carried over to the environmental arena and perhaps in part as a way to wrest environmental leadership away from the provinces). The stage does indeed seem to be set for a potential rerun of the federal-provincial pyrotechnics of the NEP. Given this, and the reality that both levels of government have the constitutional right to occupy the environmental field (as the papers in this volume by Elgie and Chalifour elaborate in detail) the initial focus of this section is to assess the likelihood of joint federal-provincial occupancy under carbon tax and C&T systems.

A convenient entry point to the discussion of federal-provincial environmental interaction is to note that, regardless of where the GHG emissions may originate within Canada, the associated negative externalities are experienced by all Canadians. Whether the externality takes the form of environmental degradation, climate change, or the usurpation of space or "quota" under a national emissions cap, provincial boundaries are largely irrelevant to the distribution of burden. There is thus little case to be made for demands that revenues from carbon pricing, whatever its form, whether CATT, carbon tax or C&T, remain within the province or territory in which they were generated. This is particularly the case with national carbon-pricing regimes, which, if effective and whether in the form of Dion's *Green Shift* or Harper's Cap & Trade, will ramify throughout the economy, burdening all Canadians. It would be quite appropriate, therefore, that part of the revenue generated be used to offset the impact of the carbon-pricing regime on low-income Canadians throughout the country, without striving to achieve a correspondence between regional revenue receipts and their subsequent disbursement. That said, a national carbon-pricing regime need not be a federal revenue grab. For example, in the case of the CATT, since it would be an indirect tax on carbon consumption, it would have to be a federal or, as is the case with the HST, at least a federally administered tax. The revenues, however, could be shared with the provinces in any manner deemed mutually acceptable, for example, on an equal per capita basis or possibly on the basis of derivation. Indeed, and following the precedent of the HST, the provinces might be encouraged to harmonize provincial CATs with a federal CATT, thereby determining the level and use of provincial carbon-pricing revenues they considered appropriate. Such revenues, together with federal carbon-pricing revenues, could be used *inter alia* to encourage climate-change-related technological development, to engage in the creation of carbon sinks or CCS, or to pursue revenue neutrality.

Matters become considerably more complicated in federal-provincial terms when the carbon tax is an upstream or origin-based tax. While the ultimate incidence of this tax may also rest with consumers, the locus of the revenue stream also moves upstream, i.e., toward where the carbon emissions are produced rather than where they are ultimately consumed. While Ottawa has the constitutional right to mount such a carbon tax (Chalifour, this volume), the reality is that it probably would not be able to monopolize the area since at least some provinces would exercise their constitutional right under s. 92A to mount their own upstream carbon taxes, given that s. 92A seems to give the provinces carte blanche to raise revenues in respect of non-renewable natural resources, forestry and electricity generation. This would likely raise a raft of efficiency and harmonization issues (both interprovincial and federal-provincial) with respect to coverage, rates, the treatment of multi-provincial enterprises, domestic and international border tax issues, etc.

Beyond these challenges, the reality will be that with an upstream tax the abatement revenues will accrue to the provinces where the emissions are generated rather than where the underlying products are consumed. While this may well be appropriate, in terms of "environmental federalism", if these provinces were to dedicate the resulting revenues toward the development of low-carbon technologies, carbon sinks and the like, it would surely wreak havoc with "fiscal federalism". To see this, suppose that the bulk of the revenues from pollution abatement were indeed to go, via upstream or origin-based emission taxes, to the energy provinces. This would mean that not only would they be receiving the huge rents/royalties from fossil energy but, as well, they would be receiving the very significant revenues from carbon-abatement taxes. Given the long-standing fiscal equity issue relating to the funding of energy royalties in the equalization program, an origin-based emissions tax the proceeds of which would accrue largely to the energy provinces would dramatically exacerbate an already acute interprovincial fiscal imbalance problem. Arguably, therefore, a nationally run, destination-based carbon tax regime seems to be the preferred policy option on this score because the revenues from carbon abatement policies will be distributed across the provinces in line with the carbon-footprint of the consumed products. If upstream taxes were to be the norm, however, then the equalization formula would probably have to be re-thought and/or redesigned. One option along these lines (Courchene 2008b) would be to bring the roughly $30 billion of federal-provincial cash transfers (i.e., CHT and CST) into play by converting them from their current equal-per-capita format to an approach where they would be subject to a clawback of, say, 25 cents for every dollar that a given province's total own-source revenues (including carbon taxes) plus equalization exceeded the all-provincial average.

Canada-U.S. Interrelationships

To round out our selected set of environmental federalism challenges, some of the features of the possible/proposed joint Canada-U.S. C&T systems merit highlighting. Turning first to the WCI regime, the overall permit limit will be the sum of the individual partner limits. Specifically, each partner state or province will be allotted its emissions limit for 2012, which will then decline in straight-line fashion through to 2020. Once distributed within each member of the partnership, the trading of permits will then determine the common carbon price among the WCI partners. (By way of a relevant aside, the E.U. ETS system also initially allocated individual limits to the participating countries, which presumably played a part in the over-issuing of permits that led to a zero carbon price. By constraining the total E.U. allotment, the most recent version of the ETS ensures a positive carbon price.) This caveat aside, the WCI individual partner emissions permits lead rather naturally to a related challenge, namely, the provision for "early reduction allowances", which would allow partners to acquire additional emissions credits for compensation, as it were, for reducing emissions prior to the start of the C&T. For example, in the Canadian context, Quebec would want additional allowances for its efforts that have brought its emissions down to levels that are within the Kyoto targets, whereas Ontario's emissions have increased relative to its 1990 emissions. Since they can be sold in the permit-trading market, these early reduction allowances are essentially money in the bank for Quebec. More intriguing, and surely of interest to Alberta, is that the WCI model also has provision for re-allocating the carbon footprint for situations where, say, energy is produced in one province/state but consumed in another, i.e., the carbon footprint embodied in exports would be assigned on the basis of the locus of consumption, as we have argued above it should be.

Other intergovernmental features of the WCI are probably less likely to be carried over to a national or Canada-U.S. C&T. For example, under WCI, provinces will obtain the revenues from auctioning permits, will have control over offsets, will have oversight over permit trading, and will have a cross-border level playing field. It is not clear whether the final (House-Senate) version of the Waxman-Markey bill will embody these features, or whether the allotments it would create will be tradable on par with those of the WCI. Both the provinces and Ottawa will be especially concerned about the possibility of protectionist spillovers from the final version of Waxman-Markey into NAFTA, compounding those related to U.S. Homeland Security.

The bottom line here is that while a Canadian C&T regime would severely challenge the design and implementation ability of our political class (and especially so if the provinces lobby for some of the WCI features to be embodied in a national C&T), the prospect of Canada buying into a U.S. C&T system after it has been wrung through the geo-economic and political rent-seeking and rent-keeping machine called Congress may be quite another matter. The upside in all of this is that Ottawa, by committing itself to link up with the U.S. C&T, will have finally jettisoned its criticized role as a Kyoto signatory but non-implementer, and will have asserted the leadership role on the environmental file that its earlier inaction had defaulted to the provinces.

CONCLUSION

Our conclusion focuses on the "federalism" part of the title rather than the "carbon pricing" part. While there has already been much attention devoted to the challenges facing federalism on the climate-change front, there is yet another perspective of the overall challenge that thus far has been ignored, or at least underplayed. Specifically, there is no equivalent on the environmental front to the more than 50-year history of federal-provincial fiscal relations dating from the inauguration of the equalization program in 1957. Fiscal federalism involves scores of meetings of federal and provincial officials each year. The processes of fiscal federalism also include a host of federal-provincial agreements on equalization, on tax-collection harmonization, on a national tax collection agency (the CRA) and even on securing the internal social, economic and fiscal unions. However, over the foreseeable future, environmental federalism will likely become every bit as important as fiscal federalism. Indeed, it may embrace key aspects of fiscal federalism. Given this, and the reality that, when compared with the fiscal-federalism infrastructure, the political and institutional machinery in the area of environmental federalism ranges from weak to non-existent, both Ottawa and the provinces (individually and/or via the Council of the Federation) need to take immediate steps to deepen the intergovernmental infrastructure relating to the substance and the processes of environmental federalism. As matters now stand, were one to call a federal-provincial meeting on climate change, it is not clear to whom one would send the invitations – to the environment ministers, to the energy/resources ministers, to the finance ministers, to the industry ministers, to the intergovernmental ministers, or perhaps even to the first ministers?

Addressing climate change is a sufficiently daunting challenge in its own right without the complication of tolerating the reality that the structures and processes of environmental federalism are in a state of disarray. Phrased differently, we will have made progress on the climate-change front when "environmental federalism" takes its rightful place in our policy vocabulary.

REFERENCES

Allan, J.R. and T.J. Courchene. 2008. "Level the greenhouse", *National Post*, 15 May, FP15.

Courchene, T.J. and J.R. Allan. 2008a. "Climate Change: The Case for a Carbon Tariff Tax", *Policy Options* (March): 59-64.

— 2008b. "Kyoto walks all over Canada: The carbon footprint belongs to the nations that produce the fossil fuels or the nations that consume them?" *The Globe and Mail*, 17 July, A13.

— 2008c. "The Provinces and Carbon Pricing: Three Inconvenient Truths", *Policy Options, Options politique* (December/January): 60-67.

Courchene, T.J. 2008a. "Climate Change, Competitiveness and Environmental Federalism: The Case for a Carbon Tax". Background Paper for Canada 2020 Address 3 June. Available as a Working Paper at http://www.irpp.org/about/index.htm.

— 2008b. "Fiscalamity! Ontario: from Heartland to Have Not", *Policy Options, Options politique* (June): 46-54.

— 2009a. "Murphy's Law Running Rampant? Gump and Gretzky to the Rescue". Presented at *Making Progress in Difficult Times*. At http://www.queensu.ca/sps/events/mpa_policy_forum/2009/documents/Couchene.pdf.

— 2009b. "Carbon Pricing as a Wicked Problem", in C. Andrews *et al.* (eds.), *Gilles Paquet: Homo Hereticus*. Ottawa: University of Ottawa Press.

Harrison, K. and L.M. Sundstrom. 2007. "The Comparative Politics of Climate Change", *Global Environmental Politics*, 7(4): 1-18.

Horne, M. 2008. "Cap-and-Trade Fact Sheet" (March). Pembina Institute. At http://pubs.pembina.org/reports/capandtrade-fs.pdf.

Hyndman, R. 2009. "The Cost of Carbon", *Alternatives Journal* 35(1).

Ismer, R. and K. Neuhoff. 2007. "Border Tax Adjustments: A Feasible Way to Support Stringent Emission Trading", Cambridge Working Papers in Economics CWPE 0409. University of Cambridge: Department of Applied Economics.

Mintz, J. and N. Olewiler. 2008. *A Simple Approach for Bettering the Environment and the Economy: Restructuring the Federal Fuel Excise Tax*, prepared for the Sustainable Prosperity Initiative. University of Ottawa: Institute of the Environment.

Mittelstaedt, M. 2009. "The high cost of blowing smoke on the high seas", *The Globe and Mail*, 1 April, A1.

O'Neil, P. 2009. "Canada snubs G8 emissions target", *National Post,* 10 July, A1.

Simpson, J., M. Jaccard, and N. Rivers. 2007. *Hot Air: Meeting Canada's Climate Change Challenge*. Toronto: McClelland Stewart.

Tamiotti *et al.* 2009. *Trade and Climate Change: A Report by the United Nations Environment Program and the World Trade Organization.*

Vidal, J. 2008. "Shipping boom fuels rising tide of global CO2 emissions", *The Guardian*, 13 February.

Weitzman, M.L. 1974. "Prices vs. Quantities", *Review of Economic Studies* 41(4): 477-491.

6

Carbon Pricing as if GHG Mitigation Matters

Rick Hyndman

Le mieux est l'ennemi du bien.

Voltaire's well known dictum – the best is the enemy of the good – is certainly applicable to public policy concerning climate change, an area where much needed action has been stalled by the advocacy of policies more costly than the public will support. Meanwhile, time continues to pass as governments debate how to commit themselves to action to reduce greenhouse gas (GHG) emissions, and scientific reports warn with increasing urgency that time is running out if we want to avoid the risk of catastrophic interference with the climate.

Economists love market solutions to problems. The damage from global GHG emissions is *the* classic example of market failure – a global external cost that emitters do not bear. Fix the failure and everything will be fine. Apply an appropriate carbon price globally and we will get the most efficient possible outcome. There are calls from environmentalists and others for the immediate adoption of very high carbon prices that will drive very costly actions by the well-off industrialized countries on their own emissions and, as well, will lend significant support to poor countries and newly industrializing countries to develop more cleanly. The problem is that voters do not like market solutions to problems in general, and oppose any policies that impact them negatively to any significant extent.

Within the United States and Canada, there is no broad support for costly actions for which the public would have to pay, and, whether the actions be taken by industry or directly by consumers, it is the public who, in the final analysis, will bear the burden. The saga of the Waxman-Markey bill in the United States provides a clear message that carbon pricing will have to start modestly, with little prospect of the carbon price rising to a level that would drive large reductions by 2020. In Canada, the National Round Table on the Environment and the Economy (NRTEE) released a report in April 2009 on its

---

The views expressed here are my own and do not represent an official CAPP position. For information on the latter, see www.capp.ca.

study of the policies necessary to meet the federal government's emission targets for 2020 and 2050.[1] NRTEE presented a scenario with \$100-150/tonne CO_2 as a feasible plan to use carbon pricing to drive the changes needed to meet the 2020 target, even with a significant contribution toward the target coming from buying foreign credits. The TD Bank commissioned a study by the Pembina Institute and the David Suzuki Foundation using modelling analysis by MK Jaccard and Associates (MKJA 2009). This study released in October, 2009 showed the need for similar cost actions, with different regional distribution of the costs and greater declines in oil and gas output resulting from their policy assumptions. It also added an alternative target even further beyond what is achievable domestically.

Both modelling exercises rely in an essential way on general equilibrium models of the Canadian economy, which reallocate Canadian capital and labour resources in response to the introduction of carbon pricing and other policies while maintaining full employment, investment levels and trade balances. For long-term analysis, those are appropriate assumptions. However, for analyzing short-term restructuring of the economy over the next 8 to 10 years, the general equilibrium assumptions miss the main challenges which governments face. Read in the context of the policy debate in the United States, both analyses are better interpreted as demonstrating the political infeasibility of relying on carbon pricing to meet near-term 2020 targets.[2]

Figure 1 below illustrates the problem. Achieving the government's 2020 target would require actions with costs up to \$100/tonne CO_2 or more. There is notional public support for the target but no real support for anything like such costs eventually falling on consumers and taxpayers. What can be achieved at a cost that is publicly acceptable falls far short of both public and government expectations.

FORMULATING NATIONAL EFFORTS ON GHG EMISSION REDUCTIONS

There is a basic question about how we arrive at and describe our national effort to reduce GHG emissions. An emission cap or target level is not a good indication of what the public will support. Setting emission-reduction objectives as a specified level of emissions in some year ignores the reality that the target Canadians will support depends on the cost of achieving it. Although most environmentalists and many policy analysts are adamant in their calls for a firm

---

[1]The Executive Summary of the NRTEE report appears as an Appendix to this volume.

[2]Some idea of the political difficulties likely associated with the NRTEE carbon-price range of \$100-150 per tonne is provided by noting that a carbon price of \$115/tonne represents a tax of \$329/tonne on coal, the market price for which is currently in the range of \$15-110/tonne.

Figure 1: Carbon-Price Gaps

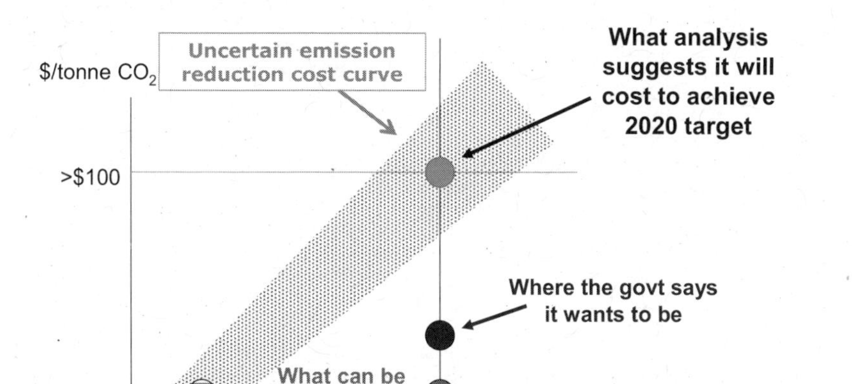

cap on emissions, any specific level would not necessarily reflect the balancing of costs and emissions that the public will actually support. The same applies to a specific price. It is more realistic to portray emission-reduction objectives as an emission-reduction demand curve, which embodies the reality that Canadians will be willing to make more reductions as the cost of achieving them declines – the lower the cost of emission reductions, the more the public will be willing to support. Figure 2 thus portrays two negatively inclined demand curves, showing the inverse relationship between the cost of emission reductions and the amount of reductions demanded. The lower of the two curves reflects the evident unwillingness of the public to bear emission-reduction costs, while the higher curve indicates the substantial carbon prices necessary to effect a 224 million tonne (mT) reduction, this being the reduction required to move from the 2020 800 mT business-as-usual (BAU) level to the 576 mT target level.

Arguably, the Canadian public's demand curve for reductions has not been explored beyond a desire to have large reductions at a low cost. Ideally, politicians would be clear about the feasible choices and explain to the electorate that the real policy choices have to be made within the set of feasible options on the emission-reduction cost curve. Unfortunately, fiction about going after big, polluting industry with no consequences to the public seems easier to sell than the fact that, one way or another, consumers and taxpayers will have to bear most of the costs of reducing emissions. The political challenge in this environment is to design efficient policy to begin action and increase the effort over time.

Figure 2: Emission-Reduction Demand Curves

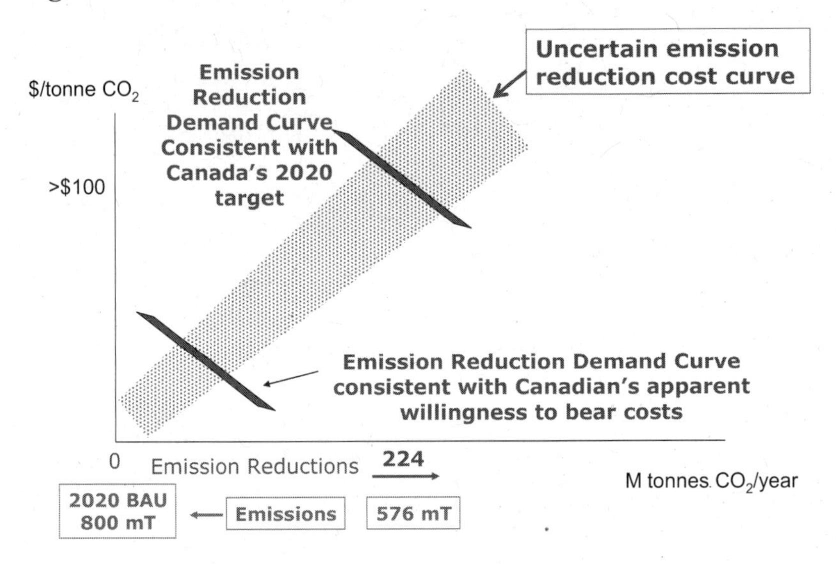

Cap and Trade: The Current Policy Favourite

What about the apparently widespread support for a limit on national emissions and a cap-and-trade carbon pricing system to achieve it? The basis of support seems to lie in four different views:

- A view of the public that cap and trade applies to industry and leaves consumers unaffected.
- A view of promoters of immediate costly actions that a commitment to significant action can only be made by focusing on outcomes to the exclusion of costs, and that cap and trade will guarantee a price as high as necessary to achieve the targets.
- A view of brokers, bankers and consultants that carbon-permit trading will create a whole new financial sector and represent a high-income industry in its own right.
- A view among some in industry that foreign credits will maintain the lowest price and cost of domestic actions.

There is frequently reference to the U.S. SO_2 cap-and-trade system as an example of success in which power-sector SO_2-emission reductions were achieved at a cost far below what was initially expected under traditional cost-of-service regulation. But the SO_2 example is not a good analogy. In that case, there was available an existing emission-control technology at a cost that society was willing to bear. Emission pricing provided an opportunity for lower-cost methods based on compliance flexibility and decision-making by plant owners. As an experiment, it was a no-risk proposition: if lower-cost opportunities failed to materialize, the default alternative was for power plants to proceed with what they would have done under regulation, namely, install SO_2 scrubbers. For CO_2 there are far fewer existing low-cost technologies than those necessary to achieve the 2020 emission objectives at a publicly acceptable cost.

The immediate, relevant policy question is how to contribute the most toward emission objectives given the current situation. For CO_2, given the public attitude to costs, the emission-cap approach is a serious distraction. As illustrated in Figure 2, the cap debate is not an informed discussion of alternatives drawn from the feasible set of cost and emission outcomes. The debate should be about types and costs of actions that have a chance of public support, and the most efficient policies to make them happen.

Unachievable Targets and International Emission Trading

In contrast with the U.S. SO_2 cap and trade example, in the case of CO_2, if low-cost reductions are not found to meet the cap – and, at this time, there is a presumption that they are simply not available [Cf. Chris Green, this volume (eds.)] – the default action to meet the cap domestically is reducing output, which is not acceptable. The alternative in an emission-trading world is to pay for reductions in other countries.

Domestic Reductions

The Kaya Identity is a useful tool for organizing the discussion of emission targets and reductions. It states as follows:

Total carbon emissions = C/E x E/GDP x GDP/Pop x Pop

where C/E is the carbon-emission intensity of energy, E/GDP is energy intensity of GDP, and Pop is population. With rising per capita output and rising population, reducing total carbon emissions requires decreases in the carbon intensity of energy supplies and in the energy intensity of GDP that more than offset rising output and population.

The United States provides a good example of the challenge of reducing emissions from 2010 to 2020 by 20 percent, the goal in the original draft of the Waxman-Markey bill. For this purpose, we can restate the Kaya identity as follows:

Total carbon emissions = CIO x GDP

where the carbon intensity of output, CIO, is given by C/E x E/GDP, and GDP by GDP/Pop x Pop. For the United States, between 1980 and 2006 the carbon intensity of output, CIO, decreased at an annual rate of 2.2 percent (a much more rapid rate of decline than the global rate of 1.3 percent). Approximately 2.0 percentage points of this annual rate were attributable to a decline in energy intensity (E/GDP), while the remaining 0.2 points were from reductions in the carbon intensity of energy (C/E). For the purpose of the following analysis, the annual rate of improvement in the carbon intensity of output has been increased by one percentage point (an increase of almost one-half in the rate of improvement) to 3.2 percent per year for the period 2010-2020. This would be a dramatic decline in the CIO.[3]

The U.S. EPA is assuming GDP will increase on average by 2.5 percent per year over the period 2010-2020, while the White House budgetary forecast is based on a 3.3 percent GDP growth rate. Clearly, at a business-as-usual (BAU) rate of improvement (i.e., decrease) in CIO, emissions would rise. Assuming the more modest of these two GDP growth rates, and an average annual rate of decrease in the CIO of 3.2 percent per year, emissions would decrease at an average annual rate of 0.7 percent. However, to meet a target of a 20 percent decrease from 2010 levels, emissions must decline by 2.2 percent, or three times as fast. Absent an acceleration in the decrease of CIO, meeting the emissions target would require sacrificing GDP growth. Just how much output would have to be sacrificed is evident in the following table, which is excerpted from the more complete calculation shown in the Appendix to this chapter.

These Kaya-identity calculations are a good complement to the general-equilibrium macro-economic analysis of what it takes to meet the 2020 targets. They indicate that, where even significant increases in the rate of improvements in the CIO are insufficient to meet the cap, GDP would have to be reduced if

Table 1: GDP Loss to Meet 2020 Emission Target

	2010	*2020*
GDP @ assumed 2.5% growth rate $trillions	$14.00	$17.92
Covered Emissions BAU m tonnes CO2	6,000	6,149
Covered Emissions target m tonnes CO2	6,000	4,800
Annual % change in CIO required by target		-4.6%
Annual % change in CIO max achievable		-3.2%
Achievable emissions	6,000	5,548
Achievable minus target emissions		748
GDP compatible with emissions target $trillions	$14.00	$15.50
% loss in GDP		-13.5%
Cumulative GDP loss $trillions		$12.62
Cost of credits @ $30/tonne $billions		$22
% GDP		0.125%
Cumulative cost of credits $billions		$129

[3]Chris Green provided the historical data and suggested the assumed rate of improvement in carbon intensity.

actual emissions in the United States were to meet the cap. Meeting the emission reductions at home (in the United States), would thus imply serious GDP costs, $2.4 trillion in 2020 ($17.92 t - $15.50 t) and cumulatively $12.6 trillion from 2011 to 2020. It seems obvious that politicians who proposed the 20 percent emission reduction target in the original Waxman-Markey bill did not have such an effect in mind. This gives an indication of the amount of economic restructuring that is assumed in general equilibrium modelling analysis, beyond what is already assumed by increasing the rate of improvement in CIO in these calculations.

In Canada's case, the gap between BAU and what is required is significantly larger. The average annual improvement in carbon intensity of GDP (CIO) from 1990 to 2007 was 1.3 percent (versus the 2.2 percent improvement in the United States from 1980 to 2006). In the MKJA study referred to above, the 2010 to 2020 projected BAU annual improvement is 0.93 percent. To reach the government's 2020 target would require an average annual decline in CIO from 2010 to 2020 of 4.4 percent. The MKJA analysis indicates this can be achieved with a carbon emissions price rising to $145/tonne in 2005 C$ terms (p. 64). Both the percentage annual improvement in CIO and the carbon-emissions price required to drive it in the models indicate the dramatic changes required in the Canadian energy system relative to both projected BAU and history. This highlights the economic and political challenges the federal and provincial governments would face in selling such a change in the direction of the economy.

FOREIGN CREDITS

The alternative to high-cost domestic actions and output reductions is reliance on foreign credits, i.e., accepting as part of the U.S. reductions emission reductions in other countries that are paid for by the United States. If these were available at, say, $30/tonne, the cost of "meeting" the target through purchases of those credits would be dramatically lower than reducing output to meet the target domestically; indeed, they would be only one-eighth of one percent of the otherwise necessary drop in GDP.[4] This is the story behind support for emission trading – the assumption that there are vast tonnes of low-cost reductions available in developing countries that developed countries can pay for to meet targets that can only be achieved domestically at much higher cost.

However, the availability of foreign credits when all major economies are contributing to global reductions will be limited. That is, if addressing global GHG emissions requires large reductions in *both* developed and other major economies, then the volume of low-cost, additional foreign reductions available to developed countries, at the moderate prices and costs of reductions being assumed, will simply not be large enough.

A more thorough analysis of the potential supply of foreign credits in a world in which China, India and other newly industrializing countries are

---

[4][($30 x 748 mT target gap) ÷ $17,952 t] x 100 = 0.125 percent.

making their own contribution to global reductions may well reveal that there will not be a large supply of reductions in developing countries for which the United States and other developed countries can pay and take credit. In that case, the price of credits and allowances in the United States would be more likely to be $50/tonne or more. An expectation of high carbon-emission prices would increase opposition to the proposed cap-and-trade system.

The MKJA analysis for Canada is more realistic about the prices required to achieve the government's emission target: $100/tonne (2005 C$), even with the purchase of 73 m tonnes of foreign credits, which represent a quarter of the reduction from BAU and half the reduction from 2010 to 2020 (pp. 20-21).

PUBLIC OPPOSITION TO HIGH CARBON-EMISSION PRICES AFFECTING THEM

A carbon charge or tax is agreed by most economists to be a better approach to pricing than cap and trade, but its simplicity and clarity facilitates focused opposition, especially in the United States. As cap and trade comes to be understood as a carbon charge levied in a new currency – CO_2 allowances – broad application of cap and trade will face similar opposition, as witnessed by the amendments and opposition to the Waxman-Markey bill, even with carbon credit prices assumed to be in the $15 to $25/tonne range in 2020.

In the face of public opposition to having to bear any significant cost of carbon emissions, what policy design can begin to address the problem and still be publicly acceptable? In other words, what is the best we can do, as an initial policy, recognizing that in the next 5-10 years, whatever one thinks is necessary, the public is not going to support incurring very large costs? There are two areas for initial focus of the policy: on the one hand, increased support for the development of low-carbon-emission technology, and, on the other, carbon pricing for large, energy-intensive industry that will guide their technology choices, while achieving this without incurring high costs that would have to be recovered from consumers. These are discussed in the following two sections.

MITIGATION IS FUNDAMENTALLY A TECHNOLOGY CHALLENGE

Slowing and reducing global GHG emissions requires action on all fronts. Broadly, this requires gains from reducing carbon and energy intensities, from halting deforestation, and from improving industrial and agriculture processes. Reducing and reversing deforestation is an urgent problem requiring its own focus. But the main challenge facing the world is the transformation of the global energy system from its overwhelming reliance on CO_2-emitting hydrocarbon supplies to low-carbon-emitting sources, while global energy demand increases at a fast pace because of newly industrializing countries. To bring about that transformation at a pace consistent with medium- and long-term emission objectives requires research, development and demonstration of a range of energy-supply and -use technologies, ones that can be deployed to meet energy needs with low- to zero-carbon emissions at a cost that most countries are prepared to incur.

Carbon pricing is an important central element of policy strategy to promote efficient, market-driven responses that will reduce emissions. A widely applied policy of a high, certain price on carbon emissions would not only drive adoption of existing lower-carbon technologies, but would also stimulate private investment in the development and deployment of new low-carbon technologies. There is, however, little prospect for high carbon prices in the near term. Without public support for very high, broadly applied carbon pricing, most of the incidence of which will fall on consumers, private investors cannot rely on an expected future carbon price that will make their risky R&D spending a good investment. Furthermore, even with a reliable, high price, we know that private decisions result in under-investment in R&D due to the inability of developers of new technology to capture all the benefits of their inventions.

Nevertheless, the world needs to get going on investment in technology research, development and demonstration now, despite the lack of public support for a price that will drive the change needed through private investment alone.

Reconciling a Slow Start on Carbon Pricing with Large Investments in Technology

The way to reconcile a publicly acceptable, low initial price on carbon emissions with the need for large investments in low-carbon-technology development is to implement the carbon price as a low charge on emissions to raise revenue for funding technology research, development and deployment (RD&D). Starting with $5/tonne CO_2 and applying this to 80 percent of emissions would raise $2.9 billion per year in Canada, and $24 billion in the United States. [$5.00/t x 6000 mt x.8= $24 billion] These amounts are not huge, but represent large increases from current support for low-carbon-technology RD&D.

It is very important to put in place the right governance of these funds. The revenue should go into a technology-development trust fund to support RD&D of transformative technologies and be managed by an independent board, with a public-interest mandate, at arm's length from governments and political interference.

Perhaps some revenue should also be diverted to provide aid to poor countries to develop cleanly, including national programs to reverse tropical deforestation.

A HIGHER PRICE SIGNAL FOR ENERGY-INTENSIVE, TRADE-EXPOSED (EITE) SECTORS

The key resistance to higher carbon pricing is coming from the public (to the extent that it would apply to their emissions), energy-intensive industries exposed to competition from countries without carbon pricing, and coal-fired power generators exposed to capital losses. For the non-EITE sectors, i.e., electricity and broad energy consumption, the free allocations under Waxman-Markey to protect consumers from price increases illustrates the political challenge of using even moderate carbon prices to drive reduced emissions. However, it is possible to begin (and, in the case of the E.U. and Alberta, continue) with a higher price of carbon applying to the EITE sectors, if the price is applied at the margin, i.e., if EITE sectors are given a free allocation tied to their output, but not their emissions. This mechanism results in a higher marginal price signal, but an average cost net of the free allocation to these sectors that is at, or below, the $5/tonne level.[5] The E.U. ETS, the system planned by Australia, the Waxman-Markey and Kerry-Boxer bills and the Alberta carbon-pricing system all do this, each in their own way for their EITE sectors. How do they work?

Facilities pay the carbon price on their emissions in excess of sector- or facility-specific performance standards. If they are better than the standard, they generate saleable credits for the difference and could even be net beneficiaries of carbon pricing. Thus, they have the full incentive of the carbon price to improve their performance, but do not pay for emissions at the performance-standard level. Expressed differently, each tonne of emission reduction they effect will reduce their charges at the high *marginal* price of carbon (or create a credit that may be sold at that high price), while leaving their *average* emission charge at a level that does not too seriously impair their competitiveness. For example, if the standard emissions intensity is 80 kg/unit of output and the charge is $15 per tonne of CO_2, a firm with an emissions intensity of 100 kg/unit of output would have an average cost per tonne of their total emissions of 0.20 x $15 = $3.00. The firm is thus called on to finance emission charges at the relatively low average rate of $3.00 per tonne, while predicating its output and investment decisions – including decisions of whether to invest in emission-reducing technologies – on the marginal charge of $20.00 per tonne. If it were deemed necessary, a higher marginal charge, say $30, could still yield a relatively low average cost if the performance-standard intensity were raised to compensate; for example, a standard of 0.90 tonne/unit of output and a $30 charge yields an average cost of 0.10 x $30 = $3.00. Over time, as the availability and cost of low-carbon-emissions technology improved, the charge could be raised and the performance-standard emissions intensity lowered.

This performance-standard approach (for EITE industries in the developed countries) is equivalent to a "free allocation" that prevents them from being put

---

[5]This may be shown symbolically by the following equation, where C_j is the charge to be paid by the jth industry or firm; O_j is the output of that industry or firm; i_j is its emissions intensity per unit of output and i_s is the performance standard intensity; and p is the price or charge per tonne of CO_2 equivalent:

$$C_j = O_j[i_j - i_s]p.$$

The product $O_j i_s$ may be thought of as a "free allocation" of emissions that may be subtracted from actual emissions, $O_j i_j$, to determine the emissions subject to the marginal charge rate p.

at a disadvantage to their competitors in the newly industrializing countries that do not face carbon pricing. It thus removes the pressure for so-called *leakage* – the shifting of investment in these sectors to those developing countries, a shift that would yield no global benefit in GHG-emission reductions.

CONCLUSION

A simple, broadly applied, significant and escalating, carbon charge harmonized across major trading countries would lead to the most efficient choices globally for reducing GHG emissions, and would provide an incentive for private investment in the development and demonstration of new, low-carbon-emission technologies.[6] Cap-and-trade systems that generate a comparable price on carbon are similar to straightforward charges on emissions but, by introducing international transfers and price volatility, more complex. In either case, there is public resistance to significant carbon prices that affect them. Given this, the major economies need to get started or step up their efforts to manage greenhouse gas emissions before such a system could be agreed upon and made acceptable to the public in many countries.

A high level of technology development and demonstration will be required to enable large scale deployment of the low-carbon-emission technologies necessary to achieve the medium- and long-term emission objectives being discussed in the UNFCCC process. Given the political constraints on imposing a large cost on carbon emissions that can be flowed through to consumers, especially in the United States, the near-term carbon-emission price signal cannot be expected to drive private investment in low-carbon technology to the level required. Therefore, in the near and medium term, there needs to be a high level of policy-supported investment in technology development and demonstration. An obvious mechanism would be to start with a broadly applied but modest carbon charge to raise the revenue for such support. This would have the added benefit of initiating broad carbon-emissions pricing as a first step toward having the price signal itself drive efficient choices for reducing emissions.

In addition to a broadly applied low-carbon price to raise revenue for low-carbon-emission technology development, a higher price incentive can be provided for large industry. Europe, Alberta and British Columbia[7] have started, and the rest of Canada, Australia and the United States are proposing to start in 2011 and 2012 with pricing emissions for large industrial emitters, while addressing the competitiveness of EITE sectors. An output-based free allocation

[6]William Nordhaus argued the case for harmonized carbon taxes yet again in a recent paper delivered at a conference in Copenhagen, March 10-12, 2009, *Economic Issues in a Designing a Global Agreement on Global Warming* (http://nordhaus.econ.yale.edu/documents/Copenhagen_052909.pdf).

[7]British Columbia has started with a broad, revenue-neutral carbon tax that does not address competitiveness of EITE sectors, though its engagement with the Western Climate Initiative discussions suggests that this might be addressed in the future.

or performance standard allows a higher price signal to drive technology choices while avoiding significant costs that would lead to trade and investment-distorting leakage in the initial global steps to reduce GHG emissions.

REFERENCES

M.K. Jaccard and Associates Inc. (MKJA). 2009. *Exploration of Two Canadian Greenhouse Gas Emissions Targets: 25% Below 1990 and 20% Below 2006 Levels by 2020*, October 18. At www.pembina.org.

APPENDIX: HYPOTHETICAL EXAMPLE OF EITE PRICING

	Pre-policy actual emissions 100kg/unit	
	Performance standard 80kg/unit with $15/tonne price	*Performance standard 90kg/unit with $30/tonne price*
Output # of units/year	1,000,000	1,000,000
Emissions per unit of output kg/unit	100	100
Annual emissions before policy, tonnes/year	100,000	100,000
Policy performance standard (free, output-based allocation per unit), kg/unit	80	90
Emissions per unit of output on which cost imposed before response to policy, kg/unit	20	10
Cost per unit of output before response to policy, $/unit	$0.30	$0.30
Cost per year before response to policy, $/year	$300,000	$300,000
Emissions/unit after policy if 5% reduction is achievable at $25/tonne, kg/unit	100	95
Emissions per unit of output on which cost imposed after response to policy, kg/unit	20	5
Cost per unit of output after policy: reduction cost + charge on emissions over standard, $/unit	$0.30	$0.275
Emissions after policy, tonnes/year	100,000	95,000
Cost per year after response to policy, $/year	$300,000	$275,000
Cost if carbon price applied to all emissions, $/year	$1,000,000	$2,850,000

IV

Federalism, Multi-Level Governance and Carbon Pricing

7

Multi-Level Governance and Carbon Pricing in Canada, the United States, and the European Union

Kathryn Harrison

INTRODUCTION

The scientific community is advocating global greenhouse-gas-emissions reductions with increasing urgency. If we are to respond in time to prevent irreversible disruption to the global climate, significant policy reforms will need to be undertaken within the context of existing political institutions. To that end, this chapter asks whether one particular institution, federalism, facilitates or deters adoption of policies to reduce greenhouse emissions, with particular attention to policy instruments that entail carbon pricing. To explore that question, the chapter employs comparison of two federations, Canada and the United States, and one quasi-federation, the European Union.

To preview the conclusions, the impact of multi-level governance in the European Union has been largely positive. Various climate-policy leaders have emerged over time among the member states, and that horizontal dynamic has been matched vertically by activism from the European Council of Ministers, Parliament, and Commission. In response, the European Union has made the greatest progress in adopting policy reforms to price carbon, most notably through its Europe-wide Emissions Trading System. In the United States, federalism also has had a positive impact in facilitating policy innovation and diffusion at the state level, albeit in the face of a policy vacuum at the national level. With respect to carbon pricing, some (though not all) state governments are collaborating to create regional emissions trading schemes. In contrast, in Canada the impact of federalism on climate policy to date has, on balance, been negative. As in the United States, there has been a dearth of action at the national level, but until quite recently Canadian provinces did not respond unilaterally to the same degree as their U.S. counterparts. Federal and provincial governments were deadlocked over how to respond to climate change for almost two decades. Provincial policy innovations have emerged since 2006, led most notably by British Columbia's adoption of a carbon tax and the commitment by BC, Manitoba, Ontario, and Quebec to join with U.S. states in emissions trading. However, those reforms have not diffused to provinces that account for half of Canada's current emissions and the majority of its projected emissions growth.

The chapter points to three factors in accounting for the very different experience of these three multi-level systems. First, greater public concern about climate change in the European Union has interacted with federal institutions, thus prompting a dynamic of vertical reinforcement in Europe not seen in either Canada or the United States. Second, differences in federal constitutions, in particular, greater central decision-making authority, has facilitated a more aggressive response by the European Union than the Canadian federal government. Finally, and arguably most important, the relatively low abatement costs anticipated by the largest states in the European Union and United States has facilitated leadership at the state level, while concentration of costs in key provinces has deterred comparable policy reforms in Canada.

INTERGOVERNMENTAL RELATIONS AND CLIMATE POLICY

Elsewhere, I have analyzed climate policy as a function of three broad factors: policy makers' own ideas, their electoral incentives, and the institutional context in which policies are adopted (Harrison and Sundstrom 2007). The first of these includes both the policy maker's perception of the problem and the policy maker's normative or ideological commitment to addressing it. Electoral incentives are a function of both public opinion and the distribution of costs and benefits within a polity, which will tend to yield a different balance of interest group pressures in different jurisdictions. Finally, the "rules of the game" – political institutions – can tip the balance in favour of some interests over others. The institution of federalism, or more generally multi-level governance, thus is just one among many factors that can be expected to influence climate policy. As always, the risk for a study that focuses on just one factor is of overstating its impact, in particular by attributing all outcomes of interest to that one variable. This chapter seeks to put federalism in perspective by drawing on insights from a larger cross-national project, which considers the impact of public opinion, interest group pressures, norms, and other political institutions (Harrison and Sundstrom, forthcoming). The relative impact of federalism will be revisited in the conclusion.

A variety of intergovernmental dynamics can emerge within systems of multi-level governance, some conducive to policy reforms to price carbon and others obstructive. One can distinguish between interprovincial or horizontal relations and federal-provincial or vertical relations. With respect to the former, the diversity of sub-national governments within a federation may facilitate policy innovation, with states or provinces serving as "laboratories of democracy" (*New State Ice Co. v. Liebmann* [1932] 311). Moreover, such innovations can spread to other states or provinces. While this dynamic is often referred to as a "race to the top", in the environmental field a "pull from the top" is much more common. Rather than seeking to *outdo* state leaders in a true race to the top (a form of reverse prisoners' dilemma), laggard states fearful of losing competitive advantage should they regulate unilaterally typically are content merely to *match* the standards set by the leaders (a form of assurance game)

(Harrison 2006). A more troubling horizontal dynamic, however, is the potential for a race to the bottom should states compete to attract investment by relaxing environmental standards, or, more plausibly, a "stuck at the bottom" dynamic in which no leaders step forward, leaving all states or provinces reluctant to tax or regulate unilaterally (Olewiler 2006).

Vertical or federal-provincial relations also offer both encouraging and troubling prospects. On one hand, a federal government can establish national standards in order to overcome a race-to-the-bottom dynamic. Federal and provincial or state governments may also coordinate their efforts to achieve complementarity, with the federal government undertaking responsibilities to which it is particularly well suited and the provinces or states undertaking those to which they are well suited. That said, agreement on which policy roles are particularly well suited to the federal or provincial governments can be elusive. On the other hand, a federal government may abdicate its role and simply "pass the buck" to the provinces, or vice versa (Harrison 1996). Moreover, even if both levels are willing to act, an expectation of federal-provincial consensus can be a recipe for deadlock rather than complementarity, a phenomenon Scharpf (1988) has referred to as a "joint-decision trap".

Finally, federal and provincial or state governments may use their authority to pursue policies that are at cross-purposes, in so doing undermining the effectiveness of the other's policies. This is not especially problematic in the case of traditional emissions regulations, since compliance with a stricter standard typically ensures compliance with a weaker standard for the same pollutant, though there may be additional costs borne by the polluter as a result of duplicative or inconsistent reporting requirements. A cap-and-trade program with grandfathering of permits presents similar questions of administrative efficiency, but also new challenges that have yet to be addressed by Canadian federal and provincial governments. For instance, will one jurisdiction accept reductions or offsets purchased by a local polluter from a polluter in another jurisdiction if the rules governing the latter are different?

While these inconsistencies create a prospect of unrealized efficiencies in trading, overlapping federal and provincial carbon taxes or auctioning of permits present more serious problems. With conventional regulation or grandfathering of tradable permits, a polluter can comply with both simply by meeting the more demanding standard. In contrast, overlapping carbon taxes or permit auctions have an additive effect: a polluter faces a much higher cost than intended by either the federal or provincial government. In that case, it is desirable that only one government would collect the tax or auction the permits. While federal and provincial governments typically are loathe to cede tax room to the other level, to the extent that future federal and provincial governments commit to *revenue-neutral* carbon taxation, the obstacles to coordination would be reduced, since neither has a financial incentive to be the tax collector. In any case, given the fate of the Liberal Party's proposed *Green Shift* in the 2008 federal election, the prospect of overlapping federal and provincial carbon taxes would appear to be moot in the foreseeable future.

The literature on policy making in federal systems is replete with examples of each of these dynamics. The question is thus under which conditions desirable intergovernmental relationships will emerge – horizontal innovation and vertical backup and coordination – and under which conditions we will see negative dynamics of provincial paralysis and federal-provincial deadlock or even obstruction. The sections that follow explore this question by comparing three "federations" which have experienced very different intergovernmental relations with respect to climate policy: the European Union, the United States, and Canada. Thereafter, the discussion will turn to variations in federal structure and background conditions that appear to explain observed differences.

MULTI-LEVEL REINFORCEMENT IN THE EUROPEAN UNION[1]

Although not a formal federation, the members of the European Union have acted in concert on climate change, both in international negotiations and in formulating E.U.-wide climate policies. The discussion here thus treats the European Union as a unit, and in particular focuses on the 15 member states that collectively negotiated and ratified the Kyoto Protocol, before the recent expansion of the European Union.

For over a decade, the European Union has acted as an international leader on climate change (Harris 2007), both calling for the deepest cuts among industrialized countries in international negotiations and moving forward with more aggressive "domestic" policy measures, most notably the E.U.-wide Emissions Trading System (ETS) launched in 2005 and covering some 10,078 individual polluters in 23 countries (European Environmental Agency 2006). The latter is particularly noteworthy since in 1997 the European Union had strongly opposed inclusion of emissions trading mechanisms in the Kyoto Protocol. The track record to date of the ETS has, however, been mixed. In the first round, the allocation of permits was left to member states with little oversight from the European Commission. Perhaps not surprisingly, many states were very generous with permits, in most cases distributing more permits than polluters even needed. When that was revealed by the release of the previous year's emissions inventory, the price of carbon plummeted to near zero. However, in the second round of trading, the Commission has played a much more activist role in questioning member states' proposed permit allocations and rejecting several states' initial submissions (Skjæseth and Wettestad 2008). With fewer permits allocated, the price of carbon has increased and round two seems likely to deliver real reductions by 2012.

In terms of the intergovernmental dynamics discussed above, horizontally the European Union saw a period of innovation and diffusion in the early 1990s. Although climate-policy making at that stage consisted more of aspirational target setting rather than concrete abatement policies (Cass 2006), in the early 1990s even embracing the science and committing to targets offered some assurance to reluctant neighbouring states that they could do the same. Schreurs

---

[1]The discussion in this section draws heavily on the work of my colleagues, Miranda Schreurs and Yves Tiberghien (2007), to whom I am most grateful for the benefit of their insights.

and Tiberghien (2007) note that leadership has shifted over time among member states, from the Netherlands and Germany in the early to mid-1990s, to the United Kingdom in the late 1990s, and more recently to France and back to Germany, often coinciding, perhaps not coincidentally, with which state held the Presidency of the Council of Ministers.

Vertically, the European Union has witnessed a dynamic that Schreurs and Tiberghien have labeled "multi-level reinforcement". The authors note:

> The open-ended and competitive governance structure of the EU in an issue of shared competence such as the global environment has created multiple and mutually-reinforcing opportunities for leadership. This suggests a kind of logic that is the reverse of that of veto points or veto players. (Schreurs and Tiberghien 2007, 24)

Although initially prompted by the leaders among the member states, actors at the E.U. level also have responded to the climate challenge with enthusiasm. Schreurs and Tiberghien (2007) argue that this is, in part, because bureaucrats within the European Commission see climate change as a unifying issue that can advance European integration. In addition to the Commission's role in creating the ETS, the E.U. climate plan includes consistent mandates for member states with respect to alternative energy and E.U.-wide plans for control of automobile and airplane emissions. The European Parliament, which includes Green Party representatives, has also been enthusiastic about expanding the E.U. role. Finally, the existence of the Council of Ministers was critical to brokering an E.U. "burden-sharing agreement" under the Kyoto Protocol, through which some member states committed to deeper cuts, thus allowing other states to continue to increase their emissions. Indeed, fewer than half of the European Union's 15 members committed to actual reductions under the Kyoto Protocol, while at the limit Greece and Portugal are allowed to increase their emissions by 25 and 27 percent respectively. Rather than mere generosity on the part of green leaders, the disparity in commitments under the burden-sharing agreement reflected pressure by some states to hold their greener neighbours to targets they had previously committed to unilaterally (Ringius 1999).

U.S. STATE ACTIVISM

The United States not only has among the highest per capita greenhouse gas emissions at 24.5 tonnes per person in 2005,[2] but was also the country responsible for the largest share, roughly one quarter, of global greenhouse gas emissions until 2008, when it was overtaken by China. Yet the U.S. federal government has done little to reduce those greenhouse gas emissions beyond modest investments in research and ineffectual voluntary programs. The United

---

[2]Calculated with emissions data as reported to UN FCCC in 2007 (http://unfccc.int/ghg_emissions_data/items/3954.php); population data from U.S. Census Bureau (http://www.census.gov/ipc/www/idbrank.html). Emissions are without LULUCF.

States also is the only advanced industrialized country not to ratify the Kyoto Protocol.

The announcement that the United States would not ratify was made by President George W. Bush in one of the first acts of his presidency, combined with the Bush White House's persistent questioning of the international scientific consensus on climate change thereafter (Mooney 2005), has led many to attribute U.S. recalcitrance on climate change to President Bush. However, it is noteworthy that U.S. climate policy was not significantly different under President Bill Clinton and his environmental-activist Vice-President Al Gore. During the Clinton administration, it was Congress that held back the White House. Democratic majorities in both Houses rebuffed a "BTU tax" proposed by the president in 2003. After the Republicans gained control of both Houses in 2004, Congress regularly attached riders to unrelated laws to preclude actions by the Environmental Protection Agency (EPA) to address climate change, some of which, like amendments to vehicle fuel economy standards, would have been good policy regardless of global warming (Lutzenhiser 2001, Skolnikoff 1999). In the lead-up to the international meeting at which the Kyoto Protocol was negotiated in 1997, the Senate drew a line in the sand by *unanimously* passing the Byrd-Hagel resolution, which stated that the Senate would not ratify any climate change treaty that included binding targets for industrialized countries unless developing countries also faced binding commitments in the same period.

By the late 1990s, the United States was arguably the world leader with respect to market-based environmental policy instruments, most notably with the SO_2 emissions trading program established by the 1990 Clean Air Act Amendments. When the Kyoto Protocol was negotiated later that year, the United States and its "umbrella group" partners (including Canada), pressed for inclusion of market-based mechanisms in order to take advantage of lower abatement costs in former Soviet and developing countries. Although the umbrella group won inclusion of three market-based approaches – international emissions trading, joint implementation, and the clean development mechanism – over the E.U.'s objections, the treaty nonetheless did not satisfy the terms of the Byrd-Hagel resolution concerning binding commitments for developing countries. In response, the White House continued to pursue "meaningful commitments" from developing countries via bilateral negotiations. When those negotiations proved futile, the administration did not even try to obtain Senate approval for the treaty, given its unambiguous position on ratification.

While United States non-ratification effectively had already been decided by the Senate, President Bush laid to rest any doubt two months after his inauguration in 2001. Thereafter, the administration not only declined to use its considerable regulatory authority under existing statutes, but also resisted a growing number of lawsuits by environmental groups and state and local governments, which sought to force the executive to undertake various measures, including regulation of greenhouse gas emissions from motor vehicles and power plants (the latter account for some 40 percent of United States emissions given the country's heavy reliance on coal for electricity). Although the Supreme Court dealt a serious blow to the administration in 2007 (*Massachusetts v. Environmental Protection Agency* 2007), rejecting EPA's claim that it did not have authority to regulate greenhouse gas emissions under the Clean Air Act, the administration dragged its heels in response, in late 2008 merely publishing for public comment a list of *possible* regulatory actions that might be undertaken.

The inauguration of President Barack Obama in 2009 already has resulted in significant U.S. climate-policy reform. Within days of the election, President-elect Obama announced, "The science is beyond dispute and the facts are clear. ... My presidency will mark a new chapter in America's leadership on climate change" (Knowlton 2008). Obama called on the now-Democratic Congress to pass legislation to establish a national cap-and-trade program. In June 2009, the House of Representatives passed the Waxman-Markey bill, which would establish a national cap-and-trade system to cut emissions to 20 percent below 2005 levels by 2020. Although it is the first time climate change legislation has advanced so far in Congress, passage of the bill by the Senate is far from a foregone conclusion, not least because avoiding the threat of a filibuster requires a threshold of 60 of 100 votes.

However, the federal government is far from the only player in U.S. climate policy. Rabe (2004; 2007; and this volume) has documented the tremendous innovation occurring in U.S. climate policy at the state level. Leading the states is California, which passed legislation in 2006 mandating a return to 1990 emissions by 2020, a bold commitment that has since been matched by several other states (Pew Center on Global Climate Change, nd). Most importantly, state governments have begun to adopt concrete measures to move toward those targets: 25 states have renewable energy mandates, 18 have proposed hard caps on emissions from industry (Barringer 2008) and 19 – comprising over half the U.S. population – have committed to more stringent motor vehicle emissions standards. The last is the result of a special provision in the Clean Air Act, which, in recognition of the historically greater air quality challenges faced in the Los Angeles basin, authorizes only the state of California to set standards for motor vehicles that depart from national standards, though only with approval of the federal EPA. However, if California is granted such a waiver, the Clean Air Act grants other states the option of matching California's standards, and a growing number have indicating their intention to do so.

In 2004, the state of California promulgated a regulation requiring that greenhouse gas emissions from individual vehicles be reduced by 30 percent by the 2016 model year. Although California petitioned EPA for a waiver in 2005, and even sued EPA in pursuit of a response, the agency did not respond until December 2007, at which time the EPA Administrator dismissed the request, the first time California has been denied a waiver in over 50 such petitions. California and other states appealed the decision in the courts. Before those cases could be resolved, newly elected President Obama directed the EPA to reconsider the Bush administration's rejection of California's request. However, even before the public comment period on EPA's proposed California waiver ended, the president announced that the administration would pass national standards to match California's standards. In making the announcement, the president was joined at the White House by supportive representatives of state governments, environmental groups and, quite remarkably, the automobile industry. Although the industry had fought California's tailpipe standards at every turn, two factors contributed to the turnaround. First, there was no question that EPA would approve California's request for a waiver, thus raising the prospect of two different emissions standards within the United States. Second, although the economic crisis strengthened other sectors' opposition to regulation of greenhouse gas emissions, the U.S. government's investments in bailing out domestic auto manufacturers gave it significant leverage over the future direction of the industry (Broder 2009. Also Broder and Maynard 2009).

In the absence of a national program to control emissions from stationary sources, state governments also are collaborating to establish regional cap-and-trade systems. The first of those was the Regional Greenhouse Gas Initiative (RGGI), through which ten north-eastern and mid-Atlantic states have agreed to cap greenhouse gas emissions from power plants at 2009 levels by 2015, and thereafter to achieve a 10 percent reduction by 2018 (RGGI, nd). RGGI permits are distributed by auction, the first of which was held in 2008. On the other coast, the Western Climate Initiative now involves seven states and four Canadian provinces (Ontario, Quebec, Manitoba, and BC), who have committed to a broader goal of a 15 percent reduction in emissions across the economy by 2020, with trading to begin in 2012.

In terms of the patterns of intergovernmental relations identified above, horizontally the states have clearly engaged in a dynamic of innovation and diffusion. In the face of federal inaction, state leaders are moving unilaterally, in so doing prompting other states to match their actions. It is not the first time that California has led the charge. Indeed, the phenomenon of environmental policy innovation by a green leader followed by diffusion to other states or countries has been dubbed "the California effect" by Vogel (1995, 259). As in the past, U.S. states have engaged in a "pull from the top", with states matching the green leaders, rather than a competitive bidding war implied by a "race to the top". While states have forged ahead on their own, two caveats are in order. First, not all states are active, and many that are selectively choose climate policies that serve other purposes (e.g., renewable portfolio standards that promote energy security). Second, and not unrelated, while willing to act unilaterally, the greenest states clearly welcome federal backup. Consistent with an assurance – rather than prisoner's dilemma game – state leaders including New York, Massachusetts, and California have sued the federal government in an effort to force national standards for various sources.

In the case of motor vehicle standards, for many years the United States experienced a joint-decision trap in which the federal government blocked state standards, though that has now given way to federal reinforcement of state leaders' efforts. In other respects, the United States has witnessed neither the positive nor negative vertical dynamics set out above. Rather, the federal government for the most part has been irrelevant to date. That said, the greater likelihood of federal action under a Democratic president and Democratic Congress will present many of the challenges noted above in terms of reconciling a national cap-and-trade program with regional programs already developed by the states. Not only will it be desirable to administratively reconcile trading rules under the federal and regional programs plans (or, as some members of Congress have proposed, for the federal government simply to pre-empt the regional programs), but the prevalence of non-revenue-neutral auctioning in the RGGI program, the WCI plan, and many congressional proposals suggests that harmonization not only will be essential, but also that it is likely to be difficult to achieve as the federal and state governments fight for revenues.

CANADA: THE JOINT DECISION TRAP

Though not a significant contributor to global emissions by virtue of its relatively small population, Canada, like the United States, has one of the most greenhouse gas-intensive economies in the world, producing 23.1 tonnes per person in 2005. While U.S. federal climate policy offers a history of persistent refusal to commit to emissions reductions, Canada's offers a series of ambitious commitments followed by inaction – and steadily increasing emissions.

In 1988, the Mulroney government committed that Canada would reduce its emissions by 20 percent by 2005. Two years later the same government set a somewhat less ambitious goal in its Green Plan of stabilization at the 1990 level by the year 2000. When the Liberal Party won the first of three parliamentary majorities under Jean Chretien's leadership in 1993, they sought to outdo their Conservative predecessors in government by proposing a 20 percent cut below 1990 levels by 2005. However, like their predecessors, the Liberals soon retreated, though in 1997 still embracing an ambitious Kyoto Protocol target of a 6 percent reduction below 1990 by 2008 to 2012. Despite United States withdrawal from the Kyoto Protocol, which prompted strong resistance from most provinces and the business community, the Chretien government unilaterally released a federal plan and ratified the Kyoto Protocol in the fall of 2002. Still ostensibly committed to the Kyoto Protocol, despite an ever increasing gap between Canada's emissions and its Kyoto target, Mr. Chretien's successor, Paul Martin, introduced his own plan, Project Green, in 2005.

With the election of a Conservative government in 2006, the Harper government announced that Canada could not meet its Kyoto Protocol target. The Conservatives introduced their own plan in the fall of 2006, promising only to eliminate Canada's emissions *growth* by 2025. When the environment soared to the top of the political agenda during 2006, however, the Harper government announced a revised target of a 20 percent reduction below 2006 levels by 2020 (3 percent below 1990 emissions) and released its second plan in less than a year, this one entitled Turning the Corner.

The past two decades have thus seen six distinct targets, each promising dramatic emissions reductions relative to gradually retreating goal posts. In the meantime, Canada's greenhouse gas emissions increased by 25 percent from 1990 to 2005. And despite at least six federal plans of action, including three proposals (one Liberal, two Conservative) for a national cap-and-trade program since 2005, Canada still does not even have *draft* national regulations, whether conventional or market-based, for a single source of greenhouse gas emissions. The situation is little different at the provincial level, though the standouts are a nominal carbon tax (less than $3 per tonne) introduced by Quebec in 2006, a not-especially-credible cap-and-trade program launched by Alberta in 2008,[3] and a more serious carbon tax in British Columbia, which took effect at $10 per tonne in 2008 with increases to $30 per tonne scheduled by 2012.

To what degree has federalism contributed to this record of inaction? Horizontally, the provinces appear to have been "stuck at the bottom" until late 2006. However, the floor of concern arguably was set globally by Canada's largest trading partner, the United States, rather than within the Canadian federation by other provinces. Consistent with that, even when the provinces reached collective agreement, as they did on a Canadian position going into the Kyoto negotiations, they agreed only to match the U.S. target (with the exception of Quebec, which advocated deeper reductions). It is, however, striking that Canadian provinces have been fixated on the U.S. floor rather than the ceiling, and thus did not engage in the innovation and diffusion dynamic evident among U.S. states years earlier. Rabe (2007) attributes the lower level of policy innovation among Canadian provinces compared to U.S. states to three factors: 1) greater administrative capacity at the sub-national level in the United States, ironically as a result of top-down federal environmental mandates to the states; 2) greater state-level experience with market-based policies, especially under the federal cap and trade program for SO2; and finally and counter-intuitively, 3) to Canada's ratification of the Kyoto Protocol. With respect to the latter, while the dismal prospects for U.S. ratification made it clear that U.S. states were on their own, Rabe argues that the Chretien government's continued support for ratification prompted Canadian provinces to delay action pending negotiation of financial compensation from the federal government.

An innovation and diffusion dynamic has emerged, however, since the resurgence of public attention to the environment in 2006. British Columbia has emerged as a provincial leader, committing to a one-third reduction of emissions below 2006 by 2020, proposing to match California's proposed automobile tailpipe standards (should they go forward), insisting on carbon-neutrality of any new electricity generation projects, and introducing a revenue-neutral carbon tax in 2008. Also during 2008, BC, Manitoba, Ontario and Quebec all joined the Western Climate Initiative and thus committed to participate in a transnational cap-and-trade program. Quebec and Manitoba also have committed to match California's auto emission standards.

As in the United States, the provincial innovation and diffusion dynamic is truncated. However, in the Canadian context, the provinces that have remained on the sidelines are especially significant. Most notable is Alberta, which accounts for roughly one third of Canada's emissions and over half of Canada's projected "business as usual" emissions growth between 2006 and 2020 (Government of Canada 2006). Alberta's own plan, released in 2008, commits

---

[3]The Alberta emissions trading program allows emitters to meet their emissions *intensity* targets by purchasing offsets that occurred several years earlier, thus raising serious questions about the additionality of claimed reductions. After all, it is difficult to argue that a reduction would not have occurred but for an agreement to purchase offsets if it already did occur years before the offset program was even anticipated. The Alberta case raises important questions about the politics of cap and trade, in particular policy makers' ability to pursue symbolic commitments via an offsets shell game.

only to a 14 percent emissions reduction relative to 2005 (18 percent *above* 1990 levels) by *2050*. Also noteworthy is the province of Ontario's repeated rejection of calls to match California's tailpipe standards (Howlett and Keenan 2008).

Although the provinces may have been "stuck at the bottom" for almost two decades, the federal government has shown little resolve to step in unilaterally. Rather, the Canadian case offers a classic example of a joint-decision trap. Given overlapping jurisdiction with respect to the environment, the Canadian federal and provincial governments have a long history of coordinating their actions – and inaction – via the Canadian Council of Ministers of the Environment, where a norm of consensus among equals prevails. However, in the case of climate change, consensus has been elusive to say the least. The federal environment minister from 1993 to 1996, Sheila Copps, later recalled, "it became clear that the rule of [federal-provincial] 'consensus' in the environmental agenda would mean moving to the lowest common denominator. There was no way that Alberta would agree to *any* reduction in fossil-fuel emissions" (Copps 2005, 93).

The federal and provincial governments did reach an agreement to match the U.S.'s position to return emissions to 1990 levels by 2010 going into the Kyoto negotiations, but provincial governments were soon outraged by the federal government's unilateral commitment in Kyoto to a 6 percent cut (even though the United States committed to a 7 percent cut). However, Prime Minister Chretien assuaged the premiers' concerns by agreeing to a joint federal-provincial process – co-chaired inauspiciously by Alberta – to develop an implementation plan. Although ongoing discussions masked federal-provincial tensions for several years, provincial opposition to ratification became more public after the U.S.'s withdrawal from the treaty was announced by President Bush in 2001. In late 2002, after five years of consultations and federal-provincial meetings, consensus was nowhere in sight when the federal government ratified the Kyoto Protocol and produced its own plan unilaterally. At the time, only Manitoba and Quebec, two provinces that stood to gain from development of their hydro-electric potential, supported ratification, though even they opposed the unilateral federal plan. Federal-provincial discussions continued bilaterally, particularly with the inclusion in the Martin government's Project Green of a "partnership fund", through which the federal government would jointly fund provincial projects. However, the Harper government cancelled the partnership fund soon after its election in 2006.

Federal-provincial deadlock was broken as both the federal and provincial governments began to act more independently in response to resurgent voter interest in the environment in 2006. The federal government has not been in a position to obstruct provincial initiatives that are more ambitious than those of the federal government, although federal ministers have vacillated between noncommittal statements that provincial initiatives such as BC's carbon tax "complement" federal plans (Bailey 2008), and more pointed criticism of provincial unilateralism (Authier 2008).

EXPLAINING DIFFERENCES

Although the European Union, United States, and Canada are all grappling with the problem of climate change within the context of multi-level political institutions, their experiences to date have been markedly different. The European Union appears to have experienced the best of federalism, combining policy innovation and diffusion at the member state level with complementary E.U.-level initiatives and reinforcement. While the United States has experienced a relatively negative vertical dynamic, horizontally state governments have responded to a federal policy vacuum with a variety of unilateral and coordinated policy innovations. In contrast, Canadian climate policy has for the most part been characterized by federal-provincial deadlock, albeit with signs of provincial innovation emerging in the last two years.

The question, then, is why similar political institutions have had such different impacts in these three systems. This section considers three explanations, two that point to the interaction of federal institutions with other variables that differ across the cases – in particular, public opinion and the regional distribution of costs – and a third that points to more subtle differences in the distribution of powers within the three federations.

As discussed by Harrison and Sundstrom (forthcoming), there is evidence that European voters, especially in Western Europe, have been more concerned about climate change and for longer than their counterparts in Canada and the United States. That has a direct impact on climate policy in providing stronger electoral incentives for policy makers at both the member state and E.U. levels to adopt measures to address climate change. However, public opinion can also interact with federal institutions, in many respects serving as a switch that shifts intergovernmental relations from one dynamic to another. While policymakers weighing supportive yet inattentive voters against a hostile business community may remain "stuck at the bottom" horizontally and engage in buck-passing vertically, those with strong electoral incentives are more likely to act unilaterally, whether at the sub-national or national level. The dynamic of "multi-level reinforcement" that has emerged in the European Union compared to the federal-provincial deadlock that has prevailed in Canada is thus consistent with how similar federal institutions have expressed differing electoral incentives in the two jurisdictions. The interaction of public opinion with federal institutions is also apparent over time within Canada, accounting for the shift from provincial inaction prior to 2006 to a dynamic of innovation and diffusion thereafter.

Second, the division of powers relevant to climate change is notably different within the three federations under study. After 1970, the United States witnessed a significant centralization of environmental policy, predicated on both federal government ownership of public lands and, more importantly, on a broad interpretation of the interstate commerce clause of the U.S. Constitution. Indeed, in the recent *Massachusetts v. EPA* decision, a dissenting Supreme Court justice complained that the majority's broad interpretation of federal authority concerning air pollution would authorize regulation of everything from "frisbees to flatulence" (*Massachusetts v. Environmental Protection Agency* 2007, 10). While the U.S. federal government has yet to exercise that authority with respect to greenhouse gas emissions, Rabe (2007) has argued that the longstanding role of state governments in fulfilling federal mandates contributed significantly to the states' administrative capacity to respond to climate change unilaterally, including their familiarity with market-based instruments.

While seemingly implausible that the confederal European Union, predicated on a principle of subsidiarity, would achieve a more centralized response to climate change than the Canadian federation, E.U. decision rules differ from those of Canada in one critical respect. In contrast to the norm of consensus that prevails within the Canadian Council of Ministers of the Environment, regulatory decisions are made by the European Union Council of Ministers via a qualified majority vote. That facilitated adoption of the E.U. Emissions Trading System, particularly in light of strong support (discussed below) from the largest member states, which carry greater weight in a qualified majority decision. Indeed, the fact that E.U. *taxation* policies, in contrast to regulation, do require unanimity explains the European Commission's greater success with the ETS than in its earlier proposal for a carbon tax.

In Canada, it is critical that 90 percent of lands remain in public hands, and that the Constitution grants ownership of Crown lands within their borders to the provinces. With respect to climate change, provincial governments thus own both the sources and sinks in question – oil, coal, and natural gas on the one hand, and forests and sites for generation of hydro-electricity on the other. Given the significance of those resources for provincial economies and provincial government revenues, a federal government proposing to regulate greenhouse gas emissions not only has to deal with provincial sensitivities concerning their own regulatory authority, but also must seek agreement with provincial governments that will, in effect, themselves bear significant costs of any federal regulation. In that context, it is hardly surprising that federal-provincial consensus has been unattainable.

The third, and arguably most important, difference among the three federations concerns the regional distribution of the costs of greenhouse gas reductions. In both the European Union and United States, the largest and wealthiest states are "green and keen", while the opposite has been the case in Canada. An often-overlooked precondition for E.U. leadership on climate change is the windfall reductions experienced by the two member states with both the largest populations and economies, Germany and the United Kingdom. The former experienced quite dramatic reductions in greenhouse gas emissions as a result of reunification and the subsequent closure of economically and environmentally inefficient enterprises in the former East Germany. For its part, a shift from reliance on coal to North Sea gas has resulted in significant emissions reductions in the United Kingdom. While neither of these windfall reductions was painless, it is nonetheless true that they would have occurred anyway. Anticipation of emissions reductions not only made it easier for these two states to lend their considerable weight (especially under qualified majority voting) to E.U. climate-policy proposals, but also facilitated the burden-sharing agreement that underpinned the E.U.'s Kyoto Protocol commitment. Indeed, with respect to the latter, Germany and the U.K. combined account for 104 percent of the E.U.'s Kyoto reduction,[4] which thus allowed other member states to continue to increase their emissions, if by less than they might have otherwise.[5]

Although the United States did not have the benefit of comparable windfall reductions, like the European Union it was fortuitous that the most populous states – California and New York – are both relatively wealthy and relatively green. California's leadership among U.S. states is not only a result of an environmentally aware electorate, but also reflects that the state is not heavily reliant on greenhouse gas-intensive industries, and thus will be less hard hit by the costs of addressing climate change. Indeed, California is offloading a share of the costs of achieving its targets through mandates concerning the greenhouse-gas intensiveness of electricity and fuels imported from out of state.

In contrast, two of the most powerful provinces within the Canadian federation, Ontario and Alberta, each stand to bear a disproportionate share of the costs of addressing climate change in Canada. While Ontario has been more activist on climate change since election of a Liberal government under Dalton McGuinty in 2003 – McGuinty's predecessor vowed that Ontario would oppose ratification of the Kyoto Protocol if it would cost "even one job" in the province (Frank 2002) – the province continues to resist efforts to strengthen emission standards for the transportation sector, which accounts for one third of Canada's emissions. Not only is auto manufacturing the most important sector in the province's economy, but Canada tends disproportionately to manufacture the least fuel-efficient vehicles within the integrated North American market. For its part, Alberta accounts for only 10 percent of the Canadian population but roughly one third of Canada's emissions and over half of its projected emissions growth. As a result, the costs of reducing Canada's greenhouse gas emissions will inevitably be borne disproportionately by Alberta, absent a massive compensation program funded by taxpayers in other provinces. Thus, while windfall reductions in the European Union are concentrated in two powerful member states, in Canada the costs of reducing greenhouses gas emissions are disproportionately concentrated in two influential provinces – provinces that to date have exercised an effective veto over measures affecting the industries that are the lifeblood of their economies.

CONCLUSION

Federalism presents both challenges to and opportunities for the policy reforms needed to achieve carbon pricing. The opportunities are most evident in the

---

[4]A 21 percent reduction of Germany's 1990 emissions yields 257.5 MT, while a 12.5 percent reduction in U.K. baseline emissions yields 97 MT. Together these exceed the E.U. 15's commitment of 341.2 MT.

[5]Harrison and Sundstrom (2007) have also argued that the fact that the European Union collectively faced a less demanding target in the Kyoto Protocol relative to the business and usual trajectory than either Canada or the United States facilitated action by the European Union.

European Union, where leadership by green states has facilitated action by laggards, and where E.U. institutions have reinforced the position of national leaders, complementing their actions with Europe-wide policies. In the United States, federalism has facilitated actions to address climate change by many, if not all, states in the absence of a federal government commitment. While innovation at the provincial level in the last two years offers promise, for the most part federalism in Canada has yielded deadlock – an archetypal example of Scharpf's "joint decision trap".

That said, it would be an exaggeration to argue that Canada's inaction to date is primarily attributable to federalism. First-past-the-post electoral systems in both Canada and the United States have also deterred action in directing policy makers' attention disproportionately to the median voter, who has for the most part been inattentive to climate change in North America. The minority of voters for whom the environment has long been a priority thus has had little impact. In contrast, systems of proportional representation in many European countries and the E.U. parliament facilitate the emergence of small parties, which provides a stronger voice for environmental voters, often through the emergence of a Green Party. Indeed, the role of Green parties in coalition governments arguably has exaggerated the influence of environmentally oriented voters in those electoral systems.

Cross-national comparison also points to the importance of public opinion. While political institutions facilitate action in some countries and exaggerate the challenges in others, comparison of public opinion over time and across countries suggests that if voters care enough, politicians can and do respond (Harrison and Sundstrom, forthcoming). Until 2006, Canadians reported that they supported action on climate change, but it was low on their priorities, often trailing issues like gas prices that belied voters' professed concern for the environment. Canadians did return their attention to the environment in 2006, prompting a variety of commitments from provincial and federal politicians alike and also shifting the dynamics of federal-provincial relations as noted above. However, that "third wave" of environmental attention subsided in late 2008,[6] not least in response to the onset of an economic recession, and Canadian politicians' enthusiasm for fighting climate change appears also to have subsided in response.

The regional distribution of costs combined with the division of powers with respect to natural resources suggest that federalism will continue to pose a challenge to Canada's ability to respond to climate change for years to come. The implication is that if we are to assume our responsibility to address a global problem to which we have contributed far more than our share, greater resolve will be needed from both the electorate and our political leaders than demonstrated by either to date.

---

[6]There were similar, short-lived bursts of public attention to the environment in 1969 and 1989.

REFERENCES

Authier, P. 2008. "Feds rap emissions accord", *Montreal Gazette*, 2 June, A1.

Bailey, I. 2008. "BC's carbon tax 'complements' Ottawa's green goals, PM says", *The Globe and Mail*, 12 March, S1.

Barringer, F. 2008. "States' battles over energy grow fiercer with U.S. in a policy gridlock", *New York Times*, 20 March.

Broder, J.M. 2009. "Obama to toughen rules on emissions and mileage", *New York Times*, 19 May.

Broder, J.M. and M. Maynard. 2009. "As political winds shift, Detroit charts new course", *New York Times*, 20 May.

Cass, L.R. 2006. *The Failures of American and European Climate Policy*. Albany: State University of New York Press.

Copps, S. 2005. *Worth Fighting For*. Toronto: McClelland and Stewart.

European Environment Agency (EEA). 2006. *Application of the Emissions Trading Directive by EU Member States*. Technical Report No. 2/2006. At http://ec.europa.eu/environment/climat/emission/index_en.htm.

Frank, S. 2002. "Where there's smoke…", *Time*, 21 October, 160(17): 47.

Government of Canada. 2006. *Canada's Emission Outlook: The Reference Case 2006*. At http://www.nrcan-rncan.gc.ca/com/resoress/publications/peo/peo-eng.php.

Harris, P.G. 2007. *Europe and Global Climate Change: Politics, Foreign Policy and Regional Cooperation*. Cheltenham, UK: Edward Elgar.

Harrison, K. and L. McIntosh Sundstrom. 2007. "The Comparative Politics of Climate Change", *Global Environmental Politics* 7(4): 1-18.

— forthcoming. *Global Commons and National Interests: The Comparative Politics of Climate Change*. Cambridge: MIT Press.

Harrison, K. 1996. *Passing the Buck: Federalism and Canadian Environmental Policy*. Vancouver: UBC Press.

— ed. 2006. *Racing to the Bottom? Provincial Interdependence in the Canadian Federation*. Vancouver: UBC Press.

Howlett, K. and G. Keenan. 2008. "Deal lets Ontario join climate-change drive", *The Globe and Mail*, 4 August. Retrieved online from http://www.theglobeandmail.com/servlet/story/RTGAM.20080804.wontclimate04/BNStory/National/home.

Knowlton, B. 2008. "Obama promises action on climate change", *New York Times*, 18 November.

Lutzenhiser, L. 2001. "The Contours of US Climate Non-Policy", *Society and Natural Resources* 14(6): 511-523.

Massachusetts v. Environmental Protection Agency, [2007] 549 U.S. 497.

Mooney, C. 2005. *The Republican War on Science*. New York: Basic Books.

New State Ice Co. v. Liebmann, [1932] 285 U.S. 262.

Olewiler, N. 2006. "Environmental Policy in Canada: Harmonized at the Bottom?" in K. Harrison (ed.), *Racing to the Bottom? Provincial Interdependence in the Canadian Federation*. Vancouver: UBC Press.

Pew Center on Global Climate Change. nd. *A Look at Emissions Targets*. At http://www.pewclimate.org/what_s_being_done/targets.

Rabe, B.G. 2004. *Statehouse and Greenhouse: The Emerging Politics of American Climate Change Policy*. Washington, DC: Brookings.

— 2007. "Beyond Kyoto: Climate Change Policy in Multilevel Governance Systems", *Governance* 20(3): 423-444.

Regional Greenhouse Gas Initiative (RGGI). nd. *About RGGI*. At http://www.rggi.org/about.

Ringius, L. 1999. "Differentiation, Leaders, and Fairness: Negotiating Climate Commitments in the European Community", *International Negotiation* 4: 133-166.

Scharpf, F.W. 1988. "The Joint-Decision Trap: Lessons from German Federalism and European Integration", *Public Administration* 66(2): 239-278.
Schreurs, M.A. and Y. Tiberghien. 2007. "Multi-Level Reinforcement: Explaining European Union Leadership in Climate Change Mitigation", *Global Environmental Politics* 7(4): 19-46.
Skjæseth, J.B. and J. Wettestad. 2008. *EU Emissions Trading: Initiation, Decision-making and Implementation*. Aldershot, Hampshire, UK: Ashgate.
Skolnikoff, E.B. 1999. "The Role of Science on Policy: The Climate Change Debate in the United States", *Environment* 41(5): 16-20 and 42-45.
Vogel, D. 1995. *Trading Up: Consumer and Environmental Regulation in a Global Economy*. Boston: Harvard University Press.

8

The Intergovernmental Dynamic of American Climate Change Policy

Barry G. Rabe

INTRODUCTION

Most scholarly and journalistic analysis presents the odyssey of climate change policy in the United States as if America was a unitary system of government. This leads to a familiar tale, whereby the federal government signed the Kyoto Protocol in 1997, spurned ratification four years later, and neither the Clinton nor the Bush administration and respective Congresses were able to agree to anything beyond climate research funding, subsidies for virtually all forms of energy, and voluntary reduction programs. In turn, the arrival of the 111th Congress and the inauguration of Barack Obama as America's 44th president may well produce major new federal climate-policy steps. At the same time, conventional analysis has assumed that climate policy would entail bargaining and implementation among nations, culminating in a world climate regime. More than a decade after the signing of Kyoto, it is increasingly evident that climate policy is proving far messier than prevailing depictions had anticipated. The Kyoto process is in tatters, attributable not only to American disengagement but also to an inability of many ratifying nations to honour their commitments. This is reflected in numerous failures to approach pledged emissions reductions, as in the Canadian and Japanese cases, or to successfully implement national or multi-national policies, as in the stumbles of the Emissions Trading Scheme in the European Union. There also continues to be enormous uncertainty about

An earlier version of this chapter appeared as "States on Steroids: The Intergovernmental Odyssey of American Climate Policy", *Review of Policy Research* 25(2) (March 2008), 105-128. My thanks to Blackwell Publishing and editor J.P. Singh for approving release of this revised version.

engagement by developing nations, at the very point where China has eclipsed the United States as the world's leading national source of GHG emissions.

But perhaps the biggest single surprise as climate policy has continued to evolve is that in the American case and many others, it is becoming increasingly evident that climate policy constitutes an issue of federalism or multi-level governance. As the emergence of California governor Arnold Schwarzenegger as a claimant to the title of "world leader" in the development of far-reaching climate policy attests, individual units across different federal or multi-level governance systems may have more in common with one another in climate policy than they have with the neighbouring units of their own federation. Indeed, one can see stronger parallels between such jurisdictions as Connecticut and Sweden, Pennsylvania and Germany, New York and New South Wales, and North Carolina and Ontario than exists across many members of the same federation. For President Obama and his counterparts in Congress, any consideration of an expanded federal role must contend with a significant sub-federal legacy of policies.

This paper will focus primarily on the American case, considering more than a decade of state and federal policy experience and attempting to distill lessons that could guide future policy development. First, it will offer an overview of American sub-federal policy development, attempting to provide a review of the tapestry of policies that have been enacted over the past decade and some of the key factors that have led to such a robust state response in the absence of federal mandates or incentives. Second, this will lead to a consideration of the divergent paths taken by the fifty states, reflected in their carbon dioxide emission trends since 1990 and varied levels of climate-policy development. This section will explore the unique contexts facing various states, particularly the differing strategic considerations for them (and for their representatives in Congress) as they consider unilateral policy steps or the possibility of federal policy in the 111th Congress and beyond. Third, the collective state experience offers some possible lessons for future policy development at either sub-federal or federal levels. In particular, we will see that there appears to be a nearly-inverse relationship between those policies that policy analysts tend to endorse as holding the greatest promise to reduce emissions in a cost-effective manner and the political feasibility of respective policy options. These patterns could offer significant lessons for the future of climate-policy development, outlining both challenges and opportunities for future policy whether enacted at the single-state, multi-state, or federal levels. Finally, we look ahead and consider alternative scenarios for future development of American climate policy, mindful of the new possibilities that coincide with Democratic Party assumption of both executive and legislative branches in 2009.

TOWARD A STATE-CENTRIC AMERICAN CLIMATE POLICY

The recent trend toward state-driven policy is hardly unprecedented in American federalism. In many instances, early state policy engagement has provided models that were ultimately embraced as national policy by the federal government. This has been evident in a range of social policy domains, including health care and education, and can either result in federal pre-emption that obliterates earlier state roles or a more collaborative system of shared governance (Teske 2004; Manna 2006). In some cases, states have taken the lead and essentially sustained policy leadership through multi-state collaboration and the absence of federal engagement. Such policy arenas as occupational licensure and regulation and oversight of organ donations have remained largely state-dominated, despite occasional federal exploration of legislation or regulation. Through early 2009, American climate policy followed the latter pattern, with prolonged federal inability to construct policy leaving substantial opportunity for state engagement and innovation. At the same time, Congress has continued to weigh a variety of policy options, many of which could ultimately encourage, constrain, or pre-empt existing state policies, including movement during 2009 toward an omnibus bill that would establish carbon emissions trading and numerous other provisions. However, the institutional impediments to any federal action remain significant, including challenges of implementation after any federal legislation was enacted, suggesting that there may well be continued state latitude to play a lead role for some time to come. In turn, this could ultimately give a number of states a strong bargaining role in any future federal policy formation or implementation, given their sunk institutional and policy investments. Many states now possess a considerable body of climate-policy expertise that may well rival or surpass federal institutions (Rabe 2004; 2007).

Many scholars scoffed at the very possibility of "bottom-up" American climate policy during the previous decade, but several factors converged to place states in increasingly central roles. First, many states that framed early policy steps that would have the effect of reducing greenhouse gases perceived the policies as being in their economic self-interest. This helps explain the ever-expanding state government interest in developing a set of technologies and skills to promote renewable energy, energy conservation, and expertise to foster a low-carbon economy. Indeed, virtually every governor has now embraced the notion of developing "home-grown" energy sources at least in part in order to foster long-term economic development. This has resulted in an active exploration of various policy tools that might achieve these goals alongside reduction of GHG emissions. Second, a growing number of states are beginning to experience significant impacts that may be attributable to climate change, whether through violent storms, forest fires, species migration, prolonged droughts, or changing vectors of disease transmission. Some of these are having the classic effect of "triggering events" that create an impetus for a policy response, however modest the climate impact that any unilateral state efforts to reduce GHG emissions may be (Repetto 2006). Third, some states have consciously chosen to be "first movers", often taking bold steps with the explicit intent of trying to take national leadership roles on climate policy. In some instances, such as California's legislation to restrict carbon emissions from vehicles and New York's efforts to establish a regional carbon-emissions trading zone in the north-eastern United States, states are also trying to establish models that will influence their neighbours to join them and position themselves to influence any future federal policy. In this regard, states are similar to corporations; some seek an early and active role, sensing potential strategic advantages over their more recalcitrant competitors (Hoffman 2006; Kamieniecki 2006). Fourth, state capitals have proven very fertile areas for the development of epistemic communities and policy networks advocating climate policy. In many instances, earlier state efforts reflected leadership from higher levels of state agencies working in environmental protection, energy or other areas relevant to climate (Rabe 2004; Montpetit 2003). These policy entrepreneurs continue to operate, but increasingly partner with other forces, such as legislators and advocacy groups, to form policy networks that build support for policy strategies that are particularly appealing to an individual state (Selin and VanDeveer 2007). Fifth, states also provide venues for alternative approaches to policy formation, including direct democracy and litigation that confronts federal institutions. Ballot propositions are proving an increasingly popular way to advance climate initiatives in cases where representative institutions stall. At the same time, the 2007 U.S. Supreme Court verdict in *Massachusetts et al. v. U.S. Environmental Protection Agency* indicates that a collective of states can wage and ultimately win an intergovernmental court battle that may serve to force a reluctant federal agency to designate carbon dioxide as an air pollutant. The decision in this case is already triggering additional multi-state efforts to use the federal courts as a venue to challenge other decisions by the private sector or federal agencies. The Bush administration decision to reject this designation was reversed in 2009 by the Obama administration, opening the way in May of that year toward a federal embrace of carbon emission standards for vehicles that was in essence first established in Sacramento in 2002.

VARIATION IN STATE EMISSION TRENDS AND POLICY ADOPTION

None of these factors converge in identical ways in differing states. Indeed, no two states have uniform profiles in terms of actual rates of GHG emissions growth or climate-policy adoption. Just as the nations of the world diverge on these dimensions, so do American states. In turn, as we shall see, the combinations of emissions growth and policy development to date vary greatly among states and this may prompt them to consider different strategic positions. This may apply to either further state-policy adoptions or to any bargaining position that they might assume in future negotiations over federal policy. Of course, the political influence of states varies enormously between the Senate, where all states retain two members regardless of population, and the House of

Representatives, where delegations from populous states such as California, Florida, and Texas have substantial power.

State Emission Trends

The range of emission trends may be particularly surprising, when weighed against the widespread reporting of national averages for emission rates. With 1990 established as a near-universal baseline internationally, American emissions increased approximately 15 percent overall from that point through 2003. This reflects steady growth throughout the 1990s, with a somewhat slower pattern in more recent years. The most recent national estimates suggest that American emissions have been essentially stable since 2003 and may have entered into decline since the onset of a major recession in 2008. Reliable state-by-state emissions data generally continues to lag behind, however. The overall pattern between 1990 and 2003 varied markedly when looking at the rates of change in the fifty states and the District of Columbia (see Table 1). One state, Delaware, is actually on track to meet the reduction targets that would have been imposed had the United States ratified Kyoto, and twelve other states have contained growth rates to single digits. These include several states, such as California, Pennsylvania, New York, and Michigan that have very large emission bases, and would rank among national leaders in global emissions were they not part of a federation. At the same time, many other states, particularly those of the Southeast and Southwest, have registered rates of emissions growth that are at least double the national average.

Such a range of emissions among sub-national units – which tends to be hidden by the tendency to focus only on national emissions trends – is not unique to the United States. Similarly in Canada – which ratified the Kyoto Protocol – the national emissions increase of 26 percent between 1990 and 2003 hid major inter-provincial disparities. Only Manitoba and Quebec approach those states toward the lower end of the growth continuum in the United States, whereas many others such as Saskatchewan, Alberta, and British Columbia, were far above the national average. Similar variability exists among the nations of the European Union, ranging from outright reduction (Germany and the United Kingdom) to major increase (Ireland and Spain). In that case, differential national reduction targets were negotiated as part of the price for ratification but many individual nations have vastly exceeded their particular targets. In each instance, political leadership of individual jurisdictions (American states, Canadian provinces, European Union member states) will be attentive to their emission patterns since 1990 and make that a factor in any intergovernmental bargaining over future emission reductions, credit for early reduction or stabilization of emissions, or selection of policy tools. In Canada, for example, the Stephen Harper government began in 2007 to explore ways to shift the baseline to a much later year, thereby trying to entice cooperation from provinces that could in effect be "forgiven" very high rates of emissions growth during the 1990s. But this faced prompt opposition from those provinces that felt they should be "rewarded" for registering lower rates of emissions growth.

Table 1: State Carbon Dioxide Emission Trends, 1990-2003*

States Below National Average				*States Above National Average*			
	1990	*2003*	*% Change*		*1990*	*2003*	*% Change*
Delaware	18	17	-5	Illinois	192	227	18
Louisiana	191	189	-1	Montana	28	33	18
District of Columbia	4	4	0	North Dakota	40	47	18
Hawaii	21	21	0	Texas	587	694	18
Connecticut	41	42	2	Kentucky	118	141	19
Michigan	180	183	2	Georgia	138	166	20
New York	208	214	3	Vermont	5	6	20
Pennsylvania	260	267	3	Maine	19	23	21
Massachusetts	83	86	4	Wisconsin	85	103	21
California	361	384	6	Iowa	63	77	22
Ohio	243	261	7	Rhode Island	9	11	22
South Dakota	12	13	8	Alabama	108	135	25
West Virginia	105	113	8	Minnesota	79	99	25
New Jersey	114	124	9	Arkansas	51	65	27
New Mexico	52	57	10	Idaho	11	14	27
Wyoming	57	63	11	Nebraska	33	42	27
Kansas	69	77	12	Alaska	34	44	29
Indiana	201	228	13	Mississippi	48	62	29
Maryland	70	79	13	Oregon	31	40	29
Washington	71	80	13	Virginia	94	121	29
Tennessee	105	121	15	Florida	186	242	30
Utah	53	61	15	North Carolina	110	144	31
Oklahoma	88	102	16	South Carolina	61	80	31
				Missouri	103	136	32
				Colorado	66	88	33
				New Hampshire	15	20	33
				Arizona	62	88	42
				Nevada	30	43	43

*Fossil Fuel Combustion, Million Metric Tons CO_2 ($MMTCO_2$). Includes emissions from commercial, electric power, industrial, and transportation sectors.

Source: U.S. Environmental Protection Agency (2007).

Policy Development Trends

The ways in which governments can enact policies that purportedly stabilize or reduce their GHG emissions is virtually limitless. Since GHG emissions emanate from essentially every sector of the economy, a vast range of policies and sectors could come into play. But this effort to measure intensity of state climate-policy development uses a measure from eight policy options that are prominently addressed either in current practice or in the scholarly literature on climate-policy options (Pew Center 2007). In Table 1, we dichotomize the states by their rate of emissions growth since 1990 and also divide them according to low (zero to one) versus high (two or more) policy adoption rates from this census of eight possibilities. These policies include: renewable electricity mandates, or portfolio standards; carbon taxes; renewable fuel mandates or equivalent programs that mandate expanded use of bio-fuels; carbon cap-and-trade programs; state-wide emissions reduction targets; mandatory reporting of carbon emissions; formal participation as a co-plaintiff in the 2007 Supreme Court case on carbon-dioxide regulation; and adoption of the carbon-emission standards for vehicles enacted by California.

This demarcation essentially divides the nation into two blocs. Twenty-two states representing about one-half of the American population have adopted two or more of these eight climate policies, indicating a considerable degree of political support for policy and formal engagement in climate-policy adoption. A few of these states, such as California, Massachusetts, Connecticut, and New York, have adopted as many as six or seven of them. In some cases, states have revisited early policies and decided to "raise the bar", elevating initial emission-reduction targets or strengthening earlier commitments to renewable energy. The remaining 28 states represent in essence the other half of the American population and have either zero or one such policy in operation, indicating less political support for policy or formal engagement in climate-policy adoption. Over the past three years, at least eight states have moved from the "low" policy cell to the "high" policy cell. That trend appears likely to continue, given the volume of activity on various climate policies in many state legislatures. At the same time, it is also possible for states to backtrack on prior commitments, as states such as Arizona and Utah wavered in 2009 on their initial commitment to join with California and other Western states (and some provinces) in the Western Climate Initiative, which would set up a regional approach to emissions reductions.

STRATEGIC CONSIDERATIONS EMERGING FROM THE INTERSECTION OF STATE EMISSIONS GROWTH AND POLICY ADOPTION

These rather fundamental differences between various states may be instructive in considering their receptivity to future policy initiatives, whether undertaken unilaterally, in concert with regional neighbours, or in response to possible future federal government actions. Figure 1 represents the convergence of these two dimensions of emissions growth and policy adoption. Each of the four quadrants reflects a different blend of emissions and policy adoption trends and includes reference to the total number of states that fall within it as well as a sample set of cases. The convergence of these factors illustrates the diverse contexts facing individual states as they contemplate future initiatives or engagement in intergovernmental bargaining as Congress explores a wide range of possible options. They further suggest that individual states may have considerable reason to view various climate initiatives in very different ways, depending upon where they stand in relation to the 1990 baseline that is used almost universally in American and international climate-policy deliberations and whether or not they have made any significant commitment to policy adoption and implementation. Just as private businesses and industries are increasingly thought to adjust their strategies based on their emissions levels and internal incentives for action or inaction (Layzer 2007, 209-210), states may face similar strategic choices and be influenced by their current context.

Figure 1: State Climate Policy Adoption and GHG Emission Trends

Emission Growth Trends (1990-2003)*

		High (>15%)	*Low (< 15%)*
*Levels of State Climate Policy Adoption***	*High (2 or more policies)*	10 States Arizona Minnesota Oregon	12 States California New Mexico Pennsylvania
	Low (0-1 policies)	22 States Alabama Florida Texas	7 States Louisiana Michigan West Virginia

* See Table 1.
**Measures the adoption of the following policies within a state: Renewable Portfolio Standard; Carbon Tax; Renewable Fuel Standard; Carbon Cap-and-Trade; State-wide Emissions Target; Mandatory Emissions Reporting; Litigation (formal support of Massachusetts in *Massachusetts v. EPA*); California vehicle emission standards.

Subsequent sections will review each of these quadrants and consider possible strategic considerations as states approach the possibility of serious federal legislative engagement and consider their desired impact on any such policy output.

Low Emissions, High Policy

States that have sustained low rates of emissions growth while pursuing significant policy adoption may be eager to exert their influence over neighbouring states and federal policy debates. They will be adamant that 1990 remain sacrosanct as the emissions baseline and insist upon maximal credit for achieving "early reductions" and for being "first movers". Pulling other states or the entire nation into their orbit is likely to maximize their leverage on overall emissions reductions. This might also serve to provide them with economic advantages, having already invested in technology and staff expertise associated with policy adoption, thereby forcing recalcitrant states to launch the process of "catch-up". States in this quadrant will be keen to make sure that any future federal policy "follows their example", both to ease transition costs and maximize credit-claiming opportunities for political leaders. One early example of the intergovernmental transfer of policy ideas involved the federal government embrace of California's approach to reducing emissions in the vehicle sector in a May 2009 decision by President Obama.

California perhaps epitomizes this quadrant, taking the long-standing term of "California effect" in American intergovernmental policy leadership to new lengths in climate change (Vogel 1995). The state has long played a pioneering role in environmental protection and other areas of policy, often stimulating cross-state diffusion and ultimate embrace at the federal level. In climate, it has literally "run the table" by adopting virtually every kind of climate policy imaginable. Politically, this allows Governor Schwarzenegger and other state leaders to claim credit for "global leadership" on climate policy, even pushing constitutional bounds in ways that allow for direct negotiation with other national heads of state or sub-national governments outside the United States (Breslau 2007; Adams 2006). The state first entered into the climate-policy arena in 1988, having already achieved one of the lowest rates of per capita GHG emissions due to major energy conservation efforts in the prior decade. But it has since followed with a veritable blizzard of climate initiatives, including the 2006 Global Warming Solutions Act that established bold state-wide reduction commitments over coming decades and set in motion a carbon cap-and-trade program with wider scope than attempted in any Western democracy to date. In turn, California state government is being reconfigured to begin to foster the inter-sectoral and inter-agency collaboration that will be necessary to secure implementation, including hiring of dozens of new state staff, although funding necessary for these tasks is in jeopardy given the state's dire fiscal deficits (Rabe 2009).

It has become increasingly clear that California intends this massive effort to achieve additional emissions stabilization to minimize any internal costs and maximize its economic development opportunities. This is reflected in a number of state policies that are designed to influence neighbours as well as national and even international policy. Evolving interpretation of California statutes to guide regulation of vehicle emissions would clearly impinge most heavily outside of the state. The state has a relatively small vehicle manufacturing sector but a good portion of it is concentrated on high-fuel-efficiency vehicles which would be boosted by regional or national adoption of the California standards (Rabe and Mundo 2007). At the same time, these policies are designed to apply to all vehicles registered in California, allowing the state to influence emissions standards for vehicles purchased outside of the state. In turn, California's evolving efforts to markedly reduce carbon emissions from utilities are being designed so that regulation would be implemented through "performance standards" that would force any utilities from other states or nations that might export electricity into the state to adhere to California standards. These approaches raise a series of political and constitutional issues, but the California case suggests that, at least in some instances, a low-emission and high-policy adoption footing can allow for simultaneous pursuit of environmental improvement and rent-seeking. In this regard, California is somewhat unique given its sheer size in terms of populace and economic heft. But one can see somewhat similar strategic thinking in play elsewhere, particularly Pennsylvania, New York, and other north-eastern states that are also trying to position themselves as models for regional and national policy adoption and to emerge as "national climate-policy leaders". It is also reflected in very recent efforts by Florida and its governor, Charles Crist, to explore the possibility of dramatic expansion of its engagement in climate policy and also attempt to assume roles of leadership among south-eastern states and perhaps nationally. All such states are on record as being concerned about climate change but want their emissions track record to be rewarded and their early steps toward policy adoption to have considerable leverage.

High Emissions, High Policy

Given other competing factors, the adoption of multiple climate policies does not guarantee their effectiveness or their ability to achieve significant emission reductions. Indeed, states such as Arizona, Minnesota, and Oregon, among others, have adopted multiple forms of climate policy, including particularly early initiatives in the latter two states that are well along into the stage of implementation. But their rate of emissions growth has remained well above national averages. States with this blend of emissions and policy development will likely approach intergovernmental negotiations from a somewhat different position. They will be more enthusiastic about modification of the 1990 baseline and seek credit for early policy initiatives even if these had little effect on reducing emissions growth. They might well seek special treatment or status for policies that were enacted more recently and are only moving into preliminary stages of implementation. This might include allowance of a two-tiered system, whereby states would be free to exceed federal minimum standards or released from adherence to any federal requirements through a waiver process.

Arizona provides an illustration of this phenomenon. It has established in recent years a series of renewable energy policies and entered into a number of collaborations with California that range from negotiation of a regional carbon cap-and-trade program to formal endorsement of its carbon emissions regulations for vehicles. But it has the second-highest increase in emissions of any American state between 1990 and 2003, with a rate of growth that is more than four times that of one of its neighbours, New Mexico. The state has experienced particularly steep emissions growth in transportation. It will likely endorse federal policies that concentrate emissions reductions in that sector, given the small presence of vehicle manufacturing in the state, as reflected in its formal adoption of the California vehicle emissions legislation. In turn, Arizona has decided to use 2000 as the baseline for its state-wide emission reduction goals in coming decades, providing a more conducive starting point for claiming any future reductions by ignoring the developments of the prior decade.

Minnesota's emissions growth is similarly dominated by the transportation sector and so it might take a position similar to that of Arizona. In turn, much of Oregon's emissions growth is in the electricity sector, due in part to diminished output from nuclear and hydro plants, which could lead it to support a different set of policies. All of these states tend to view themselves as "mini-Californias", supporting cutting-edge policy experimentation and in the vanguard of national leadership on the climate change issue. But they will want to be protected against penalty for any substantial emissions growth and be rewarded for early policy adoption in any future federal climate legislation.

High Emissions, Low Policy

More than 20 states fall into the quadrant with above-average rates of emissions growth and low levels of policy development and they are represented by more than two-fifths of the membership of the U.S. Senate, a level that is generally sufficient to block discussion on any legislative proposal. Many of these states are located in the south-eastern portions of the nation, including Alabama, Florida, and Texas. They have generally experienced steady rates of population and economic growth, tend to have expanding manufacturing industrial bases, and are heavily reliant on coal for electricity. In turn, they are generally thought to have some of the weakest potential capacity for renewable energy and have historically taken few if any steps to promote alternative sources or energy efficiency. Moreover, they tend to be among those states that receive low rankings for their levels of commitment or institutional capacity to pursue environmental protection (Resource Renewal Institute 2001).

Many of these states have been non-players in climate policy, although Texas and a few others have adopted single policies of some consequence. In some instances, these states may literally adopt policies with significant greenhouse gas reduction potential, as in the case of a renewable portfolio standard that was enacted in Texas in 1999 and expanded six years later. But this particular policy was not adopted initially for its climate-protection potential; instead, it was supported for such reasons as energy supply diversification and reduction of conventional air contaminants (Rabe 2004). Its prospects for initial passage might have been jeopardized by explicit labelling as climate policy, a pattern that has also been evident in some other states. In such instances, state proponents will sustain a "stealth" approach and emphasize other attributes, unless it proves advantageous for them to become more explicit about having taken some form of early action. Indeed, Texas's 2005 reauthorization and expansion of this program illustrates a shift, with explicit emphasis on GHG reductions being a factor contributing to the second round of legislation.

Consequently, this set of states represents a substantial area of the nation that is essentially the converse of the Northeast or Pacific West. Not only is their emissions growth high and policy adoption minimal, but they may view virtually any federal climate policy as a possible threat to their economic well-being (Rabe and Mundo 2007). They are likely to oppose any policy that would impose significant costs on them and would be particularly mindful of possible redistributive effects that could result from mandates to purchase carbon credits, offsets, or renewable-energy credits from outside their state and region. Moreover, they would have significant incentive to adjust the emissions baseline to a date well after 1990 and seek substantial federal subsidies to compensate against any possible adverse economic consequences from federal policy implementation. They are likely to be hostile to any effort to apply ambitious state policies, such as those emanating from California, on a national basis.

Ironically, many of these states may be among the most vulnerable to climate change, at least over the next few decades. Coastal states such as the Carolinas and Texas have become particularly concerned about growing risk from severe weather episodes and some significant temperature increases. Several such states are enmeshed in discussions of the future of insurance coverage for developed property, particularly from coastal property owners who are facing steep rate hikes due to increased vulnerability. This issue has begun to move climate change onto the agenda in such Gulf Coast capitals as Tallahassee, reflecting a series of initiatives in recent years that illustrate the potential for states to undertake major policy shifts in relatively short order. Each year of the past decade has seen some additional states take major policy adoption steps and this pattern could indeed continue, whether attributable to triggering events or other factors.

Low Emissions, Low Policy

The odyssey of state experience with greenhouse gas trends reveals that it is indeed possible to attain stable levels of emissions in the absence of climate policy designed to achieve these goals. In seven states – including Louisiana, Michigan, Indiana, Ohio, and West Virginia – there has been virtually no adoption of any GHG-reduction policies and yet all have emission growth rates well below the national averages between 1990 and 2003. Louisiana and Michigan emissions, for example, are largely unchanged over this period (see Table 1). However, states in this quadrant are not exactly models for effective transition to a less carbon-intensive society in that much of their stability is due to economic stagnation. In the Michigan case, an actual increase in emissions in most sectors is offset by significant outright declines in manufacturing-based emissions since 1999, reflecting the marked contraction in that sector. Given the virtual collapse of vehicle manufacturing in that state in more recent years, the Michigan decline would clearly be more dramatic if data were available through 2009. Louisiana has also undergone its own transitions and, were more recent data available, it would perhaps be evident that both states may have declined even further due to continuing economic contraction.

In some respects, this parallels the "East German model" for emissions reduction, drawing comparisons with those portions of Eastern Europe which easily met Kyoto goals through industrial collapse in the early 1990s. Any such states will approach climate policy with trepidation and will be particularly inclined to combat any policies that might further weaken vulnerable economic sectors. In Michigan, this was reflected in an aggressive but ultimately unsuccessful effort by its congressional representatives to fend off any new federal restrictions on vehicular fuel economy, on the heels of the state government's decision to formally support the federal position in opposition to Massachusetts and other states in the 2007 Supreme Court case on climate change. Indiana and Ohio also have highly vulnerable vehicle manufacturing sectors and will clearly desire to shift any climate regulations toward other sectors where they have less at risk. This may explain the strong push in these states for bio-fuel policies at the state and federal levels, as this is seen as favouring traditional American vehicle manufacturers such as General Motors which have invested in technologies to allow for expanded use of such fuels.

Consistent with the bio-fuel experience, states in this quadrant are likely to seek minimal interference with threatened industry and also to insist on very favourable financial terms to compensate them for any possible costs that might be imposed by federal policy. States of all sorts are keen to maximize federal transfer payments but the demand may be particularly great in these states given their relative economic position. Some of these, including Kansas and West Virginia, have actively called for expanded federal support to develop potential renewable energy sources that are particularly promising within their boundaries. They contend that they lack the resources to develop these on their own, given their economic circumstances, and appear particularly eager to seize maximal shares of revenues for energy development from the 2009 federal economic stimulus legislation. Relatively recent expansion of interest in possible climate-policy adoption in several of these states, for example, Michigan and Ohio, is based almost exclusively on anticipated economic development potential. At the same time, such states will want to make sure that any future policy accords them maximum "credit" for their low rates of emissions growth. Hence, the 1990 baseline will remain sacrosanct and states in this quadrant will welcome any opportunities for credit-trading programs that could deal them a favourable hand, similar to Eastern European nations and Russia which have attempted to maximize the value of their "hot air" credits.

Consequently, the combination of emission trends and policy adoption could influence future state action, whether it entails unilateral state policy or efforts to shape the direction of future federal policy. This variability may serve to complicate any consensus on future federal policy, given dramatically different strategic considerations facing different states and possibly different regions of the nation. Such conflict would also likely surface in implementation following any federal policy enactment, producing a prolonged intergovernmental bargaining process. At the same time, as we shall see, recent state experience reminds us that different types of climate policies may generate very different political responses that transcend any particular jurisdictional context. Some policies may be anathema, whereas others may have considerable appeal across a diverse range of states and regions, and this aspect of state experience may serve as a guide to likely political viability of policy options at the federal level.

CLIMATE-POLICY SELECTION: ECONOMIC DESIRABILITY VS. POLITICAL FEASIBILITY

The expanding body of state experience in climate policy may afford insight into the political prospects for future enactment of various policy instruments intended to reduce GHG emissions, whether through expanded state adoption or eventual acceptance in some form by the federal government. As discussed,

GHGs emanate from every sector of economic and social activity, opening up the possibility of an almost-infinite number of possible policy interventions. These run the gamut from more conventional command-and-control policies that emphasize rigid regulations and standards to economically based policies that allow flexibility as long as overall emission reduction goals are met. States clearly have substantial latitude to choose from this range of policy tools, and their choices offer an indicator of how the options fare when placed into a political context. Many scholars have noted a general shift in various areas of environmental protection toward the latter set of policies (i.e., towards the economically based policies), particularly in the American context (Fiorino 2006; Mazmanian and Kraft 2009). This trend is energetically embraced in much scholarship on climate policy, with widespread endorsement of policies that make use of market-based mechanisms to maximize the likelihood that any reductions will be produced in as cost-effective manner as possible (e.g., Stewart and Wiener 2003; Victor 2004; McKibbin and Wilcoxen 2002; Jaccard, Nyboer, and Sadownik 2002; Fischer and Newell 2007; Aldy and Stavins 2007; Congressional Budget Office 2007).

But the preferences of scholars, particularly economists, are not readily translated into new policy through the political process. Indeed, climate change may well illustrate the confrontation between the pecking order of policies, as endorsed by most economists and kindred scholars in other social science disciplines, and the reality of gaining political support for various policies through representative institutions stocked with officials who must contend with electoral realities. In short, those policies that tend to maintain the strongest base of support from policy analysts appear to have the greatest difficulty of being adopted by state legislators and governors. In turn, those policies that have many features of more traditional approaches that have been long criticized by policy analysts are far more successful in securing significant support from elected officials.

In some respects, as Figure 2 indicates, the relationship between the "economic desirability" and "political feasibility" of climate-policy options may be nearly inverse, based on American state experience to date. This figure indicates on the horizontal axis a measure of economic desirability as determined by an extensive review of scholarly climate-policy literature that compares various policy alternatives. This entailed a content analysis of nearly 60 scholarly books, articles, and reports published between 2000 and 2007 that formally compare the economic desirability of competing climate-policy tools. This focused primarily on leading tools under consideration for electricity generation and manufacturing, though they could easily be transferred over to other sectors. In turn, Figure 2 also considers on the vertical axis the "political feasibility" of competing tools determined through a simple measure of the number of states that have formally adopted them. This uses legislative action in 50 statehouses as a proxy measure for the political viability of alternative policies. It calculates the vertical dimension of political feasibility by a measure of the number of states that have adopted a particular policy, ranging from low (0-9 states) to medium (10-19) to high (20 and above).

Figure 2: Economic Desirability and Political Feasibility of State Climate Policy Tools

		Economic Desirability*		
		High	*Medium*	*Low*
*Political Feasibility***	High			Renewable Portfolio Standard (28)
	Medium		Cap and Trade (23)	
	Low	Carbon Tax (0)***		

*Reflected in climate policy literature review.
**Measured by number of states adopting policy: 0-9: Low
10-19: Medium
20 and above: High
***Excludes public benefits charges/social benefits funds due to modest scope.

If the experience to date of American states with climate policy is representative of other polities, it raises important questions about the political viability of those policies that might deliver emission reductions in the most cost-effective manner possible. In particular, this experience suggests that state governments are extremely reluctant to impose strategies that are explicit about any costs that will be imposed, particularly if they are likely to be evident at the point of product purchase or utility bill payment. Instead, they may have considerable incentive to produce far more complex policies, which may require greater overall costs but allow them to be less visible either by being hidden or spread out over a longer period of time. This pattern may well carry over to other polities, including those of governments in other multi-level systems such as Canada and the European Union (Rowlands 2007). It may also reflect developments of the first months of the 111th Congress, including the passage of the American Climate Energy and Security Act in June 2009 on a narrow vote in the House of Representatives.

Economic Attractiveness, Political Anathema: Pity the Carbon Tax

A review of diverse literatures on climate policy indicates a very broad consensus among scholars regarding the desirability of using carbon taxes as a central approach to climate policy. In 2007, a *Wall Street Journal* survey of leading economists showed overwhelming support for carbon taxes as the preferred tool for addressing climate change. "A tax puts pressure on the market, rather than forcing an artificial solution on it", noted one of the survey participants in a representative comment. In turn, carbon taxes have been formally endorsed by a Who's Who of very diverse economists who often agree on little. Harvard economist Gregory Mankiw, who chaired President Bush's Council of Economic Advisors between 2003 and 2005, has established a pro-carbon tax blog, known as the Pigou Club Manifesto, which has been endorsed by such diverse luminaries as Gary Becker, Martin Feldstein, Thomas Friedman, Alan Greenspan, Paul Krugman, Anthony Lake, William Nordhaus, Richard Posner, Jeffrey Sachs, Isabel Sawhill, Lawrence Summers, and Paul Volcker, among many others. As Edward Snyder, dean of the University of Chicago Business School has noted, "We need to recognize carbon is a 'bad', tax it, and let the market work" (Carbon Tax Center 2007). Even such unlikely allies as consumer activist Ralph Nader and ExxonMobil President Lee Tillerson have made favourable comments about this option in recent years.

In practice, such taxes would be based on the carbon content in respective fossil fuels, thereby establishing a higher cost for a similar unit of coal versus oil or natural gas, given the high carbon levels of the former. In theory, they give consumers incentives to use less carbon-intensive energy but do so without imposing uniform constraints on citizens or industries. In turn, it is thought that the establishment of such a tax would be relatively straightforward through expansion of existing tax code provisions, and that compliance would be high since it would be applied at a point of purchase of carbon-based energy sources. All 50 states clearly have constitutional authority to establish multiple forms of carbon taxes, as they have long used a combination of sales and excise taxes for gasoline and can use their considerable power over utility regulation to apply taxes to electricity usage. The federal government also holds vast authority to move in this direction if it were so inclined politically, and proposals for such a federal energy tax go back to the Richard Nixon-Gerald Ford era of the early 1970s.

One might anticipate that the intellectual consensus behind this option, the ever-growing need in many states for additional revenue, and the growing saliency of the climate-change issue in many states would create a groundswell of sorts behind some form of carbon taxation. But there is no evidence to suggest that any state has decided to make carbon taxes a central plank of its climate protection strategies. The lone American jurisdiction that has taken such a step is Boulder, Colorado, which enacted an explicit carbon tax through a 2006 ballot proposition that will generate revenue to help underwrite the city's climate-protection program. But Boulder appears likely to remain an anomaly. Indeed, California, the very state synonymous with an aggressive, across-the-board approach to GHG reduction, has essentially put every imaginable climate policy into play with the conspicuous exception of carbon taxation. On the same day that Boulder voters approved their carbon tax in November 2006, California voters decisively rejected Initiative 87, a ballot proposition that would have increased state-wide energy taxation as a climate-policy tool.

In the arena of gasoline, all states have maintained some form of taxation, with most of that revenue used to support highway maintenance and expansion. In 2008, the average state gas tax was 21.9 cents per gallon, which is imposed alongside a federal gasoline excise tax of 18.4 cents per gallon (Tax Policy Center 2008). The actual rates have changed only modestly over the past decade, and state tax policy analyst John Petersen has noted that they are not indexed to inflation or changes in gasoline prices. As a result, "the value of the taxes has been declining in real terms over the years" and is actually lower in price-deflated terms than it was in prior decades (Petersen 2007). Some states have explored suspending their gasoline taxes in the face of climbing prices. Consequently, there does not appear to be much political appetite for addressing this area of possible carbon tax development.

Electricity taxation is somewhat different in that 18 states have established some form of specialized taxation beyond conventional sales taxes, and at least some of the collected revenues generally are earmarked for energy efficiency programs or renewable energy development. These programs range from 0.03 to 3 mills per kWh, with one mill equivalent to one-tenth of one cent (Dernbach *et al.* 2007, 10025). These programs generate between $8 million per year in Illinois to $440 million per year in California, and the average cost per residential household across these states is quite low. They are sufficiently modest as to have little likely impact on carbon consumption, serving instead as a funding source for new energy initiatives. The most ambitious of these approximate the level of taxes imposed in Quebec but none approach the levels established in the British Columbia carbon tax initiative.

Perhaps the most revealing aspect of these policies is that they are universally not referred to as taxes in authorizing legislation or their inclusion on customer bills. Instead, they tend to be characterized by terms such as "social benefit charges" or "public benefit fees" and a number of states do not itemize them on customer electricity bills. All have been authorized through either legislative action or decisions by state public utility commissions. They have been designed to sustain a low enough level of taxation and are given a sufficiently innocuous title so as not to trigger anticipated opposition to new energy taxes. This "stealth" quality raises a number of interesting questions about future prospects for carbon taxes at either state or federal levels, underscoring the political complexities involved in being explicit about their function or setting them at levels sufficiently high to have a realistic capacity to deter energy consumption and greenhouse gas generation.

Both the executive and legislative branches have their supporters of carbon taxes. A number of prominent Obama appointees, including economic advisors like Summers and Volcker, as well as energy secretary Steven Chu, have written and spoken favourably about this option, though they have become much quieter on the subject since ascending in 2009 to federal office. In turn, a coalition of ideological opposites, ranging from Congressman Pete Stark (D-CA) to Senator Bob Corker (R-TN) have endorsed some version of a carbon tax in the 111th Congress. In fact, Stark has repeatedly introduced some version of a carbon-tax bill for nearly two decades and did so again in the early days of the 111th Congress. His most recent version looks much like the British Columbia carbon tax in its general design and gradual increase in level. Yet any carbon-tax proposal figures is likely to continue to face strong opposition, as reflected in numerous public opinion surveys, even if framed as "revenue-neutral" through some return of funding to the citizenry. In June 2009, during seven hours of House floor debate on the proposed American Climate and Energy Security Act, there was only a single reference made to a carbon tax as an alternative to the 1,300 pages of other policy approaches that were later approved by a 219-212 vote.

Economic Shortcomings, Political Attractiveness: Renewable Energy Mandates

Renewable portfolio standards (RPS) may represent the near-complete converse of carbon taxes in terms of economic desirability and political feasibility as a climate-policy option. They require that all providers of electricity within a state increase the amount of power that they derive from renewable sources over time. Most of these policies steadily increase the total percentage or volume of electricity that must come from renewable sources and establish financial penalties in the event of non-compliance. This is representative of a body of policies that follow a command-and-control pattern. Related climate-policy options include renewable fuel standards that mandate increased levels of bio-fuels and emission-control policies that mandate use of a particular technology or achievement of an identical level of emission reduction from all regulated sources.

Just as a large range of climate scholars are enamoured with the concept of carbon taxation, many view policies like RPSs with trepidation on economic grounds. Such policies are generally seen as more expensive per unit of GHG-emission reduction, in that they mandate the use of specific technologies rather than let market forces determine the most cost-effective means of reduction. This is particularly a concern as RPSs become more complex, with so-called "carve-out" provisions that require not only an overall level of renewable energy but also supplemental commitments to expand more expensive renewable sources such as solar power (Rabe and Mundo 2007). In turn, it remains very difficult to discern the actual carbon-reduction impact of RPSs, since it is not always clear which type of existing source is being supplemented and because the policy does not attempt directly to reduce demand for electricity. This issue is especially significant in those instances where the definition of renewable energy includes sources, such as biomass and animal waste, which have questionable GHG-reduction benefits and often raise other air-quality issues. Concerns about the cost-effectiveness of this tool are reflected in a number of early studies on actual RPS performance, even in cases where renewable capacity is high and overall cost is below national averages (Dobesova *et al.* 2005; Chen *et al.* 2007). These concerns are further compounded if jurisdictions adopting RPSs take steps to assure that newly mandated renewable energy is generated within their boundaries, even if that produces higher-cost electricity than through out-of-state importation. As one prominent study of competing climate-policy tools concludes, "the RPS may be one of the less efficient means of achieving greenhouse gas emission reductions. Unlike a more flexible carbon cap, it does not reward generation from non-renewable sources of low carbon power, and rewards energy conservation only very weakly" (Bushnell *et al.* 2007).

Any misgivings over RPSs from an economic standpoint have not constituted a stumbling block to their rapid adoption and diffusion. Indeed, of the eight major climate policies outlined in Figure 1, RPSs have clearly been the most popular politically. They have been approved in 28 states and the District of Columbia, and are in operation in well over half of the nation's congressional districts. RPSs have become prominent in every section of the nation except for the Southeast, although North Carolina enacted such a policy in 2007. Moreover, they are under active consideration in a number of other states and nearly half of the current RPS states have revisited their earlier enactments by setting more ambitious goals through legislative reauthorization. Many states are establishing very ambitious targets, such as 25 percent in New York by 2013, 20 percent in Colorado and New Jersey by 2020, and 18 percent in Pennsylvania by 2020.

It is not clear that states adopting RPSs have conducted systematic economic analyses or carefully assessed their capacity to reach these various targets (Chen *et al.* 2007). In turn, a number of states have faced early implementation problems, ranging from local resistance to the siting of renewable generation facilities or transmission capacity to pressures from supporters of particular renewable energy sources to receive increasingly favoured treatment in RPS implementation (Rabe and Mundo 2007). All raise added concerns over the long-term economic impact of these policies and questions of whether neighbouring states can work collaboratively to establish common renewable energy markets or instead erect barriers to discourage cross-border movement and purchase.

None of this has dampened political enthusiasm for the RPS approach, which may be attributable in part to the fact that it is commonly framed as delivering multiple benefits, only one of which is climate change. Most states enacting RPSs have characterized them as strategic investments in future technologies that could provide long-term economic benefits. In turn, renewable energy is routinely portrayed as far more labour-intensive than conventional electricity, for which imported fuel costs are high. This invariably leads to a framing of renewables as a source of within-state job creation. At the same time, various states have emphasized other co-benefits, including diminished release of conventional air contaminants through transition to new electricity sources and reduced dependence on other jurisdictions to sustain a supply of fossil fuel or uranium. Yet others have emphasized the desirability of sending early "market signals" that encourage development of energy technologies that could provide long-term benefits of accelerated energy system transformation.

Perhaps most significantly, RPSs are framed as essentially cost-free in political debates, with any added costs "passed along" to electric utilities, even though consumers will likely pay any difference for an electricity supply that has a higher level of renewable sources, whether they realize it or not. As one team of analysts of the early experience with these programs concluded, RPSs "are attractive politically because they accomplish multiple objectives with one policy, and are not perceived as a tax" despite the fact that they are likely to prove "somewhat more expensive" than more market-based strategies such as carbon taxes (Dobesova *et al.* 2005, 8583). This may explain why RPSs continue to draw broad, bipartisan support in states with every pattern of partisan control and have also been enacted via ballot initiatives in Colorado and Washington.

This political calculus that perceives an RPS as offering environmental, economic, and political benefits, however spurious in practice, may also explain why this is the one climate-policy tool that has repeatedly received serious federal legislative consideration, well before the arrival of the 111th Congress. In both 2003 and 2005, the Senate approved creation of a national RPS that would reach a 10 percent level by 2020 and allow for a two-tier system whereby states could seek higher levels through their own policies if they desired. These measures died in conference proceedings with the House of Representatives, but they remain the first instances in which either chamber voted in favour of a non-voluntary climate measure. In 2007, the House took the lead and passed a more ambitious version of a federal RPS, awaiting Senate response at year's end. The 111th Congress quickly embraced the RPS option, including a version that was inserted into the American Climate and Energy Security Act that was approved by the House in June 2009.

It is thus no surprise that while a carbon tax appears to be every bit the non-starter at the state or federal level, the prospects for continued state diffusion as well as adoption of a federal RPS remains a good deal higher. Federal RPS proponents tied their arguments closely to the 2009 economic stimulus bill, arguing that mandatory expansion of renewable energy could go hand in hand with subsidies to develop such alternatives. Extensive polling data suggests continuing popularity for this option in the United States (Borick 2010). This also appears to apply to other regulatory provisions that could influence GHG emissions, such as energy efficiency standards that have been advanced in many states and began to receive greater federal government attention in 2010.

Moderate Economic and Political Attractiveness: Carbon Cap and Trade

Emissions trading through some version of a carbon cap-and-trade system has emerged as a reasonably attractive policy option from both an economic and political perspective. Economists and policy analysts tend not to be quite as effusive about cap and trade as carbon taxes, but they do tend to characterize this approach as a very desirable alternative. Indeed, many policy analysts have championed such a policy design for many environmental problems, based in part on the extensive and near-euphoric assessment of the American sulphur-dioxide emissions-trading program that was launched under Title IV of the 1990 Clean Air Act amendments. Such a policy could theoretically be applied to specific sectors that generate carbon emissions, such as electric utilities, or an entire economic and political system. Ironically, this approach was actively pushed by the American federal government as a model for international climate policy during the negotiations that led to the Kyoto Protocol.

Cap-and-trade proponents emphasize that it injects far more flexibility into emissions reduction than conventional command-and-control approaches and holds considerable promise for achieving cost-effective reductions. Under cap and trade, an overall budget for carbon releases is established and gradually reduced over time. Once emission allowances are allocated to individual sources or jurisdictions, they are then free to negotiate transactions to allow for the most inexpensive possible reductions. These may be achieved, at least in part, through so-called offsets, such as carbon sequestration through tree planting or subterranean storage. As David Victor has noted, "Launching an emissions trading system requires creating a new form of property right – the right to emit greenhouse gases – and institutions to monitor, enforce, and secure those new property rights" (Victor 2004, xii).

This approach also has considerable political appeal, reflected initially in the adoption by the European Union of its Emissions Trading Scheme and somewhat comparable proposals in Congress. Twenty-three states have made some level of commitment to their own version of a cap-and-trade program, ten of which are working through the Regional Greenhouse Gas Initiative (RGGI) that has begun to operate a regional emissions-trading zone for utility sector emissions in the Northeast. California governor Arnold Schwarzenegger has also interpreted the state's 2006 climate legislation as allowing for development of a comprehensive cap-and-trade system. In turn, both RGGI and California are keen to expand their coverage to include as many of their neighbours as possible, and it is possible that the total number of states or Canadian provinces participating in a cap-and-trade program will expand further. A set of Midwestern states entered into a regional agreement of their own in 2008, though this has not moved beyond a very general statement of intent. Despite this flurry of activity, emissions trading does not appear to retain as strong a base of political support as such tools as renewable portfolio standards and mandatory vehicular fuel efficiency. Prior U.S. Senate votes over a national carbon-trading system received considerably fewer votes than previous proposals for a national RPS. Considerable reservations about developing such a program at the federal level have remained evident in the 111th Congress, though this approach retained general support from President Obama and his climate team as well as Democratic leaders in both chambers.

At the same time that cap and trade blends a reasonable level of economic and political appeal, its Achilles Heel may be its extreme complexity and steep implementation challenges. Whereas both carbon taxes and renewable portfolio standards are relatively straightforward policies to implement, whatever their shortcomings either politically or economically, that is simply not the case for carbon cap and trade. The early experience with this policy in the implementation stage underscores that it has features of what political scientist Charles Jones once characterized as "policy beyond capacity" (Jones 1975). This early difficulty may be exacerbated by the very weak intergovernmental institutions established to date to secure inter-jurisdictional efficacy.

The economic elegance of cap and trade quickly dissolves once one moves toward actual policy development and implementation, at least based on early experience in the United States and elsewhere. In Europe, the ETS failed to establish an institutional structure that might have allowed it to run effectively. Each member of the European Union was permitted to allocate and monitor its own emission allowances, without any overarching authority in place to assure accuracy and integrity. National compliance plans were loosely structured and repeatedly violated in implementation, with few if any consequences from the European Union or its member nations. Efforts to repair ETS after a troubled start have continued. Early North American experience with the same tool underscores these difficulties. In RGGI, multi-stage negotiations continued for more than four years before launching its initial allowance allocation in September 2008, building on a history of north-eastern regional collaboration on a wide range of environmental and energy issues.

What has emerged is a set of provisions in a treaty-like agreement, endorsed by the ten signatory states and being considered by others. However, each state must still secure formal support politically, whether through legislation or formal executive action before it can begin to move forward on implementation. In turn, many key elements of the system, such as whether emission allowances should be auctioned or distributed without charge and the methods of curbing carbon emissions from electricity generated outside of the RGGI zone, remain highly contentious. At the same time, RGGI features a dizzying array of provisions that address such issues as offsets, "early reduction credits", "triggers", and "safety valves" that will require considerable administrative sophistication and intergovernmental collaboration to sustain. Perhaps predictably, individual states and interest groups bring very different agendas to the negotiations over cap-and-trade programs, thereby weakening its economic purity. States with smaller populations or projections for higher population growth seek favoured status in the allocation of emission allowances and insist that larger neighbours pay a disproportionately large share of governance costs. States more reliant on electricity imported from outside the cap-and-trade region view import constraints differently than those states with greater energy self-sufficiency (Rabe 2008). Many of these same issues have begun to emerge in Washington as Congress has attempted to develop its own version of a cap-and-trade system on a national basis.

Perhaps the political and governance challenges of cap and trade are most evident in California. The 2006 authorizing legislation was clear about the desirability of a state-wide emissions cap but intentionally evaded the issue of whether a trading mechanism would be established because it was so divisive politically (Rabe 2009). In 2007, Schwarzenegger used his executive authority to insist on such a trading system, but this has proven extremely controversial. On the one hand, a number of industry groups suggest that such a system will be particularly disadvantageous to them. This has produced a splintering of interests and competing pressures on the California Air Resources Board to adjust any trading system to ease challenges for particular sectors. On the other hand, a range of environmental groups contend that emissions trading is a "sell-out that endorses pollution"; their reading of the 2006 statute suggests early and aggressive mandated reductions rather than a more flexible cap-and-trade system. Environmental justice advocates have further contended that any trading system will place particular disadvantages on low-income and predominantly minority communities. Many state legislators have joined this chorus in Sacramento, alleging that Schwarzenegger has exceeded his powers and should instead focus on an immediate command-and-control approach. As State Senate President Don Perata has stated, the 2006 Global Warming Solutions bill "is getting bogged down in arcane discussions over intercontinental trading schemes, 'carbon markets,' and free 'credits'. That may work for Wall Street traders and Enron economists, but it doesn't work for Californians" (*Carbon Control News*, 2007). At the same time that California is struggling to implement its own variant of a cap-and-trade system, it is also negotiating the Western Climate Initiative, a multi-state pact that generally follows the regional approach taken by RGGI. Thus far, the Governors of Arizona, Montana, New Mexico, Oregon, Utah, and Washington have signed a very general memorandum of understanding with California to begin work on the "design for a regional market-based multi-sector mechanism, such as a load-based cap and trade program". Not all of these states, however, have joined California in adopting authorizing legislation, and serious intergovernmental bargaining over the terms of a multi-state pact remain in very early stages. In turn, three states (Arizona, Montana, and Utah) gave serious thought to withdrawing from the regional agreement in 2009 and California's capacity to underwrite much of the implementation costs have come into question as the state's fiscal condition becomes more dire.

Nonetheless, as many as two dozen versions of a cap-and-trade bill were introduced into the 111th Congress and both House Speaker Nancy Pelosi and Senate President Harry Reid called for active exploration of this option in early 2009. The House and its Energy and Commerce Committee took the lead, working in close concert with a coalition of businesses and environmental groups known as the U.S. Climate Action Partnership. This produced a legislative proposal for a federal cap-and-trade system along with a federal RPS and many additional regulatory provisions. The House quickly abandoned President Obama's endorsement of an auctioning system for allowances, instead establishing a politically charged allowance allocation process that gave various industries and states highly favourable terms in exchange for supportive votes from key legislators. This led to continued expansion of the bill from its initial 624 pages to more than 1,300 pages in June 2009, with much of the growth involving specific regulatory or grant provisions designed to assemble a majority coalition. In a highly contentious and largely partisan debate, the House approved the American Clean Energy and Security Act on a 219-212 vote, where it moved to the Senate for further consideration. Political opposition to cap-and-trade legislation has remained significant and public opinion surveys suggest Americans are generally divided on the issue and unclear as to what "cap and trade" actually entails. This suggests numerous opportunities for proponents and opponents to frame public perception in the run-up to any federal decision.

LOOKING AHEAD: FROM A STATE-CENTRIC TO A FEDERAL SYSTEM?

The evolution of the state role in American climate policy suggests that it is indeed possible, at some times and in some jurisdictions, to secure a base of political support for policies that promise to make some dent in GHG emissions. Indeed, the state experience over the last decade or so offers an important set of insights and could provide a building block for new departures in this arena. This might well involve continued expansion and diversification of the state role, but could also be influential in the design of any future federal policy, whether or not that would involve formal collaboration and sharing of authority with states that have taken early action. State experience does not lend itself to easy prediction of the future, either of further state policy diffusion or eventual federal engagement. But it does suggest the possibility of three broad alternative directions over the coming years, as state policy dominance begins to confront its fit into the larger American political, economic, and policy context and given the strong indications that the Obama Administration and 111th Congress intend to take a far more active role on climate change than their federal government predecessors. Prior experience in American federalism suggests at least three distinct intergovernmental paths for American climate policy that might be pursued in the near future and beyond.

Shift Toward Top-Down

There is substantial precedent for federal government pre-emption of existing state policies. In such instances, Congress often responds to industry concerns about interstate regulatory variation and eliminates the "patchwork quilt" of policies with a uniform program (Nivola 2002; Posner 2005). The frequency with which Congress uses this tool in domestic policy has only increased in recent decades (Zimmerman 2007). Such diverse individuals as John Engler of the National Association of Manufacturers and U.S. Senator Barbara Boxer (D-CA) have referenced pre-emption as a distinct possibility, essentially wiping out existing state climate policies as part of any larger bargain to create a nation-wide policy. This would, of course, invariably raise concerns about equity among those states that have achieved low emissions growth (through whatever mechanism) and might be denied credit for their early efforts. Moreover, states that had adopted and actively implemented their own policies would argue that pre-emption was particularly unfair to them as it would invalidate their early investments. In turn, some concerns have arisen that a federal pre-emption policy of modest scope might actually achieve lower emission reductions than through the existing compilation of state policies. Nonetheless, any serious discussion of a congressionally enacted cap-and-trade program or a renewable portfolio standard increasingly turns to the possibility of federal usurpation of a policy arena heretofore developed and dominated by states. The blending of some intensive industry opposition to new federal policies and likely opposition from some set of states (and their congressional representatives) for varied reasons remains significant, however.

This may explain why the initial climate-policy steps of the incoming Obama administration have focused on administrative strategies, beginning with rapid reconsideration of the Bush Administration's 2008 rejection of declaring carbon dioxide an air pollutant. Such a reversal was signalled almost immediately by the incoming Obama climate team and suggested a willingness to push through with administrative interpretation of the Clean Air Act Amendments of 1990 to begin to impose national restrictions on GHG emissions. This step would not require new legislation and was used to pressure Congress to take significant steps, such as a federal cap-and-trade program. Such an approach would entail administrative action by the U.S. Environmental Protection Agency through an established framework of clean-air legislation that does not specifically address carbon dioxide or climate change. This work proceeded quietly in 2009 as Congress began to weigh climate-policy options, some of which involved federal pre-emption of existing state policies of cap-and-trade and energy-efficiency standards.

Continued Bottom-Up

It remains very plausible to envision a system that retains a strong bottom-up emphasis. There is no guarantee that Congress or the executive branch will reach closure on any significant climate policy in the coming years and any federal action may be confined to a specific sector or policy tool, leaving much continued room for state engagement. Federal policies may also take considerable time to be put into operation after enactment, reflecting protracted interpretation and rule development. There are numerous areas of American public policy in which nationalization has seemed inevitable but which have continued to operate with state domination (Teske 2004; World Resources Institute 2007). The recent patterns of diffusion, proliferation, and regionalization in state climate policy seem very likely to continue short of federal pre-emption. This will be reflected in expanded adoption of policies already operational in multiple states and the growing pattern of multi-state negotiation once neighbouring states establish similar or identical policies. It is increasingly possible to envision "climate-policy regions" whereby two or more states join common cause, building on early movement in this direction and perhaps working further on a collaborative basis with Canadian provinces or Mexican states (Selin and VanDeveer 2009). It is conceivable that the United States could even set a national cap of sorts and simply allocate overall allowances or reduction requirements state by state, then allowing for interstate bargaining over the mechanics of reduction. This would follow the model of the European Union and was reflected to some extent in the first major legislation of the Obama era likely to influence American GHG emissions, namely the 2009 economic stimulus plan. This initiative allocated billions of dollars to a wide range of renewable-energy and energy-efficiency projects, as well as other "climate friendly" initiatives, but gives enormous latitude to states and localities in determining how to use these funds. Some of the allowance allocations for cap and trade under the American Climate Energy and Security Act appear to have followed this format, such as in the anticipated oversight role of state public utility commissions which monitor electricity generating facilities within their boundaries.

Toward Collaborative Federalism

It is at least possible to envision an American climate policy that builds on the respective strengths of both state and federal governments and engages in active policy learning across governmental levels. As discussed, many states have developed considerable climate-policy expertise and may remain particularly well equipped to target areas of "low-hanging fruit", namely low cost emission-reduction opportunities unique to their state. At the same time, the federal government retains the ability to develop consistent rules and incentives on a national scale and, of course, the constitutional authority to work collaboratively with other nations. Perhaps the United States could evolve into a multi-level climate-governance system, consistent with practice in other areas of environmental protection (Scheberle 2004). One such option is a two-tiered mechanism whereby, unlike pre-emption, the federal government would establish a national minimum but states would be free to retain or develop policies that were more ambitious. Such a policy could be crafted so as not to penalize states for early reductions and could also be designed to reward such actions. It remains entirely possible that climate policy will follow an iterative path for some time, with at least some states continuing to play a role of policy innovator, and thereby influence various rounds of federal policy. Recent experience in American intergovernmental relations finds few examples of such collaborative federalism in environmental protection, energy, and virtually any other sphere of domestic policy over the past decade (Conlan and Dinan 2007), though the construction of the economic stimulus plan could indicate a new direction as well as President Obama's active engagement with state governors and large-city mayors in negotiating implementation and accountability plans.

The resurgent interest in climate change reflected in both President Obama and the 111th Congress raises the serious possibility that a new course in federal climate legislation could be established at some point during the next few years. Proposals in both the House and the Senate have involved renewable-energy mandates (for both electricity and fuel) and a cap-and-trade system, while generally confining carbon taxes to the fringes. High initial expectations for early federal government action on a major climate bill are, of course, complicated by the dire state of the American economy and the reluctance to require any cost-imposition strategies during severe recession. Significant interstate and cross-regional strains have emerged in both the House and Senate as proposals have moved forward. Nonetheless, the possibility of federal-level action atop a growing tapestry of state and regional policies offers unique opportunities and challenges. Even the scores of hearings on climate change in the 110th and 111th Congresses demonstrate little serious effort at policy learning, with most discussion of intergovernmental lessons involving brief presentations by high-profile governors or periodic congressional threats to overturn pre-existing state policies as part of a larger bargain with organized interests. Moreover, it also remains clear that Congress is badly fragmented and faces steep hurdles before producing anything other than piecemeal legislation, with unified Democratic Party control not a guarantor of significant policy outputs on climate or other domestic issues (Mann and Ornstein 2006). Any new federal legislation that might emerge is also likely to face protracted litigation and an extended period of policy interpretation and development. As a result, a continued period of bottom-up policy development amid some forms of federal policy expansion may be the most likely near-term outcome. In turn, continuing intergovernmental conflict (whether state-to-state or state-to-federal) is likely to persist for many years in this area whether or not significant federal legislation is forthcoming.

Ironically, the American case may have some striking parallels with climate policy in other federal or multi-level governance systems, whether or not they have ratified the Kyoto Protocol (Rabe 2007). Ratifying parties such as Canada and the European Union have struggled mightily not only to meet their reduction targets but also to strike an effective balance of authority between central governments and their constituent units. Canadian provinces and E.U. nations vary by emission trend and level of policy engagement in ways that are highly analogous to the range of American state responses discussed herein. At the same time, Australia bears especially striking resemblance to the United States, given its rejection of Kyoto under Liberal Party rule, prolonged federal-level policy inertia, a flurry of policy development in some but not all of its states, and expanded federal attention to climate policy following the election of new leaders in November 2007. All of this suggests that climate policy can no longer be framed as the exclusive province of international relations and instead must also be acknowledged as an enduring challenge for multi-level governance. Even a dramatic election result, as in the American case in 2008, is unlikely to change this factor of climate policy.

REFERENCES

Adams, L. 2006. "California Leading the Fight Against Global Warming", *ECOStates* (Summer): 14-16.

Aldy, J.E. and R.N. Stavins, eds. 2007. *Architectures for Agreement: Addressing Global Climate Change in the Post-Kyoto World*. Washington, DC: Resources for the Future.

Breslau, K. 2007. "The Green Giant", *Newsweek* (16 April): 51-60.

Borick, C. 2010. "American Public Opinion and Climate Change", in B.G. Rabe (ed.), *Greenhouse Governance: Addressing American Climate Change Policy*. Washington, DC: Brookings Institution.

Bushnell, J., *et al.* 2007. *California's Greenhouse Gas Policies: Local Solutions to a Growing Problem?* Berkeley: University of California Energy Institute.

Carbon Control News. 2007. "Major Bank Pushes Credit Trading 'Interoperability' Among GHG Systems", unsigned article in *Carbon Control News* (10 April).

Carbon Tax Center. 2007. www.carbontax.org/who-supports/.

Chen, C., *et al.* 2007. *Weighing the Costs and Benefits of Renewables Portfolio Standards: A Comparative Analysis of State-Level Policy Impact Projections.* Berkeley: Lawrence Berkeley National Laboratory.

Congressional Budget Office. 2007. *Trade-Offs in Allocating Allowances for CO2 Emissions.* Washington, DC: Congressional Budget Office.

Conlan, T. and J. Dinan. 2007. "Federalism, the Bush Administration, and the Transformation of American Conservatism", *Publius: The Journal of Federalism* 37(3): 279-303.

Dernbach, J. and the Widener University Law School Seminar on Energy Efficiency. 2007. "Stabilizing and then Reducing US Energy Consumption: Legal and Policy Tools for Efficiency and Conservation", *Environmental Law Reporter* 37 (January): 10003-10031.

Dobesova, K., *et al.* 2005. "Are Renewable Portfolio Standards Cost-Effective Emission Abatement Policy?" *Environmental Science and Technology* 39(22): 8578-8583.

Fiorino, D.J. 2006. *The New Environmental Regulation.* Cambridge, MA: MIT Press.

Fischer, C. and R.G. Newell. 2007. "Environmental and Technology Policies for Climate Mitigation", Resources for the Future Discussion Paper 04-05.

Hoffman, A.J. 2006. *Getting Ahead of the Curve: Corporate Strategies That Address Climate Change.* Arlington: Pew Center on Global Climate Change.

Jaccard, M., J. Nyboer, and B. Sadownik. 2002. *The Cost of Climate Policy.* Vancouver: University of British Columbia Press.

Jones, C.O. 1975. *Clean Air: The Policies and Politics of Pollution Control.* Pittsburgh: University of Pittsburgh Press.

Kamieniecki, S. 2006. *Corporate America and Environmental Policy.* Stanford: Stanford University Press.

Layzer, J.A. 2007. "Deep Freeze: How Business Has Shaped the Global Warming Debate in Congress", in M.E. Kraft and S. Kamieniecki (eds.), *Business and Environmental Policy.* Cambridge, MA: MIT Press.

Mann, T.E. and N.J. Ornstein. 2006. *The Broken Branch: How Congress is Failing America and How to Get it Back on Track.* New York: Cambridge University Press.

Manna, P. 2006. *School's In: Federalism and the National Education Agenda.* Washington, DC: Georgetown University Press.

Mazmanian, D.A. and M.E. Kraft, eds. 2009. *Toward Sustainable Communities,* revised edition. Cambridge, MA: MIT Press.

McKibbin, W.J. and P.J. Wilcoxen. 2002. *Climate Change Policy after Kyoto: Blueprint for a Realistic Approach.* Washington, DC: Brookings Institution.

Montpetit, E. 2003. *Misplaced Distrust: Policy Networks and the Environment in France, the United States, and Canada.* Vancouver: University of British Columbia Press.

Nivola, P.S. 2002. *Tense Commandments: Federal Prescriptions and City Problems.* Washington, DC: Brookings Institution.

Petersen, J.E. 2007. "Fueling a Tax", *Governing* (March): 66.

Pew Center on Global Climate Change. 2007. *Climate Change 101: Understanding and Responding to Global Climate Change.* Available at http://www.pewclimate.org/global-warming-basics/climate_change_101.

Posner, P.L. 2005. "The Politics of Pre-emption: Prospects for the States", *PS: Political Science and Politics* 38(3): 371-374.

Rabe, B.G. 2004. *Statehouse and Greenhouse: The Emerging Politics of American Climate Change Policy.* Washington, DC: Brookings Institution.

— 2007. "Beyond Kyoto: Climate Change Policy in Multilevel Governance Systems", *Governance: An International Journal of Policy, Administration, and Institutions* 20(3): 423-444.

— 2008. "Regionalism and Global Climate Change Policy: Revisiting Multi-State Collaboration as an Intergovernmental Management Tool", in T. Conlan and P.

Posner (eds.), *Intergovernmental Management in the 21st Century*. Washington, DC: Brookings Institution.

— 2009. "Governing the Climate from Sacramento", in D.F. Kettl and S. Goldsmith (eds.), *Unlocking the Power of Networks*. Washington, DC: Brookings Institution.

Rabe, B.G. and P.A. Mundo. 2007. "Business Influence in State-Level Environmental Policy", in M.E. Kraft and S. Kamieniecki (eds.), *Business and Environmental Policy*. Cambridge, MA: MIT Press.

Repetto, R., ed. 2006. *Punctuated Equilibrium and the Dynamics of US Environmental Policy*. New Haven: Yale University Press.

Resource Renewal Institute. 2001. *State of the States*. San Francisco: Resource Renewal Institute.

Rowlands, I.H. 2007. "The Development of Renewable Electricity Policy in the Province of Ontario: The Influence of Ideas and Timing", *Review of Policy Research* 24(3): 185-207.

Scheberle, D. 2004. *Federalism and Environmental Policy: Trust and the Politics of Implementation*. Washington, DC: Georgetown University Press.

Selin, H. and S. VanDeveer. 2007. "Political Science and Prediction: What's Next for US Climate Change Policy?" *Review of Policy Research* 24(1): 1-27.

Selin, H. and S. VanDeever, eds. 2009. *Changing Climates in North American Politics*. Cambridge, MA: MIT Press.

Stewart, R.B. and J.B. Wiener. 2003. *Reconstructing Climate Policy*. Washington, DC: American Enterprise Institute.

Tax Policy Center. 2008. www.taxpolicycenter.org.

Teske, P. 2004. *Regulation in the States*. Washington, DC: Brookings Institution.

U.S. Environmental Protection Agency. 2007. *The Inventory of U.S. Greenhouse Gas Emissions and Sinks*. Washington, DC: U.S. Environmental Protection Agency.

Victor, D.G. 2004. *The Collapse of the Kyoto Protocol and the Struggle to Stop Global Warming*, revised edition. Princeton: Princeton University Press.

Vogel, D. 1995. *Trading Up: Consumer and Environmental Regulation in a Global Economy*. Cambridge: Harvard University Press.

World Resources Institute. 2007. *Climate Policy in the State Laboratory: How States Influence Federal Regulation and the Implications for Climate Change Policy in the US*. Washington, DC: World Resources Institute.

Zimmerman, J.F. 2007. "Congressional Pre-emption During the George W. Bush Administration", *Publius: The Journal of Federalism* 37(3): 432-452.

V

Carbon Pricing: Constitutional and Institutional Perspectives

9

Carbon Emissions Trading and the Constitution

Stewart Elgie

INTRODUCTION

Combating climate change will require governments to use a range of regulatory tools to achieve significant reductions in emissions of greenhouse gases (GHGs), particularly carbon.[1] Of these tools, cap and trade is rapidly emerging as the main instrument for reducing *industrial* carbon emissions. Europe has had a carbon market in effect since 2005 (EU ETS System). Globally, carbon trading exceeded $60 billion in 2007 – and is likely to grow substantially once the United States and other countries come on board (World Bank 2007).

Canada's federal government has released a proposed framework for a national carbon emissions trading system, slated to come into effect in 2010 (Environment Canada 2007). Alberta already has in place a provincial emissions trading system (*Climate Change and Emissions Management Act* 2003), and several other provinces have announced their intention to develop their own systems, as part of the Western Climate Initiative (2007).

Emissions trading is a "next generation" approach to environmental regulation. These new systems will mark Canada's first experience with emissions trading of any kind, other than small scale or pilot projects (Elgie 2007, 246). As such, there are very real, unanswered questions about the respective constitutional powers of federal and provincial governments in this area.

The aim of this paper is to analyze the constitutional limits on federal and provincial powers to enact cap-and-trade legislation for carbon emissions. It begins with a brief overview of emissions trading. Then it assesses the federal government's powers, followed by an assessment of provinces' powers.

The conclusion reached is that a federal carbon emissions trading system will test uncharted constitutional waters and require the courts to extend existing constitutional law doctrine – which they are likely to do if the law is carefully drafted to minimize provincial intrusion. The provinces' authority is less clear. Assuming they have the authority to regulate carbon emissions (which is far from clear), provinces very likely have the authority to establish *intra*-provincial

---

[1]Carbon dioxide (CO_2) is the main GHG, and for simplicity will be referred to as "carbon" hereinafter.

carbon trading regimes; however, their authority to address extra-provincial trading is suspect. If provinces were to act through a coordinated multi-jurisdictional approach, courts may look somewhat more favourably on their authority, particularly if the federal government does not actually come forward with national legislation.

EMISSIONS TRADING OVERVIEW

Emissions trading is attractive because it can allow firms to achieve emissions reductions at a lower cost than through a traditional "command and control" approach. The theory is simple. If firm A can reduce its emissions less expensively than firm B, then B will pay A to achieve some or all of its required emissions reductions. The end result is the same total reductions, but at a lower cost. In addition to these economic benefits, cap and trade can also lead to environmental gains. By reducing the cost per tonne of emission reductions, it allows governments to set more ambitious reduction targets. In other words, some of the cost savings can be invested into greater emissions reductions.

These benefits have been realized in other types of emissions trading systems. The best known example is the U.S. Acid Rain Program. Its trading system allowed SO_2 reduction targets to be met at about 40 percent lower cost than through a traditional regulatory approach (Carlson *et al.* 2000, 1292-1326; Ellerman *et al.* 2000, 280-296). These cost savings led the U.S. government to set 25 percent more ambitious reduction targets (Environmental Defense 2000).

While emissions trading may not be effective for all types of pollution, carbon is well-suited to an emissions trading approaching, particularly because its impacts are global, not local – so it does not matter where the carbon is emitted (IPCC 2001, 247).

FEDERAL CONSTITUTIONAL POWER

Canada's Constitution demarcates areas of federal and provincial legislative authority (*Constitution Act, 1867*). To pass legislation, the federal or provincial government must show that the law's subject matter addresses one of the powers on its list. The Constitution says little about either government's power to legislate over environmental protection (not a major concern in 1867), so such authority normally must be found by extrapolating from the powers that *are* specified.

Federal powers are found principally in section 91 of the Constitution. They include specific powers, such as over fisheries, interprovincial works and navigation, as well as more general powers, such as over criminal law, trade and commerce, and a general residual power (discussed below). A detailed analysis of federal powers to legislate over carbon emissions trading is beyond the scope of this paper, but can be found elsewhere (Elgie 2008). Instead, I will cut to the constitutional chase and highlight the essential questions, and where I predict the courts would likely fall.

Environmental regulation only really began in early 1970s in Canada. In a handful of cases since that time the courts have begun to flesh out the scope of federal powers to legislate over the environment. In general, they have been trying to strike a balance between two competing goals: i) the need for national standards to address environmental problems that cross borders (which most do); and ii) not defining federal environmental powers so broadly as to allow for significant intrusion into provinces' economic or environmental authority (Elgie 2008).

In general, the courts have bounded federal environmental powers in two main ways: first, they have imposed limits on the *breadth of subjects* that Parliament may address (under the POGG power), and second, they have imposed limits on the *depth of tools* that Parliament feds may use (under the Criminal power) (Elgie 2008). The upshot is that the federal government has been given strong authority to address environmental problems that are specific and well bounded, such as marine pollution, and limited tools (mainly prohibitions) to address broader environmental problems, such as toxic pollution.

The courts so far have sketched the broad parameters of these powers. However, many important questions remain unanswered. One such question is the authority to regulate over carbon emissions and trading.

Although the precise scope of federal power over this subject cannot be known until it is tested in court, it is possible to make some educated guesses based on the existing case law. The two main federal powers for regulating carbon emissions and trading are POGG and Criminal Law.

POGG

The preamble to section 91 gives the federal government general power "to make Laws for the *Peace, Order, and good Government* of Canada, in relation to all Matters not coming within the Classes of Subjects by this Act assigned exclusively to the Legislatures of the Provinces ...". The POGG power, as it is known, has been interpreted as a residual power. In particular, it allows Parliament to address subject matters of "national concern" that were not specifically articulated in the Constitution.

A subject matter will be found to be a matter of national concern under POGG if it meets the following three-part test, established by the Supreme Court of Canada in *R. v. Crown Zellerbach* (1988):

> i) *The subject is "single, distinct and indivisible".* The boundaries of environmental jurisdiction cannot be defined so broadly as to give the federal government sweeping power to intrude into provincial areas of economic and environmental authority. For example, the courts have ruled that "pollution" is too broad to qualify as a subject matter under POGG. Rather, Parliament must distill environmental problems down into more compartmentalized, bounded topics. For example, marine pollution qualifies, as does interprovincial water pollution (*Crown Zellerbach* 1988, 542-546).

Greenhouse gas pollution would very likely meet this test. It is bounded; the *United Nations Framework Convention on Climate Change* (1994) identifies just six specific GHGs – carbon being the main one. The problem cannot be further subdivided; all six must be addressed to control climate change. And all have an international impact.

ii) *A province's failure to effectively deal with the problem would lead to significant extra-provincial consequences* (known as the "provincial inability" test). Carbon pollution almost certainly meets this test too. The impacts of carbon emissions (and all GHGs) are global. A tonne emitted in Winnipeg has as much effect on India's climate as on Manitoba's (IPCC 2001, 247). The impacts of a province's failure to effectively regulate carbon emissions would be primarily extra-provincial in nature.

iii) *Conferring federal authority over the subject would have a scale of impact on provincial jurisdiction that is reconcilable with the constitutional distribution of powers.* This would be the most difficult issue in litigation over a federal carbon cap-and-trade law. Federal regulation of carbon would affect many areas of provincial economic activity – and of peoples' lives. It could have broad impacts, such as increasing fuel prices (and thus transportation, home heating, etc.), significantly increasing the cost of coal power (a major source of energy in several provinces), or significantly affecting certain industries (e.g., cement, oil and gas, etc.).

Here is where the first big unanswered constitutional question is encountered: in deciding what degree of impact on provincial jurisdiction is acceptable, will the courts take into account the magnitude and significance of the problem being addressed? This question does not appear to have been addressed in prior cases.

It seems reasonable to argue that, although addressing climate change will inevitably involve significant impacts at the provincial level (as well the local, national and international ones), those impacts flow from the gravity and scope of the problem, not the level of government doing the regulating. Fossil fuels – the cause of climate change – are the engine of modern economies. By the same token, effective national and international measures will confer significant *benefits* on provinces (by avoiding the consequences of dangerous climate change) – benefits that could not be achieved by provincial action alone. Does the magnitude of the climate change problem, and the impacts it could have on provinces, justify somewhat greater provincial interference by federal legislation? At this point, we simply don't know.

If the federal government *was* authorized to address carbon emissions under POGG, it would get broad power to do so. That is the nature of POGG; it confers plenary power over a subject (*Crown Zellerbach* 1988, 426-427). Therefore, federal legislation almost certainly could address emissions trading, and could include provincially regulated activities such as forestry, agriculture and energy production. However, federal power would not be unlimited. For example, if it reached too deeply into provincial economic matters – for example, by prescribing the use of specific industrial technologies, or certain types of forest practices – that may cross the line.

Criminal Law

The Criminal Law power, in section 91(27) of the Constitution, allows Parliament to address wrongful conduct in society. The Supreme Court in *A-G Canada v. Hydro-Quebec* (1997) opened the door to federal environmental regulation under Criminal power.

For legislation to be upheld as Criminal Law, it must pass two tests (*Hydro-Quebec* 1997):

> i) *Have a valid Criminal Law purpose.* Traditionally the list of acceptable public purposes included "health, safety and morality". In *Hydro-Quebec* (para. 123-127), the Supreme Court ruled that protection of the environment has become sufficiently important that it is now a valid Criminal Law purpose. That case dealt with the control of toxic substances, but it is highly likely that addressing climate change would also meet this test.
>
> ii) *The legislation must be mainly* ***prohibitory*** *as opposed to regulatory in its approach.* This element of the test will be the difficult one for cap-and-trade legislation to meet.

The courts have said that a statute need not rely *exclusively* on prohibitions to fall under the Criminal Power. They will allow some regulatory elements, like: use of permits, qualified prohibitions, and setting standards by regulation. But if a statute becomes too regulatory in nature, it won't be deemed criminal (Hogg 2002, c.18.10).

The Canadian Environmental Protection Act (CEPA) barely passed this test, with the court splitting 5:4 on this issue in *Hydro-Quebec*. Based on that decision, one could predict that federal legislation controlling carbon emissions under CEPA, or a CEPA-like approach, would likely be upheld. And that is what federal government is currently proposing – to regulate carbon emissions under CEPA (Environment Canada 2007).

The big question is whether federal carbon regulation can extend to emissions trading under the Criminal Law power. The *Hydro-Quebec* decision never addressed the issue of emissions trading (it did not arise in that case). Certainly there is a legitimate argument that allowing trading of pollution rights is a regulatory, not prohibitory, approach. It is not the kind of approach one could envision falling under the Criminal Code (imagine trading units of blood alcohol with other drivers).

On the other side, there is a plausible argument that environmental legislation is not like other types of criminal legislation. It doesn't deal with *individual* morality, safety or health, like most other "criminal" laws. It deals with collective behaviour that exceeds ecological limits. The climate does not care how much individuals emit; all that matters is collective emissions. Emissions trading simply allows for a reallocation of firms' pollution limits within an overall threshold. One could argue that, having recognized the environment is a valid Criminal Law purpose, the courts may need to adapt the existing constitutional test to reflect the distinct nature of environmental problems.

It is hard to predict which way the courts would go on this issue, since it is venturing into new ground. Peter Hogg, arguably Canada's top constitutional expert, has written that he thinks emissions trading likely would be upheld under the Criminal power, albeit with only limited explanation (Hogg 2008). I am not quite as convinced. I would put the odds at around 50 percent. The courts may prefer to look for another head of federal power under which to uphold emissions trading, in an effort to stay true to the prohibitory foundation of the Criminal Law power.

If, as the above analysis suggests, it is far from certain that Parliament can implement carbon emissions trading under its two main heads of environmental power, what other options does it have? At least two other heads of power appear to be plausible candidates: the Trade and Commerce power, and the power to implement treaties. Each of these options raises complex legal questions. For reasons of space, they are briefly summarized below, but have been explored more fully elsewhere (Elgie 2008).

Trade and Commerce

The federal government, like the provinces, does not have to rely on just one head of constitutional power to pass legislation; it can use two or more powers to support different parts of a statute. Thus it might be possible to combine the Criminal power (to address carbon emission restrictions) with the Trade and Commerce power (to address emissions trading). The question is, would emissions trading be upheld under Trade and Commerce (hereinafter T & C)?

There are two arms to the T & C power (Hogg 2002, c. 20):

> i) *International and interprovincial trade:* This prong allows the federal government to regulate activities involving interprovincial or international T & C. Such federal laws cannot apply to purely *intra*-provincial activities, except perhaps as a merely incidental effect of regulation aimed at *extra*-provincial trade (Hogg 2002, c. 20.2(b)). A federal carbon trading law would not be focused just on extra-provincial trading (or at least it is hard to imagine it being so limited, since that would exclude huge parts of the market[2]). Therefore, it would be hard pressed to qualify under this prong of the T & C power, at least as traditionally applied.[3]

---

[2]For example, Alberta and Ontario each contribute more than 25 percent of Canada's total carbon emissions, and firms in those provinces would be denied access to those trading internal markets if only *inter*-provincial trading were allowed.

[3]There is a novel argument that emissions trading is not like traditional commercial activity. Its only purpose would be to achieve compliance with the emission requirements of a federal statute. It is not an activity that a province *could* regulate, even its intra-provincial aspects. Therefore, the traditional judicial restriction on the T & C power (that

ii) *General regulation of trade affecting the whole country*: This prong applies when Parliament is not addressing a particular sector or commodity, but is regulating economic activity generally. The best-known example is the federal *Competition Act,* which regulates anti-competitive practices of all businesses, and has been upheld under the "general" T & C power.

The courts have set out a five-part test for legislation to qualify under this power (*General Motors* 1989, 674-683). This test is quite similar to "national concern" test under the POGG power, except that it does not include the requirement that the law have "limited impact on provincial jurisdiction" – which, as noted above, would be the hardest part of the POGG test for a carbon trading law to meet.

The application of this five-part test to federal carbon-trading legislation has been discussed elsewhere, and will not be repeated here (Elgie 2008). However, the Supreme Court of Canada has distilled these five criteria down to their "common theme": that the problem "cannot be effectively regulated unless it is regulated nationally" (*General Motors* 1989, 680). Carbon emissions trading very likely meets this standard. It is hard to imagine emissions trading happening at a national scale, let alone an international scale, through the combined actions of ten provinces alone. Alberta, for one, has refused to participate in any national scheme (Greenberg and Fekete 2007). In fact, as will be discussed below, it is highly questionable whether provinces even have the power to regulate extra-provincial carbon emissions trading.

My conclusion, therefore, is that there is a fairly strong argument for upholding a federal carbon trading law under T & C power, if it is not upheld under Criminal Law. But given the unexplored nature of this area (T & C has never before been applied to the environment) this conclusion must be tempered with some caution. Therefore, it is worth exploring the other main constitutional option.

Treaty Implementing Power

One of the great unanswered questions of Canadian constitutional law is whether the federal government has the power to implement international treaties that Canada signs.

In 1937, back when the British Privy Council was Canada's highest appeal court, it ruled that the federal government could sign treaties, but had no special power to implement them (*Labour Conventions* 1937). Implementation authority had to be found in the existing lists of federal and provincial constitutional powers. The upshot, in that case, was that Parliament had no authority to implement an international treaty on labour standards, since that was normally an area of provincial authority.

---

it cannot address intra-provincial activities) should not apply. This argument is canvassed in Elgie (2008, 112-115).

The *Labour Conventions* decision has been widely criticized by academics (Elgie 2008). Moreover, since 1937, several Supreme Court judges have indicated that the issue of a federal treaty implementing power should be revisited (*Francis* 1956; Rand 1960; *MacDonald* 1977; *Schneider* 1982). But in those 70 years there hasn't yet been a case that has squarely raised this issue – largely because the federal government has been very reluctant to pass legislation that relies on a treaty implementing power.

It seems quite possible that federal climate change legislation may open this 70-year-old Pandora's box. This issue would arise if Parliament passes such legislation and indicates that at least one of its purposes is to satisfy (or partially satisfy) Canada's international treaty commitments – a likely scenario for carbon legislation given that Canada is a party to the UNFCC and Kyoto Protocol (even though Canada isn't meeting its Kyoto target, it has remained a party to the treaty).

There are several strong arguments for why the courts should uphold a federal treaty implementing power, if and when the issue eventually arises:

i) When Canada was formed, the British government signed our treaties, and the Constitution (s. 132) said that Canada's federal government had the power to implement all treaties. It seems illogical that, once Canada gained the power to sign its own treaties in 1926, the federal government would lose the power to implement treaties.

ii) Canada's federal government has the weakest power of any federal government in the world to implement treaties. Other federations, like the United States, Australia, and Switzerland (with notoriously weak federal powers) give their federal governments authority to implement treaties (Opeskin 1996). In the case of the United States and Australia, courts have read this power into their constitutions (*Missouri* 1920; *Tasmania Dam* 1983). Germany appears to be the only other federation that does not give its federal government full treaty implementing power, but it still gives broader powers than Canada does (Opeskin 1996).

iii) A number of authorities have concluded that the lack of a federal treaty implementing power "has impaired Canada's capacity to play a full role in international affairs" (Hogg 2002, c. 11.5(c)), although the degree of impairment is a matter of debate (Elgie 2008, 96-98).

The main counter-argument is that an unconstrained federal treaty implementing power could swamp provincial jurisdiction (Lederman 1981, 357; Hogg 2002, c. 11.5(c)).[4] Because of that concern, it is likely that the courts would put some kind of boundaries on a federal treaty power, if they did recognize it.

---

[4]Note that this does not appear to have been the case in the United States and Australia, despite broad federal treaty powers (see authorities cited in Elgie 2008, note 151).

Several authors have proposed methods for bounding such a power. Hogg, for example, suggests that Parliament should have full power to implement treaties dealing with international or trans-border issues, but not ones dealing primarily with domestic matters, such as minimum domestic standards for human rights or labour (Hogg 2002, c. 11.5(c)). This approach has some merit, although it could raise difficulties in delineating between treaties addressing "international" and "domestic" matters (many treaties have elements of both).[5]

I propose a different approach. Parliament could be given the power to implement *all* treaties, but when doing so means straying outside traditional areas of federal authority it would be restricted to simply implementing the treaty's requirements, nothing more. In other words, Parliament would be on a short leash when stepping beyond its traditional constitutional turf (Elgie 2008). This approach strikes a balance between allowing Canada to acquit itself on the international stage, while minimizing the potential for intrusion into provincial powers. It also avoids the need for judicial interpretation about which treaty matters the federal government can and cannot implement.

In any event, if either one of these approaches were adopted to contain Parliament's treaty implementing power, federal legislation on carbon emissions control and trading would very likely pass constitutional muster. Both matters – the control of carbon emissions and emissions trading – are explicitly called for in Kyoto, and both address a problem that is clearly international in nature (climate change).

Thus, if Canada's Supreme Court were to recognize a federal treaty-implementing power (and there are many signs that it would), it would provide a solid basis for a federal carbon trading regime – and obviously would have broad significance for other issues as well.

To sum up, federal legislation to regulate carbon emissions and trading would test the current boundaries of federal constitutional powers. Under any of the four powers reviewed, it would require the courts to answer questions that have not yet been answered – in some cases very significant questions. Up to now, Canada's courts have been able to skirt around the hard questions about the federal government's environmental powers; they have given answers that sufficed for the statute at issue, but which left larger questions unanswered (Elgie 2008, 79-81, 120-126). Climate change legislation is likely to force these hard questions onto the front burner. Its implications – both ecological and economic – are far reaching. It seems clear that national measures, as part of a larger global effort, are needed to address the problem – and in particular to put a price on carbon. Canada's courts will have to decide if our federal government has such powers. My view is they probably will say yes, provided the federal law is drafted to minimize unnecessary intrusion into provincial powers.

---

[5]Hogg's proposed approach is based on categorizing the *effect* of the treaty, and can lead to difficult line drawing. A more effective approach may be to focus instead on the *primary purpose* of the treaty (see Elgie 2008, at 100-101).

PROVINCIAL POWERS

The next question is to what extent provinces have the power to legislate over carbon emissions trading.

At the outset, it should be noted that the federal government's apparent authority to legislate on this subject does not necessarily preclude valid provincial legislation. Canada's Constitution allows for a good deal of federal-provincial overlap (Hogg 2002, c. 15.5(c)). This is particularly common in the environment field. As long as a government (federal or provincial) is addressing a subject matter within its constitutional powers, the existence of overlap is allowed. It is only when a federal and provincial law are in actual *conflict* – when they require companies to do inconsistent things – that the federal law prevails (Hogg 2002, c. 16.2, 16.3).

The issue of provincial power is particularly important for addressing climate change, since some provinces are moving faster than the federal government. Alberta already has legislation in place to regulate carbon emissions and trading, albeit using an intensity-based approach (Alberta *CCEMA* 2003). BC and Quebec have put in place carbon taxes (British Columbia 2008; Quebec 2008). And four provinces have joined with western U.S. states in committing to develop a carbon cap-and-trade scheme (Western Climate Initiative 2007).

At present, these provincial efforts set the bar for GHG regulation in Canada. But are they on solid constitutional ground? How far can provinces go in regulating carbon emissions and trading?

The conventional wisdom is that provinces have broad powers to regulate polluting emissions within their borders (Hogg 2002, c. 30.7(c)). That may be true for most types of problems, but it is less clear that this is the case for GHGs.

Before turning to the question of emissions trading, let us first look briefly at provincial authority to regulate GHG emissions in general, since authority over trading – if it exists – likely derives from the power to regulate GHG emissions.

Can Provinces Regulate GHG Emissions?

Provinces have broad authority to regulate pollution and other types of environmental problems within their borders. Their authority to do so derives mainly from their authority to regulate "property and civil rights" under section 92(13) of the Constitution. Hogg states that s. 92(13) "authorizes the regulation of land use and most aspects of mining, manufacturing and other business activity, including the regulation of emissions that could pollute the environment", citing as authority the case of *R. v. Lake Ontario Cement* (1973), which upheld a provincial pollution control statute (Hogg 2002, c.30.7(c)). Provinces also derive environmental authority from a variety of other powers, particularly section 92(16), "matters of a merely local and private nature in the province".

Based on this line of authority, Bankes and Lucas (2004) conclude that there is "a strong likelihood that [Alberta's *Climate Change and Emissions Management Act*] is constitutionally valid", including its provisions regulating GHG emissions and trading. The authors conclude that "the central feature of [Alberta's Act] appears to be GHG emission reduction to protect the Alberta environment", and that this purpose falls within provincial constitutional power, particularly under s. 92(13). Similarly, Hsu and Elliot (2008) conclude that proposed legislation in BC to control GHG emissions is "primarily directed" at protecting "the environment of the province", and thus likely falls within provincial constitutional power. In both cases, the authors' conclusions are based on the well-established principle that provinces have authority to regulate polluting emissions within their borders.

However, it is not so obvious that this principle applies to GHG emissions. GHGs are a global atmospheric pollutant. Once emitted, they migrate to the atmosphere, where they diffuse and mix with other GHG emissions, resulting in a build up of pollutants that affects the global climate (IPCC 2001, 138-140). The GHGs emitted in a Canadian province have virtually no direct affect on that province – a point not addressed in either of the above-noted articles. They simply contribute to a global effect. And their contribution is very small. Even the highest-emitting provinces, Ontario and Alberta, contribute less than 1 percent of total global GHG emissions.

Almost all other types of air pollution have at least *some* meaningful direct effect on the province, since much (or all) of their impact occurs locally or regionally. GHGs are different; their effect will be felt almost entirely outside the province.

In the *Lake Ontario Cement* case, referenced above, the Court upheld the pollution control provisions of Ontario's *Environmental Protection Act* as constitutionally valid. The section in question, section 14(1), prohibited the "discharge [of] a contaminant ... into the natural environment that ... causes or is likely to cause impairment of the quality of the natural environment..." The Act makes it clear that "natural environment means the air, land and water ... *of the Province of Ontario*". GHGs may not meet this test, for the reasons explained above. Climate change is certainly "impairing the quality of the natural environment in Ontario" (and it will get a lot worse), but the GHGs *discharged in Ontario* make only a tiny contribution to the problem.

It is an interesting constitutional question whether a province can regulate emissions that will have very little direct effect on the province – over 99 percent of their impact is global. It is certainly arguable whether such legislation can be said to be aimed at benefiting the province, or addressing matters of provincial concern.

Provinces do have broad authority to regulate business and commercial activity with their borders (*Insurance Reference*). But that power is not unlimited. For example, they cannot regulate aspects of a business's activities that fall within federal jurisdiction, such as the use of radioactive substances or the manufacture of illicit drugs (Hogg 2002, c. 23.2). Provincial laws may *incidentally* address matters outside their jurisdiction, if done as part of achieving a valid provincial objective (*Global Securities Corp.*). However, it is questionable whether a province could regulate business activity primarily for the purpose of achieving an international or extra-provincial objective, such as controlling exports or preventing investment in certain countries – or combating global climate change. Certainly municipal governments, which are creatures of a province, cannot do so (*Shell Canada Products*). The fact that a province's GHG emissions will have *some* impact on that province, albeit very small, may provide enough of a connection to support provincial emissions controls, but it is far from clear that a court would take this view.[6]

Given this uncertainty, a province would be well advised to refer to factors other than just environmental protection as the purposes for any legislation to control GHGs. For example, it could also refer to provincial economic objectives, such as the need to "position the province's citizens and businesses to compete in a carbon-emissions-constrained future", as Alberta has done in the preamble to its *CCEMA*.

In addition, provinces no doubt have broad authority to address many aspects of GHG emissions through other provincial powers, including electricity generation, transportation, the construction of buildings and homes (energy efficiency), forestry and agriculture, etc. – all of which are grounded in clear provincial powers. The question is whether they also have the authority to legislate controls on industrial GHG emissions.

From a pragmatic perspective, it makes sense to allow provinces to regulate GHGs, since all the other emissions coming out the same smokestacks are provincially regulated. This is particularly so in a situation where there is no federal legislation in place (as at present). No doubt a court would be influenced by such considerations. However, the conceptual basis for provincial power to regulate GHGs is not as obvious as many seem to think it is.

Assuming that provinces *do* have the authority to regulate GHG emissions, let us now consider their authority to regulate emissions trading.

Can Provinces Regulate Extra-Provincial Emissions Trading?

There seems little doubt that provinces have the authority to regulate emissions trading within their borders as part of an overall GHG regulatory regime, as Alberta has done in the *CCEMA*. Provinces have broad constitutional authority to regulate intra-provincial commercial transactions (*Shannon* 1938; Hogg 2002, c. 21.9(c)).

The more difficult question is whether a provincial scheme that included *inter*-provincial (or international) emissions trading would be seen as constitutionally valid. As with federal emissions trading, such a provincial regime could be characterized, for constitutional purposes, as relating either to environmental or commercial objectives.

Seeking to justify a provincial emissions trading regime as relating to provincial environmental objectives seems difficult. Cross-border emissions trading means a company in one province will be allowed to emit more GHGs if it pays someone in another province (or country) to emit less. Overall, emissions

---

[6]For a general discussion on this issue see Edinger (1982).

trading is environmentally neutral – it simply redistributes emissions within an overall cap. On its face, it is hard to see how a province would make a convincing case that such trading is meant to achieve a provincial environmental objective. In fact, emissions trading could lead to an *increase* in provincial emissions – if a province is a net buyer of GHG emission credits.

Could a province justify the regulation of extra-provincial emissions trading under its powers over commercial activity, which arise under mainly the Property and Civil Rights power? This question is hard to answer in the absence of specific legislation, since this is a murky area of the law and much would turn on the actual details of the statute. However, some general principles can be discussed.

It is well established constitutional law that provinces may not directly regulate inter-provincial commercial activity. They may enact regimes whose main aim is to regulate intra-provincial commercial activity, and that have some incidental effect on extra-provincial trade, such as establishing provincial marketing boards that raise the cost of exported goods (*Carnation* 1968). But they cannot directly regulate extra-provincial trade (*Lawson* 1931). Indeed, they likely cannot even regulate intra-provincial commercial activity if the main purpose of doing so is to affect extra-provincial trade (*Central Canada Potash* 1979).

In the case of emissions trading, it is likely that most trading will occur across provincial borders. Even the highest emitting provinces (Ontario and Alberta) each account for less than 30 percent of Canada's total GHG emissions; this means that for a provincial company, more than 70 percent of its GHG trading market will be outside the province (and more if international trading is allowed).

Therefore, viewed through the lens of regulating commercial activity, it seems a hard stretch to argue that provinces can regulate extra-provincial GHG emissions trading. To be clear, there is a strong argument that a province could regulate the actual emission trading *transaction*. Provinces have broad authority to regulate securities transactions within their borders even if it involves extra-provincial companies or traders (*Gregory & Co.* 1961), and there seems little doubt provinces could do the same for emissions trading transactions. This would involve regulating *how* the trading takes place – e.g., ensuring the integrity of transactions.

What is questionable is where a province would get the constitutional authority to authorize extra-provincial emissions trading in the first place. Recall that emissions trading is not like most other commercial transactions, where there is an existing commodity that people want to trade. Emissions trading is a creature of statute; it exists because of a requirement to meet legislated GHG emission reduction requirements (leaving aside the small voluntary market). So the real question, to be precise, is not whether provinces may *regulate* emissions trading transactions, but where they get the constitutional authority to *authorize* GHG emissions trading.

That being said, this is a largely unexplored area of the law. Much would turn on how a province drafted its legislation. In the author's view, a province's best chance of success would be to draft legislation directed at achieving both an environmental and economic objective. For example, a statute's preamble could state that that emissions trading reduces the costs of emission reduction, and that these lower costs will allow for greater emission reductions and improved economic outcomes. By indicating that its objective is to achieve both economic (lower costs) and environmental (lower emissions) benefits for the province, such legislation may stand a greater chance of being upheld than if it rests on just environmental objectives.

In addition, if a group of provinces acted in concert to create a reciprocal trading regime, their chances of constitutional success may be increased. The Supreme Court has at times shown a willingness to stretch the boundaries of constitutional power somewhat when multiple jurisdictions co-operate through parallel legislation to set up interprovincial trading or marketing schemes (*Re: Agricultural Products Marketing Act* 1978; Hogg 2002, c. 20.2(b)). Even with such a joint approach, though, provincial regulation of extra-provincial emissions trading would be on questionable constitutional ground.

Lastly, it should be pointed out that the validity of provincial emissions trading regulations would depend, to some extent, on what the federal government does in this area. Under Canadian constitutional law, provincial legislation will be struck down if found to be in conflict with, or incompatible with, federal legislation on the same topic, under the doctrine of "paramountcy". Such conflict would arise if it were impossible to comply with both the federal and provincial regimes, or if the provincial regime frustrated the objectives of the federal one (Hogg 2002, c. 17). It is impossible to say in advance whether such conflict would arise over GHG emissions trading; it would depend on the scope and design of the laws in question. But one could easily imagine a scenario in which such conflict could arise – e.g., where firms were faced with trading the same emission reduction under different rules to meet different federal and provincial GHG targets. In such case, the federal regime would prevail. On the other hand, the two levels of governments could enact coordinated legislation meant to integrate federal and provincial GHG trading regimes across the country, as has been done for some marketing boards (*Re: Agricultural Products Marketing Act* 1978).

CONCLUSION

In sum, governments across Canada are proposing cap-and-trade legislation to address GHG emissions, largely because trading can significantly lower the costs of emissions reductions. This paper reviews the constitutional power of federal and provincial governments to address GHG emissions trading. Since neither the Constitution nor the courts have addressed this question, one must extrapolate from existing cases and powers.

It seems likely that the federal government has the power to regulate GHG emissions trading across the country. This authority most likely stems from a combination of the Criminal and Trade & Commerce powers, although the POGG or treaty implementing power (if it is found to exist) may also support it. In each case, it would require a modest expansion of the scope of these powers by the courts to accommodate the far-reaching effects of GHG regulation. However, there appear to be sound reasons for the courts to uphold federal power in this area, particularly because GHGs have a global impact, and can only be effectively addressed through coordinated global action – which requires national leadership.

That being said, much would turn on the way in which the legislation was drafted. A federal law could be on shaky constitutional ground if it reached too far into areas of provincial jurisdiction, for example by prescribing certain types of industrial or forest practices that must be followed.

For provinces, the first question is whether they even have the constitutional power to regulate GHG emissions – which is uncertain, given that the impacts are global. Assuming provinces have this power, they almost certainly have the authority to regulate emissions trading within the province, as an element of local commerce. However, their authority over extra-provincial carbon trading is doubtful, since inter-provincial trade is an area of exclusive federal constitutional power. A provincial carbon trading regime would have the best chance of being upheld if it were done for economic purposes (cost saving), not just environmental ones, and if were part of a coordinated multi-jurisdictional approach – which, in the end, is also likely to be the best approach from a policy perspective.

REFERENCES

A.-G. Canada v. A.-G. Alberta [1916] 1 A.C. 589 [*Insurance Reference*].

A-G Canada v. A-G Ontario [1937] A.C. 326 [*Labour Conventions*].

A-G Canada. v. Hydro-Quebec et al. [1997] 3 S.C.R. 213.

Alberta, *Climate Change and Emissions Management Act*, S.A. 2003, c. C-16.7 [Alberta CCEMA].

Canada. 1867. *Constitution Act* (UK), 30 & 31 Vict., c. 3, reprinted R.S.C. 1985, App. II, No. 5.

Carlson, C.P., D. Burtraw, M. Cropper, and K. Palmer. 2000. "SO2 Control by Electric Utilities: What are the Gains from Trade?" *Journal of Political Economy* 108(6).

Carnation Company Limited v. Quebec Agricultural Marketing Board [1968] S.C.R. 238.

Central Canada Potash v. Government of Saskatchewan [1979] 1 S.C.R. 42.

Edinger, E. 1982. "Territorial Limitations on Provincial Powers", *Ottawa Law Review* 57: 14.

Elgie, S. 2007. "Carbon Offset Trading: A Leaky Sieve or Smart Step?" *Journal of Environmental Law & Practice* 17(3): 235 (July).

— 2008. "Kyoto, the Constitution and Carbon Trading: Waking a Sleeping BNA Bear", *Review of Constitutional Studies* 13: 1.

Ellerman, A.D., P.L. Joskow, R. Schmalensee, J.-P. Montero, and E. Bailey. 2000. *Markets for Clean Air: The US Acid Rain Program*. Cambridge, UK: Cambridge University Press.

Environment Canada. 2007. *A Climate Change Plan for the Purposes of the Kyoto Protocol Implementation Act 2007*. Ottawa: Environment Canada. At http://www.ec.gc.ca/doc/ed-es/p_123/CC_Plan_2007_e.pdf.

Environmental Defense Fund. 2000. *Annual Report*. New York.

Francis v. The Queen [1956] S.C.R. 618.

General Motors v. City National Leasing [1989] 1 S.C.R. 641.

Global Securities Corp. v. British Columbia (Securities Commission) [2000] 1 S.C.R. 494.

Government of British Columbia. Ministry of Finance. 2008. *Budget Backgrounder: BC's Revenue-neutral Carbon Tax*. At http://www.bcbudget.gov.bc.ca/2008/backgrounders/backgrounder_carbon_tax.htm.

Government of Quebec. 2008. *Action Plan on Climate Change: 2006-2012*. At http://www.mddep.gouv.qc.ca/changements/plan_action/2006-2012_en.pdf.

Greenberg, L. and J. Fekete. 2007. "Premiers Split on Climate Change", *CanWest News Service*, 10 August. Accessed at http://www.canada.com/victoriatimescolonist/news/story.html?id=7ac41ae9-de84-473d-a185-89190bcc771c&k=34336.

Gregory & Co. Inc. v. Quebec Securities Commission [1961] S.C.R. 584.

Hogg, P. 2002. *Constitutional Law of Canada*, student ed. Scarborough: Carswell.

— 2008. "A Question of Parliamentary Power: Criminal Law and the Control of Greenhouse Gas Emissions". C.D. Howe Institute Backgrounder No. 114 (August).

Hsu, S-L. and R. Elliot. 2008. "Regulating Greenhouse Gases in Canada: Constitutional and Policy Dimensions", Working Paper. 8 September. At SSRN: http://ssrn.com/abstract=1265365.

Intergovernmental Panel on Climate Change (IPCC). 2001. *Contribution of Working Group I (Science) to the Third Assessment Report of the United Nations Intergovernmental Panel on Climate Change: Summary for Policy Makers*. c. 4 [IPCC, *Working Group I, 2001 Report (Summary)*]. Cambridge, UK: Cambridge University Press.

Lawson v. Interior Tree Fruit and Vegetable Committee of Direction [1931] S.C.R. 357.

Lederman, W.R. 1981. *Continuing Canadian Constitutional Dilemmas*. Toronto: Butterworths.

MacDonald v. Vapor Canada Ltd. [1977] 2 S.C.R. 134.

Missouri v. Holland, 252 U.S. 416 (1920).

Opeskin, B. 1996. "Federal States in the International Legal Order", *Netherlands International Law Review* 43: 353.

Re: Agricultural Products Marketing Act [1978] 2 S.C.R. 1198.

R. v. Crown Zellerbach [1988] 1 S.C.R. 401.

R. v. Lake Ontario Cement Ltd. [1973] 2 O.R. 247, 35 D.L.R. (3d) 109.

Rand, I. 1960. "Some Aspects of Canadian Constitutionalism", *The Canadian Bar Review* 38: 135.

Schneider v. The Queen [1982] 2 S.C.R. 112.

Shell Canada Products Ltd. v. Vancouver [1994] 1 S.C.R. 231.

The Commonwealth v. Tasmania [1983] 158 C.L.R. 1 [*"Tasmania Dam"*].

United Nations Framework Convention on Climate Change. 1994. 1771 U.N.T.S. 107. 21 March.

Western Climate Initiative. 2007. At http://www.westernclimateinitiative.org.

World Bank. 2007. *State and Trends of the Carbon Market 2007*. Washington DC: World Bank.

10

The Constitutional Authority to Levy Carbon Taxes

Nathalie J. Chalifour

INTRODUCTION

Debates about the choice of carbon pricing instruments cover a wide range of issues, from environmental effectiveness and efficiency to administrative ease. However, the question of which level of government has the requisite constitutional authority to implement carbon pricing, and in particular carbon taxation, is often forgotten in these debates. This question is of fundamental importance, since governments will not only want to avoid investing resources implementing a policy that may later be judged *ultra vires* but also and more importantly will aspire to the vast revenues that may be at stake.

Accordingly, the role of this paper is threefold: i) to analyze the federal and/or provincial governments' constitutional authority to implement carbon taxes; ii) to draw upon this constitutional analysis to highlight those design features of a carbon tax that might render it *intra vires* of the implementing jurisdiction; and iii) in light of the above, to evaluate the constitutionality of the Quebec and BC carbon tax regimes. Toward these ends, the paper is organized as follows. The next section examines the federal government's authority to institute a carbon tax, highlighting the design features that would be important to ensure that such a tax would fall within its constitutional authority. The federal heads of power addressed in this context are the taxation power (s. 91(3)), the criminal law power (s. 91(27)), the trade and commerce power (s. 91(2)) and Peace, Order and Good Government (the preamble to s. 91). The third section then repeats this exercise for the provincial governments and carbon taxation, where the highlighted heads of power are the provincial taxing power (s. 92(2)), the natural resources powers (s. 92A), the licensing power (s. 92(9)) and property and civil rights (s. 92(13)). The penultimate section focuses on the

---

This paper is based on, and with permission reproduces portions of, my earlier publication (Chaifour 2008).

likely constitutional underpinning for the Quebec and BC carbon tax regimes. A brief conclusion completes the paper.

THE FEDERAL GOVERNMENT AND CARBON TAXATION

Canadian courts have established a two-step analytical framework for evaluating the jurisdictional validity of legislation under the Constitution. The analysis requires first determining the "pith and substance" of the legislation in question, i.e., identifying the "true meaning or essential character" of the law or policy. To do this, courts will look at the text of the legislation, the social and economic purposes for which the legislation was enacted, the circumstances surrounding the enactment of the legislation, the legal effects of the legislation and, sometimes, the practical or anticipated effects of the legislation. The second step involves determining whether the legislation as construed by the "pith and substance" test falls under one of the allocated heads of power in sections 91 and 92 of the *Constitution Act, 1867* or the federal Peace Order and Good Government (POGG) power.

Constituent with this framework, this section examines four potential sources of federal authority for carbon taxes, namely the federal taxing power, the criminal power, the trade and commerce power, and POGG.

The Taxation Power: S. 91(3)

The federal taxation power gives the federal government authority to legislate in relation to "the raising of Money by any Mode or System of Taxation". This has been interpreted to confer a broad power to apply direct and indirect taxes and is the source of justification for a variety of federal tax regimes, from income tax to goods and services tax (*Winterhaven Stables Ltd. v. Canada,* 1988). No matter how broad the power, however, justifying a carbon tax under the taxation power would require clearing two important hurdles. First, the federal government would have the challenge of convincing the courts that the measure truly had revenue-raising as a dominant purpose. No matter how broad the taxation power, courts have repeatedly held that the purpose of taxes is to raise general revenue for public purposes, unconnected to a regulatory scheme (*Connaught* 2008). Second, it would need to avoid coming within the scope of an important limitation created by s. 125 of the *Constitution Act, 1867.*[1]

It would be difficult for the government to demonstrate that a carbon tax had revenue-raising as its *dominant* purpose. Carbon taxes have been discussed and recommended over many years primarily as a mechanism for reducing GHG

[1]Section 125 states that "[n]o Lands or Property belonging to Canada or any Province shall be liable to Taxation". This provision applies only to the taxation power. Thus federal legislation justified under another head of power, such as trade and commerce, would not be subject to s. 125.

emissions, by influencing the price of carbon in a way that reduces demand for products that release GHG emissions. It is difficult to find a policy document or report discussing carbon taxes that does not identify climate change as the motivational force behind such a measure. Conversely, there are few, if any, policy documents that suggest carbon taxes primarily as a means of raising revenue. While it is clear that carbon taxes would generate revenue, which could be used in a variety of ways, court interpretations of this power suggest that the federal government would have to clearly articulate revenue-raising as a priority objective for a carbon tax, and would have to carefully design the enabling legislation to demonstrate that the primary focus was revenue rather than climate change policy. Courts recognize that it can be difficult to find a single "pith and substance" of tax measures, in part because "a fiscal instrument may be chosen precisely because it can kill two birds with one stone, regulating the industry while raising revenue" (*Re Goods and Services Tax, 1992*). It is easy to imagine a carbon tax measure as having this double aspect, meaning that it could be justified under more than one head of power. However, this double aspect does not eliminate the requirement for there to be a clear and dominant revenue-raising purpose to justify a carbon tax under the taxation power.

In the event that the federal government did design a carbon tax measure with revenue-raising as a dominant purpose, it would still have to consider s. 125. This provision has posed difficulties for federal tax measures in the past. In *Reference Re Proposed Federal Tax on Exported Natural Gas* (1982) the province of Alberta successfully argued that a federal tax on natural gas that the province had extracted from its own Crown lands was invalid under s. 125. The Court concluded that the natural gas subject to the tax was the property of the provincial Crown, and thus should be exempt from federal taxation under s. 125.

Of course, there are many federal taxation measures that do not trigger s. 125 problems. For example, the federal government has long imposed taxes on gasoline and diesel. The tax is imposed at the point of import or manufacture or production in Canada, payable at the time of delivery to the purchaser. Manufacturers and importers are licensed and required to pay the tax under the *Excise Tax Act.* While these taxes relate to natural resources which are inevitably located in the provinces, the taxes have not been challenged under s. 125, likely because the products they tax (gas and diesel) are the product of natural resources *after processing*, rather than the resources prior to exploitation. The main impact of s. 125 on these taxes is that the province is not responsible for paying the tax when it is the manufacturer or importer of gas or diesel.

Should the government want to use the taxation power to justify a carbon tax, it would want to design the measure in a way that created a perceptible gap between the subject of the tax (i.e., carbon, or more correctly, carbon emissions) and provincial property (the source of the carbon) to reduce risk of conflict with s. 125. Perhaps this would argue for a tax that focused on actual emissions rather than the carbon content of fuels at source.

In summary, justifying a carbon tax under solely the federal taxation power would require designing the measure so that its pith and substance was primarily revenue-raising, which I think would be quite difficult given the policy drivers for carbon taxes. Unless the courts take an unprecedented turn in their "pith and substance" analysis (and perhaps they would be inspired to do so in light of the "new social reality" of climate change), the federal government would at best be able to demonstrate revenue-raising as *one* purpose for a carbon tax, rather than the dominant purpose, which would require relying upon another source of authority. As discussed below, there is sufficient scope for this.

The Criminal Law Power: S. 91(27)

While criminal law is not an obvious choice for carbon taxation, it turns out to be a relevant power for environmental legislation and thus, possibly, carbon taxes. There are two requisite criteria for a law to fall within the federal criminal law power: i) a valid criminal purpose; and, ii) a prohibition accompanied by a penalty (Hogg 2007, 18-12). The criminal law power has been interpreted as a broad "plenary" power (*R. v. Hydro-Quebec*, 1997, para. 119). The main limitation applied to this broad power (beyond the requirement to meet the two criteria identified above and the rights guaranteed by the Charter) is that "the power cannot be employed colourably" (*R. v. Hydro-Quebec*, 1997, para.121). That is, the power cannot be used in a way that would invade unreasonably areas of exclusive provincial jurisdiction.

In terms of the first requirement, the valid criminal purpose has been traditionally interpreted as promoting "public peace, order, security, health and morality". However, in an important interpretation of this power for environmental purposes, the Supreme Court unanimously agreed that environmental protection can also be a valid criminal purpose (*R. v. Hydro-Quebec,* 1997). The more challenging criterion for environmental law is the requirement of a prohibition, since much of environmental law involves complex regulation. However, in the *Hydro-Quebec* case, the Court found that the impugned provisions were aimed at environmental protection and were not an indirect attempt to interfere with the province's jurisdiction over property and civil rights (i.e., they were not "colourable" as defined above). As per Lamer J:

> The use of the federal criminal law power in no way precludes the provinces from exercising their extensive powers under s. 92 to regulate and control the pollution of the environment either independently or to supplement federal action (*R. v. Hydro-Quebec*, 1997, para. 131)

Given this *Hydro-Quebec* decision, the federal government opted to regulate GHG emissions through the 1999 *Canadian Environmental Protection Act* (CEPA), relying on the criminal law power as a justification for the regulation. Arguably, the regulations fulfill a valid criminal law purpose, which has been held to include environmental protection, and they entail a prohibition associated with a penalty, since the regulations proscribe a penalty for going beyond the permitted release of GHG emissions.

While I believe that the regulation of GHG emissions through CEPA is justifiable under the criminal law power, the question at hand is whether a carbon tax could also fit within the scope of this power. It seems unlikely that a *stand-alone* carbon tax would fall within the criminal law power, as it is far from a prohibition coupled with a penalty. Designing the tax as part of the GHG regulatory scheme would mean using the tax as the prohibition, i.e., for going beyond regulatory thresholds of GHG emissions. While this would likely bring the tax within the constitutional authority of the criminal law power, as part of the GHG regulatory scheme, this design would be undesirable as it would seriously limit the range of behaviour that such taxes would normally target.[2] A broad interpretation of the criminal law power in *RJR-MacDonald* leaves open the possibility that a tax designed as a price on GHG-emitting fuels could be justified as part of the regulatory scheme.[3] The argument would be that since Parliament has the authority to legislate under the criminal law power with respect to protection of the environment, which includes climate change, it follows that Parliament may use a variety of means (from GHG regulations to carbon pricing mechanisms, including carbon taxes) to achieve that objective.

In sum, I believe it would be possible for a carbon tax to be justified under the criminal law power, but it would require the Court to broaden its interpretation of the prohibition and penalty requirement or, more likely, to view the tax as the means of achieving a legitimate criminal law purpose – namely reduction of GHG emissions. It remains to be seen whether the courts will be willing to stretch the criminal law power to this extent. Again, one hopes they might be inspired to do so given the importance of dealing with climate change and the need to find room for environmental policy instruments in a Constitution that was drafted at a time when environmental issues were unknown.

The Trade and Commerce Power: S. 91(2)

Under s. 91(2), the federal government has power over interprovincial and international trade and commerce. The courts have traditionally interpreted the federal trade and commerce power in a limited way because of its potential overlap with the provincial power over property and civil rights. The trade and commerce power justifies legislation addressing: i) interprovincial and international trade and commerce, and ii) general trade and commerce affecting the whole country (*Citizen's Insurance Co. v. Parsons*, 1880).

---

[2]Ideally, a carbon tax would be an economy-wide measure designed to influence price signals of products that emit GHGs, and not limited to penalizing emissions beyond a given regulatory threshold.

[3]The Court in *RJR-MacDonald* held that the tobacco advertising legislation was a valid exercise of the criminal law power, as protection of human health (through reduction of tobacco consumption) was a valid criminal law purpose and the legislation created a prohibition coupled with a penalty. The Court noted that Parliament has used a variety of means to reduce tobacco consumption, including taxes on cigarettes. However, it also held that the means selected to achieve the purpose were not relevant to the jurisdiction issue: "…once it is accepted that Parliament may validly legislate under the criminal law power with respect to the manufacture and sale of tobacco products, it logically follows that Parliament may also validly legislate under that power to prohibit the advertisement of tobacco products and sales of products without health warnings". *See RJR-MacDonald Inc. v. Canada*, 1995, para. 33.

The federal government's authority to legislate with respect to interprovincial and international trade is logical, since provinces could not adequately regulate trade outside of their borders. Relatedly, trade that remains intra-provincial is outside of federal authority unless justified elsewhere. The power to legislate over general trade and commerce, which is not limited to extra- or interprovincial trade, has until recently been interpreted narrowly. However, the Supreme Court appears to have expanded the scope of this power when it upheld the constitutional validity of the federal *Competition Act* (*General Motors of Canada Ltd. (GM) v. City National Leasing*, 1989). Specifically, in the *GM* case, the Supreme Court established a five-part test for determining whether legislation falls within the second branch of the trade and commerce power: i) the impugned legislation must be part of a general regulatory scheme; ii) the scheme must be monitored by the continuing oversight of a regulatory agency; iii) the legislation must be concerned with trade as a whole rather than with a particular industry; iv) the legislation should be of a nature that the provinces jointly or severally would be constitutionally incapable of enacting; and v) the failure to include one or more provinces or localities in a legislative scheme would jeopardize the successful operation of the scheme in other parts of the country.

The first criterion requires the measure to be designed as part of a general regulatory scheme. Considering the question of when a measure forms part of a regulatory scheme, the Court in *Reference Re Proposed Federal Tax on Exported Natural Gas* (1982) commented that regulation suggests "…a restraint upon or a channeling of economic behaviour in pursuit of policy goals". The Court recognized that a general tax on natural gas would change price signals for natural gas and could have the effect of discouraging natural gas production or consumption. It also acknowledged that there could be valid policy reasons for such discouragement and that excise taxes have been justified on such grounds apart from revenue generation. Based on an analysis of the legislation, the Court in this case found that the purpose of the gas tax was to raise revenue rather than to regulate by changing price signals. However, the commentary by the court suggests that taxes clearly designed to change price signals as part of a general regulatory scheme, rather than simply raise revenue, could satisfy this first leg of the five-part test.

Whether the scheme would be considered to be monitored by the continuing oversight of a regulatory agency (criterion 2) would depend very much on the design of the measure. It is certainly possible to imagine a carbon tax that was integrated as part of the regulatory scheme for GHG emissions being monitored by relevant tax authorities or those responsible for the GHG regulatory scheme.

Could the measure be justified as being concerned with trade as a whole (criterion 3)? Hogg (2007, 21-9) has pointed out that this branch of the federal trade and commerce power has been used to justify federal regulation of competition, but that regulation of wages and prices remains within provincial power under property and civil rights. Given the importance to the effectiveness of the measure of establishing a carbon price across the economy, it might be possible to distinguish previous cases and successfully argue that a carbon tax would be concerned with trade as a whole, rather than a particular industry. It would be important in this case to design the carbon tax to apply as widely as possible, not only to a given industry (such as oil and gas producers), but also to other emitters of GHG emissions (such as the heating and transportation industries and agricultural producers). The broader the measure's application, the more likely it could be considered as being concerned with trade as a whole. Indeed, in the earlier referenced *General Motors* case the Court relied upon commentary by Hogg and Grover (1976, 199-200) that argued significant regulation of the Canadian economy had to be done nationally:

> It is surely obvious that major regulation of the Canadian economy has to be national. Goods and services, and the cash or credit which purchases them, flow freely from one part of the country to another without regard for provincial boundaries. Indeed, a basic concept of the federation is that it must be an economic union. An over-all national policy is the key to efficiency in the production of goods and services... Any attempt to achieve an optimal distribution of economic activity must transcend provincial boundaries, for, in many respects, Canada is one huge marketplace...

All of these points in favour of a national approach to regulate competition could be used to justify a federal approach to carbon taxation under the general trade and commerce power. Would a carbon tax be major regulation? Some would argue yes. Changing the price signal of carbon would be ineffective unless done nationally, given the fluidity of the markets of the goods and services to which such a tax would apply, and such a change could surely be considered "major".

The importance of a national approach to carbon pricing would in turn lend support to the fourth and fifth criteria, which require demonstrating that the carbon tax is of a nature that the provinces jointly or severally would be constitutionally incapable of enacting and that the failure to include one or more provinces or localities in a legislative scheme would jeopardize the successful operation of the scheme in other parts of the country. In the *GM* case, the court relied upon its characterization of competition as a national issue to justify these two additional criteria.

By way of summary, while it would be novel to use the trade and commerce power to justify environmental legislation, there is a reasonable chance that a carbon tax designed as part of a GHG emissions regulatory scheme could fit within this power.

Peace, Order and Good Government (POGG)

Section 91 of the *Constitution Act, 1867* reserves a residual power for the federal government, stating that Parliament may "make Laws for the Peace, Order and good Government of Canada, in relation to all Matters not coming within the Classes of Subjects by this Act assigned exclusively to Legislatures of the Provinces". This residual power has been classified into three branches: i) responding to national emergencies, ii) filling gaps in the Constitution, and iii) areas of national concern. Since authority to legislate environmental matters has been mainly justified under the national concern branch (Hogg 2007, 17-5), I will focus on this branch.

In order to determine whether the federal government has jurisdiction under the national concern branch of POGG, the Supreme Court in *R. v. Crown Zellerbach* (1988, para. 33) stated the following:

> For a matter to qualify as a matter of national concern, it must have a singleness, distinctiveness and indivisibility that clearly distinguishes it from matters of provincial concern and a scale of impact on provincial jurisdiction that is reconcilable with the fundamental distribution of legislative power under the Constitution. In determining whether a matter has attained the required degree of singleness, distinctiveness and indivisibility that clearly distinguishes it from matters of provincial concern it is relevant to consider what would be the effect on extra-provincial interests of a provincial failure to deal effectively with the control or regulation of the intra-provincial aspects of the matter.

One of the deciding factors for the majority in the *Zellerbach* case was the "extra-provincial interests" test. The majority found that the failure of one province to protect its waters could lead to the pollution of other provincial waters as well as the high seas under federal jurisdiction. This justified federal intervention. Peter Hogg (2007, 17-15) has suggested that the extra-provincial interests test may be the determinative factor in finding authority under the national concern branch of POGG:

> The most important element of national concern is a need for one national law which cannot realistically be satisfied by cooperative provincial action because the failure of one province to cooperate would carry with it adverse consequences for the residents of other provinces.

Would a federal carbon tax have the requisite singleness, distinctiveness and indivisibility to clearly distinguish it from matters of provincial concern? The answer, in my opinion, is yes. The impact of GHG emissions on the atmosphere is the same regardless of where the emissions come from. This gives GHG emissions much more singleness, distinctiveness and indivisibility than something like the management of a local watershed. Bryce and Stevens (2000, 12) point to the following three points about GHG emissions that strengthen the argument for federal jurisdiction under the national interest: i) GHG emissions are a trans-boundary type of pollution; ii) the adverse effects of GHG emissions are predominantly extra-provincial in nature and international; and iii) Canada's GHG emission reduction commitments will be in jeopardy if a province refuses to participate in any initiatives. Barton (2002, para. 33) adds that "[o]nce emitted, greenhouse gases can have long lifespans in the atmosphere. The GHG emissions from any province will, during their lifespan in the atmosphere, contribute to climate change outside that province and outside Canada".

The courts might draw an analogy with the regulation of nuclear energy which has been justified as within federal jurisdiction under POGG. In *Ontario Hydro v. Ontario (Labour Relations Board)* (1993, para. 84), the Court said of atomic energy that "[i]t is predominantly extra-provincial and international in character and implications, and possesses sufficiently distinct and separate characteristics to make it subject to Parliament's residual power". The Court then ruled that "a provincial failure to sufficiently regulate nuclear energy could result in the risk of a human health and environmental catastrophe of extra-provincial and international implications" (Barton 2002, para. 25). Applying this reasoning to carbon taxes, one could argue that carbon taxes are one of the most economically efficient and likely effective means of reducing GHG emissions (and thus addressing climate change), which could argue in favour of a national interest justification.

I will now examine the heads of power that could support provincial carbon tax regimes.

THE PROVINCES AND CARBON TAXATION

The Provincial Taxation Power: S. 92(2)

Section 92(2) of the *Constitution Act, 1867* gives the provinces authority to make laws relating to "direct taxation within the province in order to the raising of a revenue for provincial purposes". The provinces' jurisdiction to tax under s. 92(2) is limited in three ways: 1) the tax must be direct, 2) it must be within the province, and 3) it must be for provincial purposes (ibid). Peter Hogg notes that the third limitation has turned out to have little relevance and that the second limitation is relevant, but easy to apply (Hogg 2007, 31-3). The limitation over direct taxes is important and can be more difficult to ascertain. The well accepted definitions of direct versus indirect taxes were articulated by John Stuart Mill a century ago as follows, and continue to be used by the courts:

> A direct tax is one which is demanded from the very person who it is intended or desired should pay it. Indirect taxes are those which are demanded from one person in the expectation and intention that he shall indemnify himself at the expense of another (Mill 1884).

To find authority for a carbon tax under s. 92(2), the provinces would need to design the measure as a direct (downstream) tax applicable only within the province, with the dominant purpose of revenue-raising. The provinces have long justified a variety of fuel taxes under the direct taxation power. Provincial fuel taxes have remained within provincial jurisdiction by being imposed at the final point of sale, namely on purchasers of gasoline. As such, they have been considered direct taxes within the authority of s. 92(2). The provinces could try to broaden the scope of fuel taxes (or increase rates) in order to target carbon. As long as the tax was imposed at the point of purchase by consumers, for instance at the pump or by utilities selling electricity, this downstream tax would be characterized as a direct tax because the consumer is the last purchaser of the good or service and cannot pass on the price.

A carbon tax on the content of fossil fuels would be considered an "indirect" tax because the entity releasing the pollutant would incur the cost of paying the tax, but would indemnify itself at the expense of another. In an upstream tax, producers or importers would easily pass the cost of the tax onto their distributors or customers. It would be solely within the authority of the federal government to implement such a tax unless it was carefully designed to fall within s. 92A or 92(9), heads of power that are addressed below in turn.

To fall within s. 92(2), a direct provincial tax would also have to be in "pith and substance" a tax, rather than a regulatory charge. I return to this distinction, which has been a source of contention in numerous decisions, in the later section dealing with the BC carbon tax. Provincial taxation powers are also subject to the limitation of s. 125 of the *Constitution Act, 1867.* Just as a federal tax cannot be imposed upon provincial property, a provincial tax cannot be imposed on federal property.

Natural Resources Taxation Power: S. 92A

Section 92A(4) gives the provinces power to tax natural resources both directly and indirectly:

> In each province, the legislature may make laws in relation to the raising of money by any mode or system of taxation in respect of: (a) non-renewable natural resources and forestry resources in the province and the primary production therefrom, and (b) sites and facilities in the province for the generation of electrical energy and the production therefrom.

This section means that since 1982 indirect resource taxation is now a field of concurrent, and potentially overlapping, federal-provincial jurisdiction. Therefore, a tax on the primary production of such resources – which would be indirect – is now *intra vires* of the province. This power has made it easier for provinces to derive revenues from oil, gas, minerals and other natural resources within provincial boundaries.

In order for a provincial carbon tax to fit within the scope of s. 92A(4), the tax would need to be applied to non-renewable natural resources "in the province" and relate to the "primary production therefrom". Thus, the tax would need to be tied to resource production in order to be valid. However, unlike s. 92(2), the tax could be either direct or indirect. Given the need to tie the tax to resource production, the measure would need to be applied to the exploitation or production of fossil fuels themselves rather than to emissions of CO_2 or CO_2 equivalents. While fossil fuels are clearly natural resources, it is certainly not clear whether the CO_2 emitted from the burning of fossil fuels or other GHG emissions could be characterized as natural resources.

Perhaps most importantly for both provincial taxation powers (92(2) and 92A(4)), the greatest hurdle would be convincing a court that the pith and substance of a provincial carbon tax was revenue-raising. While raising revenue might be one of a number of motivating factors, it would be difficult to identify it as the principle motivator given the political context of climate change which is clearly motivating policy action relating to carbon.

Licensing Power: S. 92(9)

The provinces have another revenue-raising power in addition to the two taxation powers discussed above. Section 92(9) authorizes the provinces to legislate in relation to "shop, saloon, tavern, auctioneer, and other licenses in order to the raising of a revenue for provincial, local, or municipal purposes". Cases relating to s. 92(9) have often turned on the question of whether a charge is a regulatory fee or levy justifiable under this section, or a tax (*Allard Contractors v. Coquitlam,* 1993).

The Supreme Court's most recent pronouncement on the definition of taxes was made in *620 Connaught Ltd. v. Canada (Attorney General)* (2008), in which Rothstein J. summarized the key characteristics of a tax as follows: i) compulsory and enforceable by law; ii) imposed pursuant to the authority of the legislature; iii) levied by a public body; and iv) intended for a public purpose (ibid., 7). Given that these four characteristics could be applied to most government levies, the courts must determine whether these are the dominant characteristics of the levy – which would argue for a tax – or whether they are only incidental – which would argue for a regulatory charge (ibid., para. 23).

The Court in *Westbank* added a fifth criterion, which has become determinative in deciding whether the first four characteristics are dominant or incidental. This fifth criterion is that a charge will be considered a tax (rather than a regulatory charge) if it is "unconnected to any form of a regulatory scheme"; Rothstein J. states that even if the levy has all the other indicia of a tax, it will be a regulatory charge if it is connected to a regulatory scheme (*620 Connaught Ltd. v. Canada*, 2008, para. 24). I will refer to this test to determine whether a measure constitutes a tax as the *Five Westbank Tax Criteria.*

Determining whether a charge is connected to a regulatory scheme is not a simple matter. Rothstein J. summarizes the two-step approach used in *Westbank*: "The first step is to identify the existence of a relevant regulatory scheme … the second step is to find a relationship between the charge and the scheme itself" (*620 Connaught Ltd. v. Canada*, 2008, para. 25 & 27 citing *Westbank*). I will refer to this as the *Westbank Two-Part Connection Test.* To identify the existence of a relevant regulatory scheme, the Court in *Westbank* offered the following four indicia. "Is there: (1) a complete, complex and detailed code of regulation; (2) a regulatory purpose which seeks to affect some behaviour; (3) the presence of actual or properly estimated costs of the regulation; (4) a relationship between the person being regulated and the regulation, where the person being regulated either benefits from, or causes the need for, the regulation?" (ibid. at para. 24 citing *Westbank*). I will refer to this test, which is used to determine whether there exists a regulatory scheme, as the *Four Westbank Regulatory Scheme Criteria.*

With respect to the second part of the *Westbank Two-Part Connection Test*, the Court is looking for a relationship between the fees paid and the regulatory scheme (*620 Connaught Ltd. v. Canada*, para. 44). In *Westbank*, the Court stated that a fee will be considered connected to a regulatory scheme "… when the revenues are tied to the costs of the regulatory scheme, or where the charges themselves have a regulatory purpose, such as the regulation of certain behavior" (*Westbank First Nation v. British Columbia Hydro and Power Authority*, 1999, para. 44).

To bring a carbon tax within the scope of the licensing power, a province would need to design the charge as part of a comprehensive code of GHG emissions regulation. It would need to state a regulatory purpose that targeted entities to reduce their GHG emissions. Tying the charge to the actual or estimated costs of the regulation might be a bit more difficult, since a carbon tax proposal would theoretically be aimed at cost internalization rather than the costs of regulation. I think it would be worthwhile arguing before the courts that the third criterion should be broadened to include cost internalization as a motivator for the regulation as well as defraying of costs. However, should a given province not wish to risk courts refusing this argument, a measure could be designed to finance the overall climate change regulation for the province. The fourth criterion involves demonstrating a relationship between the person being regulated and the regulation, where the person being regulated either benefits from, or causes the need for, the regulation. This threshold would not be difficult to satisfy given that the reason for a carbon tax stems from the need to reduce GHG emissions, the motivator for the regulatory scheme.

With regard to the second part of the *Westbank Two-Part Connection Test*, it is the regulatory purpose of the charge connecting it to a regulatory scheme that would be of interest in the case of a carbon charge. An analogy may be drawn to the *Johnnie Walker* case, which examined customs duties that were aimed at encouraging the importation of certain products while discouraging the importation of others. These duties were the method chosen of advancing the regulatory purpose of changing import and export levels:

> Where a charge itself is the mechanism for advancing a regulatory purpose, such as a charge that encourages or discourages certain types of behaviour, or where a charge is "ancillary or adhesive to a regulatory scheme" which may be used to defray the costs of that scheme, then they will usually be applicable to the other order of government[4] (*Westbank First Nation v. British Columbia Hydro and Power Authority*, 1999, para. 32).

Property and Civil Rights in the Province: S. 92(13)

A province's power over property and civil rights gives it wide authority to regulate industrial activities within the province, including the activities of its oil and gas, hydroelectric and other energy generating industries. This power is complemented by the provincial jurisdiction over natural resources under s. 92A, giving the provinces a large degree of control over energy policy. Provincial authority over energy policy is subject to certain restrictions. As noted earlier, the federal government has declared nuclear power to be within federal control. And, when legislation affects works and undertakings that have interprovincial or extra-provincial characteristics, federal authority becomes relevant.

---

[4]The "other order of government" refers to the provincial government.

The property and civil rights power has been used to justify provincial jurisdiction over many aspects of environmental regulation, including land use within a province and business activity. According to one author, a province's broad jurisdiction over property and civil rights in the province translates into a strong provincial interest in any measures to promote the reduction of greenhouse gas emissions (Stockdale 2004). This argument is further supported by Hogg (2007, 30-24) who states that "[t]he power over property and civil rights authorizes the regulation of land use and most aspects of mining, manufacturing and other business activity, including the regulation of emissions that could pollute the environment".

Could the property and civil rights power justify a carbon tax? Possibly, if the measure were characterized as part of the province's regulatory agenda pertaining to pollution control. This power has been interpreted broadly and been the justification for a number of provincial pollution measures. However, because the measure is in fact a charge rather than a traditional regulation, the courts would most likely seek justification under one of the revenue-related heads of power.

Conclusion

In conclusion, the most likely source of authority for a provincial carbon tax is the licensing power. The broad property and civil rights power, which is the source of justification for the majority of provincial environmental regulation, could be helpful. And, depending on whether the main purpose of the tax could be construed as revenue-raising, the two provincial taxation powers could be a source.

CASE STUDIES: QUEBEC AND BRITISH COLUMBIA

Quebec's Carbon Tax

The province of Quebec was the first Canadian jurisdiction to implement a carbon tax. The measure is part of the province's 2006-2012 climate change plan, which has as its objective the reduction of Quebec's GHG emissions by 10 million tons annually. Part of the climate change plan includes an annual duty on fuels that emit GHG emissions. Set annually by regulation, the duties are paid into a provincial Green Fund which is used in part to finance climate change mitigation and adaptation projects. The duty is payable by natural gas distributors, fuel distributors and any person or partnership bringing fuel to Quebec for the production of electricity. The rate and method of calculation is established by the *Régie de l'énergie* in regulations based on the CO_2 emissions generated by the combustion of natural gas and fuel.

The media has called Quebec's initiative a carbon tax, but the Quebec enabling legislation and regulations consistently refer to it as a "redevance annuelle" (annual duty). The naming and characterization of the measure was undoubtedly done very consciously by the government to strengthen the case for the measure being *intra vires* of the province.

Quebec's overarching *Energy Strategy Act* appears clearly aimed at developing a modern, sustainable energy policy for the province of Quebec. As part of this policy, the levy is aimed at reducing GHG emissions and financing the Green Fund. Given the context of the measure within the province's efforts to develop a sustainable energy policy, as well as the emphasis on sustainability, the development of new technologies and the levy's role in helping to reduce GHG emissions, the levy could be characterized as having as its dominant purpose the implementation of a climate change policy which is founded on the modernization of the province's energy policy. Although difficult to separate from the climate change context, the levy could also be characterized as an instrument designed to raise revenue for the Green Fund.

Given the nature of the measure as a charge, the province needs to find justification for the measure within one of the powers that permits revenue-raising. The strongest justification is under s. 92(9). Indeed, as will be shown, there are clear signs that the province intentionally designed the carbon tax to fit within this authority. In terms of the earlier-mentioned *Westbank Regulatory Scheme Criteria*, Quebec's carbon levy easily satisfies the second and fourth requirements since the charge is conceived at least in part to encourage the province's energy sector to shift towards more renewable energy resources (by raising the price of carbon). Similarly, there is a clear relationship between the entities regulated (energy distributors within the province) and the *Quebec Energy Strategy Act*, since the legislation is meant to transition the industry towards greater energy efficiency and more production of renewable resources.

The first criterion (a regulatory scheme) is a bit more difficult (see section Licensing Power, s. 92(9), above). Quebec's carbon levy has been introduced as part of the *Quebec Energy Strategy Act.* This legislation aims to modernize the province's energy policy in a way that improves its sustainability and reduces GHG emissions, i.e., by offering modifications to a number of pieces of existing legislation relating to energy policy. There is a strong argument that the changes the Act brings to the existing energy strategy and legislation are significant enough to satisfy this first *Westbank* criterion. Specifically, the revised energy strategy incorporates a range of new approaches to the province's energy regulation, from requiring electric power and natural gas distributors to prepare comprehensive energy efficiency and new technology plans to granting expanded powers to two regulatory agencies involved in the management of energy resources.[5] The mechanics of the change (via modification of several acts versus one new law) should not be determinative in deciding whether there is a complete code of regulation.

It is likely that the province designed the charge in part to satisfy the third *Westbank Regulatory Scheme Criteria* criterion. The *Régie* will be responsible for calculating the amount of the levy based on CO_2 emissions, but also based on the financial needs of the Green Fund – which will finance the

[5]The agencies affected are the Régie de l'énergie and l'Agence de l'efficacité énergétique.

implementation of the Climate Change Plan. By designing the carbon levy as a mechanism to finance the implementation of the Climate Change Plan, the province has established a sound link between the charge and implementation of the overall energy strategy, which should be sufficient to satisfy the third *Westbank* criterion.

The second part of the *Westbank Two-Part Connection Test* (see section Licensing Power, s. 92(9), above) could be satisfied by tying the revenues from the fees to the costs of administering the regulation or by the charges having a regulatory purpose, such as incenting behaviour. This test should be met fairly, given that the charge is meant to influence behaviour in a way that reduces carbon consumption.

In sum, Quebec's carbon charge is very likely *intra vires* of the province, with the strongest source of jurisdictional authority to be found within s. 92(9).

The British Columbia Carbon Tax

British Columbia's *Carbon Tax Act* (Bill 37) became law in May 2008. The Act imposes a tax on the purchase of a broad range of fuels, including gasoline, diesel, natural gas and coal. It also creates an administrative system for the collection of the taxes that mirrors that of the province's existing fuel taxes.

One of the more novel features of the Act is a series of requirements which helps to ensure that the revenue generated by the taxes will be used to reduce other taxes – in other words, to ensure the carbon tax is revenue neutral. The Act pursues this objective by requiring the minister of finance to prepare a carbon tax plan that estimates the revenue raised by the tax and identifies how the revenues will be used to reduce other taxes.

In contrast to Quebec, which created a midstream tax, BC designed its carbon tax as a downstream or consumption tax, applicable to purchasers (and importers) of the affected fuels and combustibles. While many factors undoubtedly played a role in this instrument design, one influential factor may have been the Supreme Court's decision in *Air Canada v. British Columbia* (1989). It was in this case that the Supreme Court judged BC's gasoline taxes *intra vires* of the province under s. 92(2). Modelling the carbon tax after the gasoline tax could lend support to an argument that the carbon tax is justifiable under the province's revenue-raising power.

Of course, as discussed earlier, the determinative factor in assessing jurisdiction is identifying the dominant purpose of the provision. Bill 37 does not include a purpose clause. The Budget Speech (2008, 2) introduces the carbon tax in its section entitled "Action on the Environment", and explains the intention of the tax as being to "put a price on carbon emitting fuels in British Columbia". The speech goes on to explain that putting a price on carbon will create new incentives to "change the habits that created global warming in the first place" (ibid). When introducing the Bill into the legislature, Finance Minister Carole Taylor placed the tax squarely in the context of climate change:

> Bill 37 introduces a groundbreaking revenue-neutral carbon tax that will encourage all British Columbia families and businesses to lower their carbon

> footprint and will help meet our goal of reducing emissions by 33 percent by 2020 (Hansard).

The context within which the carbon tax was introduced, along with the statements in the budget and in the bill's introduction, make it clear that the bill's purpose is related to the environmental goal of reducing carbon emissions. The province did not institute the measure in order to generate revenue. It is true that the measure will generate revenue and that the revenue will be used to reduce other taxes. However, the carbon tax was not motivated by a desire to raise revenue in order to reduce other taxes. As such, it is not a foregone conclusion that the province could rely upon s. 92(2) to justify the measure.

In order to fit within the scope of s. 92(2) (see section The Provincial Taxation Power, s. 92(2), above), the measure needs to satisfy the criteria for a tax, and that tax needs to be a direct tax. The latter criterion is easily met, given the measure's design as a downstream tax. With respect to whether the measure is a tax … [t]he tax is not primarily for raising revenue, nor is the revenue for general purposes. However, applying the *Five Westbank Tax Criteria* (see section Licensing Power, s. 92(9), above), the carbon tax could fit within the definition.

The fifth of the *Five Westbank Tax Criteria* requires a bit more thought. BC's carbon tax is part of an overall climate change plan that includes several different objectives and several pieces of legislation. The overall goal is to reduce the province's GHG emissions by at least 33 percent below 2007 levels by 2020, and by 80 percent below 2007 levels by 2050 (*Greenhouse Gas Reduction Targets Act*). The province is pursuing this goal through a number of means, including not only the *Carbon Tax Act,* but also the establishment of a cap-and-trade system for GHG emissions (*Greenhouse Gas Reduction (Cap and Trade) Act*). The province's approach to climate change also includes a variety of other measures, such as requirements for the government to achieve carbon neutrality by 2012 and investments in more efficient transportation (*Greenhouse Gas Reduction Targets Act*).

Does this package constitute a regulatory scheme to which the tax is connected? Following the reasoning in *Connaught* one could argue that the whole of the province's climate change package should be considered part of the regulatory scheme, or at least all of the regulatory components (which includes three pieces of legislation).

To determine whether this package constitutes a regulatory scheme to which the tax is connected, a Court would apply the *Four Westbank Regulatory Scheme Criteria* (see section Licensing Power, s. 92(9), above). As with the Quebec charge, it is easy to satisfy the second and fourth criteria. The climate change package clearly has a regulatory purpose – to reduce GHG emissions in order to reduce the impacts of climate change – which seeks to affect some behaviour, notably that of individuals and entities that emit GHGs. There is certainly a relationship between the individuals and entities being regulated (those whose purposes will now be subject to the carbon tax) and the climate change regulation. The emitters are causing the need for the climate change policy measures.

Is the BC climate change package a complete, complex and detailed code of regulation? The province has an ambitious approach to climate change that includes targets and the introduction of two economic instruments (tax and trading systems), along with a variety of other measures. Each instrument requires a detailed code of regulation. For instance, the *Carbon Tax Act* requires identification of all the different fuels to which the tax will apply, establishing the rates for each of these fuels, correlating these rates to the overall emissions reduction targets, establishing a system for administration and collection of the tax and for enforcement. In my opinion, this is a sufficiently complete, complex and detailed code of regulation to satisfy the first criterion.

With regard to the third criterion, one of the central features of the *Carbon Tax Act* is its revenue neutrality, and achieving this requires the minister of finance to estimate the revenue raised by the tax and identify how the revenues will be used to reduce other taxes (*Carbon Tax Act*, ss. 3 & 4). While these estimates are not the same as the costs of implementing the regulation, I would argue they are sufficient to satisfy the third criterion.

The next issue would be to determine whether the carbon tax is connected to this regulatory scheme. It would be difficult to justify a finding that the carbon tax is not connected to the province's regulatory scheme for climate change.

This clear connection between the regulatory scheme and the carbon tax leads to the conclusion that the tax is not in fact a tax, but a regulatory charge, and thus not justifiable under s. 92(2). The Supreme Court has made this distinction between taxes and regulatory charges quite clear, for instance, stating that "...the federal government imposes a levy primarily for regulatory purposes, or as necessarily incidental to a broader regulatory scheme ... then the levy is not in pith and substance 'taxation' and s. 125 does not apply" (*Reference Re Proposed Federal Tax on Exported Natural Gas*, 1982, 1070). In spite of the connection between the provincial climate change scheme and the carbon tax, it would be possible to argue that the carbon tax could have been implemented on its own, and that as such it is not *necessarily incidental* to the climate change package. However, the Courts have tended to find the connection even where the measure in question could have stood alone.[6]

Assuming the Courts maintain a narrow view of s. 92(2) and considered the carbon tax to be outside of its scope, I believe s. 92(9) would be available as a justification for the measure. Determining whether a measure can be justified under s. 92(9) requires applying the *Westbank Two-Part Connection Test.* We already established that the first part of this test should be met by applying the *Four Westbank Regulatory Scheme Criteria* outlined above. With regard to the second part, the courts could simply determine that the charges have a regulatory purpose, which is to influence the behaviour of individuals and entities to reduce their GHG emissions (*Westbank First Nation v. British Columbia Hydro and Power Authority*, 1999 para. 44). As in the *Johnnie Walker* case, where customs duties were used to discourage the importation of certain

---

[6]For instance, the regulation of alcoholic beverages in *Connaught* could have stood alone, but it was part of the broader parks administrative scheme and thus connected as per the Court's interpretation.

products, the carbon tax is meant to discourage emissions of GHGs, and the tax was one of the means chosen to advance the regulatory purpose of reducing GHG emissions (*Westbank First Nation v. British Columbia Hydro and Power Authority*, para. 29). Since the *Westbank Two-Part Connection Test* would, in my view, be satisfied by the BC carbon tax, the measure should be considered *intra vires* the province under s. 92(9).

There is also a reasonable argument to be made that the carbon tax could be justified under the property and civil rights power. The courts have generously interpreted s. 92(13) to include the regulation of business activities, including energy production and distribution. This is the power under which the provinces could probably justify regulations relating to GHG emissions within the province. It would thus be fair, in my opinion, to include within this power legislation that changes the price of carbon within the province. Changing the price signals to better internalize the environmental costs of carbon is arguably part of the province's regulation of all activities, including businesses. One could make an analogy to regulations relating to pollution, which have been held to fall within this power. Relying on s. 92(13) would require the courts to avoid seeking authority under a revenue-raising power and instead to view the measure as a regulation of business (and other) activity in the province through a price correction mechanism.

CONCLUSION

While there are innumerable considerations involved in the selection and design of policy instruments to address climate change, jurisdictional authority is a critical factor in Canada. This paper has shown that both the federal and provincial governments have jurisdiction to implement carbon taxes, as long as they are carefully designed to fit within the appropriate powers. However, it has also shown that the federal and provincial taxation powers – which are often the first to come to mind as possible justifications – are not the optimal sources of authority for a carbon tax. Federally, I have argued that carbon taxes would find their strongest source of authority under the national concern branch of the POGG power, with possible justification under the criminal law and trade and commerce powers depending on design and, of course, court interpretation of those powers. The taxation power is a possible source, but least likely of those analyzed. Provincially, I have argued that the power to charge license fees offers the best source of authority, though there may be room to find authority within the property and civil rights and possibly the taxation powers. And indeed, examining the Quebec and BC carbon tax measures showed that they are best justified under the licensing power (and were probably designed with this in mind).

REFERENCES

620 Connaught Ltd. v. Canada (Attorney General) [2008] S.C.C. 7.

Act respecting the Ministère du Développement durable, de l'environnement et des Parcs, R.S.Q. c. M 15.2.1.

Air Canada v. British Columbia [1989] 1 S.C.R. 1191.

Allard Contractors v. Coquitlam [1993] 4 S.C.R. 371.

Barton, P. 2002. "Economic Instruments and the Kyoto Protocol: Can Parliament Implement Emissions Trading without Provincial Co-Operation?" *Alberta Law Review* 40: 417.

British Columbia Legislature. 2008. Hansard 31(5) at 11610 (28 April). At http://www.leg.bc.ca/HANSARD/38th4th/h80428p.htm#11610.

Bryce, J.T. and J. Stevens. 2000. "Legal Issues Arising from Canada's Constitution", in J. Kowalski (ed.), *Climate Change Handbook for Agriculture 2000.* Centre for Studies in Agriculture, Law and the Environment, University of Saskatchewan. At http://www.csale.usask.ca/PDFDocuments/cchLegal.pdf.

Citizens' Insurance Co. v. Parsons [1880] 4 S.C.R. 215.

Chalifour, N.J. 2008. "Making Federalism Work for Climate Change: Canada's Division of Powers over Carbon Taxes", *National Journal of Constitutional Law* 22: 119.

Constitution Act, 1867 (UK), 30 & 31 Vict., c. 3, reprinted in R.C.S. 1985, App. II, No. 5 and *Constitution Act, 1982.*

Excise Tax Act, R.S.C. [1985] c. E-15.

General Motors of Canada Ltd. v. City National Leasing [1989] 1 S.C.R. 641, 58 D.L.R. (4th) 255 [cited to S.C.R.].

Government of British Columbia. 2008. *Budget 2008 Speech.* At http://www.bcbudget.gov.bc.ca/ 2008/speech/2008_Budget_Speech.pdf.

— 2008. Bill 37, *The Carbon Tax Act.* At http://www.leg.bc.ca/38th4th/1st_read/gov37-1.htm.

Hogg, P.W. 2007. *Constitutional Law of Canada*, 5th ed., looseleaf. Scarborough: Thomson Carswell.

Hogg, P. and W. Grover. 1976. "The Constitutionality of the Competition Bill", *Canadian Business Law Journal* 1: 197.

Mill, J.S. 1884. *Principles of Political Economy*. Book V., ch. 3. New York: D. Appleton and Company.

Ontario Hydro v. Ontario (Labour Relations Board) [1993] 3 S.C.R. 327.

R. v. Crown Zellerbach [1988] 1 S.C.R. 401.

R. v. Hydro-Quebec [1997] 3 S.C.R. 213.

Re Goods and Services Tax (GST) [1992] 2.S.C.R. 445, 94 D.L.R. (4th) 51.

Reference Re Proposed Federal Tax on Exported Natural Gas [1982] 1 S.C.R. 1004.

RJR-MacDonald Inc. v. Canada (Attorney General) [1995] 3 S.C.R. 199.

Section 2, *Règlement relative à la redevance annuelle au Fonds vert,* Loi sur la Régie de l'énergie (L.R.Q., c. R-6.01, r.0.2.3.1).

Stockdale, C. 2004. "The Constitutional Implications of Implementing Kyoto", Centre for Studies in Agriculture, Law and the Environment, University of Saskatchewan. (Unpub). At http://www.csale.usask.ca/PDFDocuments/constitutionImplication. pdf.

Westbank First Nation v. British Columbia Hydro and Power Authority [1999] 3 S.C.R. 134, 176 D.L.R. (4th) 276.

Winterhaven Stables Ltd. v. Canada (Attorney General) [1988] A.J. No. 924.

11

Carbon Pricing, the WTO and the Canadian Constitution

Andrew Green

INTRODUCTION

The World Trade Organization (WTO) rules can be viewed as almost a quasi-constitutional set of constraints on the climate change policies of domestic governments. They raise a common set of concerns to the federalism debates in Canada. What are the fundamental rules governing measures that governments can take to address climate change? What is the role of tribunals in interpreting or giving substantive content to the scope of limits on governments? Can the fundamental rules be changed? Whose values should be considered in making determinations about the legitimacy of particular policies? What is the role of multinational environmental agreements in defining the scope of governments' powers? This paper explores some of these similarities and differences between the Canadian Constitution and WTO agreements and their implications for carbon pricing policies in Canada.

The overlap of the WTO and climate change policies has become a growth industry both academically and in policy circles. In part this growth may be because individual or groups of countries have implemented or are in the process of implementing more apparently stringent climate change policies. These policies tend to increase the costs of domestic firms or importers or both, thus impacting international trade. This trade impact raises concerns about competitiveness of domestic industries in relation to industries in countries without stringent climate policies. Relatedly, it may reduce the environmental effectiveness of climate policies if there is "leakage" of production to countries without stringent policies. Countries may take unilateral action to attempt to overcome these competitiveness and leakage concerns such as through taxes on imports from countries with weak climate policies.[1]

---

[1]For example, the Leiberman-Warner *Climate Security Act of 2008* (s. 2191) proposed an emissions trading scheme with imports from countries without equivalent climate policies having to bear essentially a tariff prior to entering the United States. Similarly, the more recently proposed *American Clean Energy and Security Act of 2009* (the Waxman-Markey Bill) (H.R. 2454) may lead to requirements on importers of a wide

Trade rules are also being analyzed in the climate change debate because of their potential to overcome some of the deficiencies of the current international climate regime. As is discussed further in the next section of this paper, a main concern with international efforts to address climate change is that countries may free-ride – that is, continue to obtain the benefits of greenhouse gas (GHG) emissions while allowing others to make the emission reductions necessary to stave off dangerous climate change. The current Kyoto Protocol framework does little to either induce countries to enter into the agreement or to enforce compliance with commitments. As a result, there are a number of large emitters that did not sign onto the Kyoto Protocol (notably, of course, the United States) and others such as Canada have no hope of meeting their commitments. As in the Montreal Protocol on ozone-depleting substances, trade measures could be used to induce both participation and compliance with the next iteration of an international climate agreement (Barrett 2007).

WTO rules constrain both the trade impacts of domestic measures and the ability of countries to use border measures (such as a tax on imports). There are parallels between these WTO rules and the rules governing federalism under the Canadian Constitution. In fact, the WTO has been argued to have elements of a constitutional framework (Trachtman 2006, 623). Without entering into the debate about whether the WTO actually constitutes a constitution, there are interesting comparisons to be made between the WTO rules and the institutions on the one hand and the Canadian Constitution on the other. Bodansky (forthcoming) argues there are several key (though possibly not necessary) elements to a constitution. First, it must constrain government rather than private action, which is true of both the federalism provisions of the Canadian Constitution and the WTO agreements. WTO agreements place limits on governments' ability to use both domestic (internal) measures and measures imposed at the border. These limits are intended to constrain protectionist policies by governments – that is, policies which favour domestic industry at the expense of imports.

Second, the rules are generally entrenched in the sense of being more difficult to change than general legislation. The Canadian Constitution is obviously entrenched in this sense, having special amendment rules that do not apply to regular laws.[2] Similarly, the WTO agreements can essentially only be changed through its consensus decision-making process. This process has led to protracted negotiations over changes to the substantive rules.

Third, constitutional rules in general are superior to ordinary legislation. For example, any provincial or federal legislation which does not accord with the division of powers set out in the Constitution is not valid. WTO rules also have a form of superiority. Any domestic legislation which is found not to comply with

---

range of goods to purchase "international reserve allowances" for the emissions associated with their production. However, these allowances would not appear to be applicable until 2017.

[2] *Constitution Act, 1982*, ss. 38-49 (essentially requiring constitutional amendments to be approved by the House of Commons, the Senate and a 2/3 majority of provinces with at least 50 percent of the population).

WTO agreements must be amended so as to comply with WTO commitments. If it is not, the complaining party or parties can seek permission to impose sanctions on the non-complying party.[3]

Fourth, in addition to substantive limits, constitutions may define basic institutions and decision-making processes. The Canadian Constitution creates the basic democratic institutions in Canada including the court system. Similarly, the WTO agreements specify legislative rules and, importantly for our discussion, also create a tribunal system. Under this system, complaints by one member about measures taken by another member are heard by a dispute panel. The parties may appeal the decision of the dispute panel to a permanent Appellate Body.

Finally, constitutional rules are intended to exist for a long time and therefore tend to be more general than ordinary legislation. While some of the provisions on federalism in the Canadian Constitution are very specific, others such as the federal residual power to make laws for the Peace Order and Good Government of Canada are broadly interpreted to cover any matters that are not specifically enumerated in the Constitution (Hsu and Elliot 2009, 54). The WTO rules are similarly broad, in part because the rules are meant to persist and there are so many different measures that governments could take that could impact trade.[4] For example, GATT Article III, one of the core substantive rules in the GATT, relates to national treatment – requiring broadly that a member's tax or regulatory measures not discriminate against imports in favour of domestic products. This rule is tremendously flexible in order to take account of the myriad of different forms of taxes and regulations that governments may take.

In what follows, the focus is on some aspects of these constitution-like features of WTO agreements to discuss how WTO rules impact carbon pricing by the Canadian federal government and the provinces. It examines these issues in the context of border measures (such as border tax adjustments). Border measures have become particularly controversial in recent years as countries threaten to impose such measures to overcome free rider problems underlying climate change.[5]

Before getting into the core questions, the next section briefly discusses why climate change is so difficult for the WTO. The article then addresses three questions. The third section discusses what instruments are permissible under

---

[3]The principal requirements in the event of a challenge to a WTO measure of one member by another are set out in the WTO's Dispute Settlement Understanding which was adopted as part of the Uruguay Round (for an overview, see Trebilcock and Howse 2005).

[4]Horn, Maggi, and Staiger (2006). Note that the incomplete contract view tends to focus on the manner in which these gaps are created and may be filled and the potential for parties to find efficient solutions to trade disputes. For the opposite view that WTO agreements are mandatory and that they must be complied with in all cases, see Jackson (2004, 98).

[5]See footnote 1 for proposed U.S. legislation that includes provisions that would impose costs on imports from countries that do not have similar greenhouse gas emissions abatement measures to those in the United States.

WTO rules. It argues that, as with the division of powers provisions of the Canadian Constitution, the WTO rules limit the measures governments can take and tend to most clearly favour tax measures. The fourth section then turns to who is covered by the rules. The Canadian Constitution covers both the federal and provincial governments. The WTO rules, on the other hand, only directly cover the federal government, with the federal government then under an obligation to ensure provincial governments comply with these rules. Further, member governments are not covered equally by WTO rules because of the nature of the dispute settlement process and in particular the sanctions permissible under WTO rules. Finally, the fifth section raises the issue of who gets to decide which measures are permissible and the nature of institutional competence of the international tribunals of the WTO.

WHY IS CLIMATE CHANGE SO HARD FOR THE WTO?

Addressing climate change is an additive global public goods problem (Barrett 2007; Stern 2007). It will depend on the aggregate efforts of those who emit GHGs. Being a public good, however, means that there is an incentive for actors to free ride. Countries, and individuals, may be inclined to continue to receive the benefits of GHG emissions while allowing others to bear the costs of reducing emissions so as to stave off climate change. If each country free-rides in whole or in part, there is too much of the GHG emitting activities and climate change continues to be a problem.

This free riding could manifest itself in three broad fashions. First, a country could take no action on reducing GHG emissions. It could, for example, not participate in any international climate agreement and not take any action to reduce GHG emissions. Alternatively, it could sign onto an international climate agreement and not comply with the commitments it took on under the agreement. The Kyoto Protocol provides examples of both. The U.S. federal government did not ratify the Protocol and has yet to take significant action on climate change. Canada signed on but will not meet its commitments.

Second, and perhaps worse, countries could claim they are taking action on climate change but in fact be taking solely actions that favour their domestic industry. For example, a country may claim to be adopting standards to reduce GHG emissions but in reality the standards do nothing but create a barrier to entry to imports. This type of action is worse than no action because there are no benefits in terms of climate change and there is the economic cost of protectionist measures. Protectionist measures favour domestic industry such as by raising the costs of imports through a tariff or a standard that imposes added costs on imports but not domestic producers. Such measures tend to benefit domestic producers but harm both importers and domestic consumers (who face higher costs). Domestically, the harm to consumers in general outweighs both the benefit to producers and any revenue the government receives from the measure – such as tariff revenue (Krugman and Obstfeld 2005).

Finally, a country could take a measure that does address climate change but does so in a manner that restricts international trade more than is necessary. For example, a country may put in place a requirement to use a minimum percentage of ethanol in gasoline but then also require that the ethanol be domestically produced. The measure may (arguably) help address climate change but does so in a manner that privileges domestic production with its attendant economic, and potentially environmental, costs. In such a case, social welfare is not as high as it could be because, while there is an environmental benefit, it is in some cases not as high as would be the case if a different measure was used and because there is the economic cost of protectionism.

Trade measures and trade rules can aid in addressing climate change both by inducing action to address climate change and by reducing the ability of countries to take protectionist measures in the guise of taking action on climate change. In terms of inducing action, trade measures could take three related forms: carrots, sticks, and measures that reduce disincentives to action. An example of a "carrot" is a country giving a trade preference (such as reduced tariffs) to imports from a country that is taking action on climate change. A "stick" could take the form of a punitive tariff or ban on imports from countries that are not participating in the effort to address climate change (such as by not signing onto or complying with an international climate change agreement). Measures that reduce disincentives to climate policies are similar to "sticks", but are not necessarily punitive in nature. For example, a country may wish to address the political obstacles to climate policies by offsetting the cost of domestic policies at the border such as by imposing the same cost on imports from countries that do not have equivalent policies (Epps and Green 2008).

Trade rules can also be used to reduce the ability of countries to use protectionist policies. In fact, this purpose is central to the WTO. The WTO was initially focused mainly on reducing tariffs. As average tariffs have decreased, the importance of WTO rules on non-tariff barriers such as the national treatment principle has increased (Trebilcock and Howse 2005). These rules include the national treatment principle, which prohibits states from using taxes or regulations in a manner that discriminates against imports. There are exceptions to these rules – not for protectionist policies but for policies that are aimed at legitimate ends that happen to have protectionist impacts. Article XX of GATT permits measures that are "necessary to protect human, animal or plant life or health" (Article XX(b)) or are "relating to the conservation of exhaustible natural resources if such measures are made effective in conjunction with restrictions on domestic production and consumption" (Article XX(g)). In either case, the measure must not be applied in manner that is "a means of arbitrary discrimination between countries where the same conditions prevail or a disguised restriction on international trade". As a result, even with legitimate policies, there is a further check that they are not applied in a protectionist fashion.

Sorting out legitimate domestic policies from protectionist policies has been one of the more controversial functions of the WTO's dispute settlement institutions. It is particularly difficult for the WTO in the context of climate change for a number of reasons. First, even assuming everyone in the world had the same preferences for environmental protection, the costs and benefits of climate policies vary across countries. Some countries are likely to suffer extreme effects (such as many developing countries) while for others the near term impacts are less clear. If a country's climate measures are challenged as protectionist, it will claim these measures are part of its optimal policy package. A WTO panel or the Appellate Body will have difficulty sorting out whether the policy is within a reasonable set of policies if it is hard to determine the costs and benefits of different policies for the particular country.

Second, not all countries have the same preferences for environmental protection. Climate change is at its core an ethical issue. Given that the effects of climate change will vary across countries, how much does each individual owe to people in other countries?[6] Further, the main impacts of climate change are unlikely to be felt until many years into the future. How much do individuals today owe to future generations who will suffer harm from climate change? How much do we value the environment in and of itself? The preferences of the citizens of different countries will vary on such issues. In a challenge to a country's climate policies, WTO panels or the Appellate Body will be faced with the difficult (and potentially impossible) task of analyzing the preferences of the citizens of the country imposing the measure. Do they really care about the environment as much as their government says they do? What if the government says the population does not seem to care but it should and therefore argues that it is attempting to exhibit leadership?

Finally, tests such as whether the measure is "necessary" under Article XX(b) depend on information on climate change and the efficacy of different measures that is in general not available. There is considerable uncertainty not about whether climate change is occurring but about related issues including how quickly cuts have to be made to significantly reduce the risk of catastrophic climate change, which particular instruments work best to foster reductions in GHG emissions, and how instruments will work when used together. In the presence of such uncertainty, the WTO will be faced with the question of how much it should defer to the policy determinations of the domestic governments. Higher levels of deference exhibit sensitivity to a lack of expertise and information on the part of the WTO but reduce its ability to actually police protectionism.[7]

As a result, climate change suffers from free-rider problems. It is rife with opportunities for countries to benefit domestic industries by either not taking action or by instituting protectionist policies while appearing to be addressing climate change. The trade measures and WTO rules potentially can aid with these problems by providing tools to increase participation and reduce

---

[6]Climate change is likely to have differential effects both over time and across countries and the appropriate policy choice will involve ethical judgments about these distributional effects (Stern 2007, XV). Stern used a particular discount rate to account for intergenerational equity, acknowledging this choice was based on an ethical determination. For criticisms of the discount rate chosen by Stern, see for example, Weitzman (2007) and Nordhaus (2007).

[7]For the connection between deference and policing for protectionism in the context of scientific determinations, see Sykes (2002) and Howse (2000).

protectionism. The difficulty is that the nature of climate change makes these tasks difficult. The next section will focus on some of these issues in discussing how WTO rules limit choices governments can make for pricing carbon.

TAXES, EMISSIONS TRADING AND BORDER TAX ADJUSTMENTS

WTO Rules as Constraints

The Canadian Constitution limits the options available to the federal or provincial governments through establishing a basic division of powers across levels of government (Elgie and Chalifour, this volume; Hogg 2008; Hsu and Elliot 2009). Provinces appear to have the power to put in place certain taxes or trading regimes within their jurisdiction. The federal government, on the other hand, can enact a carbon tax but there is some uncertainty about its ability to put in place a nationwide emissions trading program. Further, the head of power under which the government acts may constrain the form that the trading program may take.

Similarly, WTO rules make some instrument choices impermissible or limit the form they can take. Consider carbon taxes or emission trading schemes. Both types of instruments are subject to the national treatment provisions of GATT (Article III). For taxes, where there is a competitive relationship between domestically produced goods and imports, the tax burdens must be similar.[8] For example, suppose Canada put in place a tax based on the level of emissions from cars that appears to fall more heavily on cars imported from the European Union because of the type of cars made there. The European Union may challenge this tax under the national treatment provision. A WTO dispute panel would first have to determine whether there was a sufficiently competitive relationship between the products. Among other things, it would typically examine whether there are physical similarities between the domestic and imported goods, whether the end uses are the same and whether consumers distinguish between them (Trebilcock and Giri 2005). The first two would likely be relatively easy in the case of the cars. Whether consumers distinguish between emissions levels of cars is more difficult and may require extensive economic analysis. Assuming there was a sufficiently competitive relationship, the panel would then have to assess the tax burden on domestic goods and imports to see if there is a difference.

This paper focuses on another type of instrument – border tax adjustments (BTAs) – as they have been the source of much controversy of late. BTAs may be used to attempt to overcome the political disincentives to putting in place

[8]GATT Article III sets out two tests for tax measures. If the domestic good and the import are "like" products (a very close relationship) then the tax burdens must be identical. If they are "directly competitive and substitutable" (not as quite as close a relationship), the tax burdens must be similar.

climate policies and to provide an inducement to other countries to take action. They do so by reducing the competitive disadvantage for industries in countries with strict climate policies. BTAs can be placed on either imports or exports. BTAs on imports are taxes on imports from countries with less stringent climate policies. BTAs on exports are rebates of or exemptions from taxes the domestic producers paid under climate policies. In either case, the general principle is that the BTA cannot exceed the level of tax paid if the good were bound for domestic consumption.[9]

Taxes, Trading and BTAs

How do the WTO rules interact with the ability of Canadian governments to price carbon? Take first the choice between carbon taxes and emissions trading. As the name implies, border tax adjustments may be used to adjust for the competitive impacts of taxes. However, even in the case of taxes, not all taxes are open for adjustment. BTAs on both exports and imports are limited to "indirect" taxes – that is, taxes levied on products rather than producers.[10] In principle, most carbon taxes would fit within this description, or could be framed as such.

However, whether BTAs can be used for emissions trading programs is more controversial. For BTAs on imports, the BTA can only offset an "internal tax or other charge". The question then is whether the emissions trading program can be considered an "other charge". There is not much WTO case law on the nature of "other charge".[11] It will depend on the nature of the trading scheme. If the permits are auctioned or firms are required to purchase permits over an allocated level, a panel may view the requirement to purchase a permit as being in the nature of a "charge". If the permits are given away for free, the issue is even more uncertain. A panel could view the provision of permits as a form of subsidy to the recipients as opposed to a charge. Whether panels will find BTAs can be used for emissions trading schemes is therefore uncertain.[12]

While the status of BTAs on imports to offset the costs of an emissions trading program is uncertain, the United States is currently considering such a measure. A recent set of U.S. federal bills proposed an emissions trading

[9]BTAs on imports cannot exceed the level of tax paid by "like" goods bound for domestic consumption (GATT, Article II.2) and BTAs on exports cannot exceed the level of the tax if the good were bound for domestic consumption (GATT Article Ad XVI.4 and Agreement on Subsidies and Countervailing Measures Article 1.1(ii) (footnote)).

[10]Report of the Working Party on Border Tax Adjustments, BSID 18S/97, adopted 2 December 1970.

[11]A recent WTO decision touched on the issue. The panel examined a measure that was part of a tax administration scheme and found the measures to be "charges" on the basis that "other charge" includes measures that impose a pecuniary burden and creates liability to pay money. Panel Report, *Argentina – Leather* (2000).

[12]Pauwelyn (2007) outlines arguments relating to BTAs for emissions trading programs and argues that BTAs could be used for emissions trading schemes.

program that would require imports from countries without any similar climate policy to bear costs that are similar to those faced by U.S. firms (Leiberman-Warner 2008; Waxman-Markey 2009). They have gone a step further and proposed that these costs take the form of a requirement for importers to purchase emissions units, rather pay a specific tax. This emissions purchasing requirement adds a significant number of additional questions, not the least of which is whether a requirement to purchase allowances constitutes a relevant tax or charge that can be imposed at the border.

BTAs relating to emissions trading programs seem even less likely in the case of exports. BTAs on exports can offset a "duty or tax". While an emissions trading program could be seen as a charge, it seems less likely to fit within the apparently narrower terms "duty or tax".

BTAs and PPMs

WTO rules therefore may limit governments' ability to use BTAs for emissions trading programs, although they are in general possible for carbon taxes. Constitutional rules may not only determine when a particular instrument may be used but also the form it may take. Under the Canadian Constitution, for example, if the federal government wishes to base its policy on the criminal law power, the policy must meet certain criteria. The policy must be for a public purpose and it must take the form of a prohibition backed by a penalty (see *R. v. Hydro Quebec*, [1997] 3 S.C.R. 213; and the papers in this volume by Elgie and by Chalifour). These criteria limit how the measure can be framed.

Similarly, WTO rules limit the form of BTAs. In the climate change context, a key question is whether BTAs can pertain not only to the characteristics of products themselves but also how the products are made. For example, the Canadian government could (subject to the limits discussed previously) use a BTA to offset a tax on emissions from vehicles. However, what about a tax on emissions from the production of steel? Assuming the steel itself is no different depending on the carbon emitted in its production, could a BTA be placed on imports from countries which do not tax or otherwise regulate emissions from steel production? This issue of non-product related "process and production methods" (PPMs) has been a continual source of controversy in the trade and environment area (as well as in related areas such as the overlap of trade and labour issues).

The PPM issue plays out differently for BTAs on imports and on exports. For BTAs on imports, the issue in part will depend on whether products that are otherwise identical become different because of how they were made. This issue is important because BTAs on imports can only be placed "in respect of the like domestic products" and "an article from which the imported product has been manufactured or produced" (GATT, Article II.2(a)). In terms of "like domestic products", any such BTA must be consistent with the national treatment provisions of GATT (ibid). As noted previously, the national treatment provisions require that if there is a sufficiently competitive relationship between the products (that is, they are "like"), then they cannot be taxed differently. The question will be whether if domestically produced steel is taxed in accordance with its GHG emissions, this tax can then be imposed on imported steel. The steel may face different tax burdens because of different processes. Governments for the exporters would argue that such a tax difference is inappropriate as there is nothing different about the steel itself. If consumers actually differentiate between the products because of how they were made, the argument goes, the products could be taken to be "unlike". Absent such environmental concern, it is unclear whether such a tax on PPMs could be offset through a BTA.

BTAs on imports may also be imposed in respect of "an article from which the imported product has been manufactured or produced", i.e., in terms of the second part of GATT Article II.2(a). This appears to provide more scope for basing BTAs on PPMs of imports such as for taxes on carbon used but not directly present in steel. Unfortunately, the WTO dispute settlement body has not clearly settled the issue of whether any such article must be present in the imported product or could be used up in the production.[13] The United States has used such BTAs in the past to offset the costs of policies limiting ozone depleting substances (ODS). The United States charged imports of ODS a tax equal to a domestic tax and rebated the tax on exports. The tax related to both the ODS as a product as well as products containing or produced with them. The tax was never challenged at the WTO (Barrett 2007; Pauwelyn 2007).

BTAs on exports face slightly different constraints. The concern is that the BTA not constitute an illegal subsidy for the exported product. A BTA on exports may be used to rebate taxes where the amount rebated or exempted for the exported product is the same as the amount levied on "like" products bound for domestic consumption. If, for example, the Canadian government imposed a tax based on the GHG emissions from domestically produced cars, they could likely use a BTA on cars that are exported to offset the competitive effects of the tax. However, the question, as before, is whether such a BTA could be used to offset the costs of a tax on the emissions from or energy used in the production of a product (such as steel).

The relevant WTO rules in this case are found in the Subsidies and Countervailing Measures (SCM) Agreement. It allows members to exempt or remit "prior stage indirect cumulative taxes" on inputs consumed in the production of the good. It defines "inputs consumed" as including energy, fuels and oils used in the production process. These provisions appear to provide greater scope for BTAs on exports in relation to PPMs, although there remains debate about whether "prior stage indirect cumulative taxes" actually encompasses taxes on inputs or emissions (Epps and Green 2008).

[13]Panel Report, *U.S. – Taxes* (1987). The panel found that BTAs on chemicals contained in products were permissible but was not clear on the issue of inputs that are not physically incorporated in the product.

Applying the Measure: Constraints under Article XX

So far we have seen that WTO rules potentially constrain governments' use of measures in the carbon pricing area. BTAs can be used to offset the costs of taxes but only indirect taxes. It is not clear whether they can also be imposed to address the costs of an emissions trading program. However, while they may not meet these substantive rules, the country imposing the measure could still attempt to save the measure under GATT Article XX. However, Article XX does not apply to the SCM Agreement and the SCM Agreement does not currently have any provision that allows exceptions. As a result, only BTAs on imports could be saved under Article XX and not BTAs on exports.

The two exceptions that seem most relevant to climate policies are Article XX(b) and (g). Under Article XX(b), countries can take measures that are "necessary to protect human, animal or plant life or health" (Article XX(b)). The difficulty here stems from the "necessity" test. It is in essence a form of balancing of costs and benefits, although with a twist. The Appellate Body has stated that determining whether a measure is necessary

> … involves in every case a process of weighing and balancing a series of factors which prominently include the contribution made by the compliance measure to the enforcement of the law or regulation at issue, the importance of the common interests or values protected by that law or regulation, and the accompanying impact of the law or regulation on imports or exports.[14]

The determination therefore depends on at least three factors: the importance of the objective; the measure's contribution to the objective; and the trade impact of the measure. At the same time the Appellate Body held that each country has the right to set its own public health or environmental objective and the level of protection related to that objective.[15] The combination of the right of each member to set its own level of protection and this balancing test seems to imply that panels are not to assess the level of benefits as in a strict cost-benefit or balancing test. Instead, panels should use the level of importance as decided by the member as a "margin of appreciation" in assessing the relationship

[14] Appellate Body, *Korea – Measures Affecting Imports of Fresh, Chilled and Frozen Beef*, WT/DS161.WT/DS169/AB/R, 10 January 2001, at para. 164. In its most recent decision in this area, the Appellate Body adopted this test, stating that in determining "necessity", "a panel must assess all the relevant factors, particularly the extent of the contribution to the achievement of a measure's objective and its trade restrictiveness, in the light of the importance of the interests or values at stake". Appellate Body, *Brazil – Measures Affecting Imports of Retreated Tyres*, WT/DS332/AB/R, 3 December 2007, at para. 156.

[15] Appellate Body, *United States – Standards for Reformulated and Conventional Gasoline*, WT/DS2/AB/R, adopted 20 May 1996 and Appellate Body, *European Communities – Measures Affecting Asbestos and Asbestos-Containing Products*, WT/DS135/AB/R, adopted 5 April 2001, at para. 168.

between how environmentally effective the measure is and how severe are the effects on trade.[16]

Even if the measure passes this test, panels compare the measure against other alternatives that the member could have taken.[17] Panels examine three factors in assessing whether the member should have taken an alternative measure to meet its objective. First, the alternative must "preserve for the responding Member its right to achieve its desired level of protection with respect to the objective pursued".[18] Second, the alternative must be less trade restrictive than the impugned measure (Appellate Body, *Brazil – Tyres* 2007, para. 156). Third, even if the measure provides the same benefit and is less trade restrictive than the impugned measure, it must also be "reasonably available". The Appellate Body stated that "an alternative measure may be found not to be 'reasonably available' … where it is merely theoretical in nature, for instance, where the responding Member is not capable of taking it, or where the measure imposes an undue burden on that Member, such as prohibitive costs or substantial technical difficulties" (Appellate Body, *US-Gambling* 2005, para. 308).

While Article XX(b) demands a form of balancing and examination of alternatives, Article XX(g) may be somewhat easier to satisfy. Article XX(g) permits countries to adopt measures "relating to the conservation of exhaustible natural resources if such measures are made effective in conjunction with restrictions on domestic production and consumption". The Appellate Body has interpreted "related to" to mean "reasonably related to the ends".[19] The AB found that this latter test involved an examination of the "general design and structure" of the measure and its relationship to the objective (ibid.; Charnovitz 2007, 701). However, the Appellate Body also noted that the measure was "not disproportionately wide in its scope and reach in relation to the policy objective" (Appellate Body, *U.S. – Shrimp I* 1998, para. 141). The Appellate Body did not expand on the "disproportionate" test but it appears to permit examination of the

---

[16]Regan (2007) at 356 arguing that the Appellate Body did not write the Korea-Beef judgment in terms of the importance of the goal providing a margin of error but instead as a cost-benefit test which they never actually apply. On the other hand, Sykes (2003) argues that the Appellate Body has created a "crude cost-benefit analysis" with the importance of the interests standing in for the cost of error – that is, the more important the goal, the more the Appellate Body will defer to the member in the analysis because the cost of improperly finding that a measure does not fall within Article XX(b) is so high.

[17]Appellate Body, *Brazil – Measures Affecting Imports of Retreated Tyres,* WT/DS332/AB/R, 3 December 2007, at para. 156. The complaining party must identify possible alternatives and the responding party has the opportunity to show that these are not reasonable.

[18]Appellate Body, *United States – Measures Affecting the Cross-Border Supply of Gambling and Betting Services* WT/DS285/AB/R, circulated on 7 April 2005.

[19]Appellate Body, *United States – Import Prohibition of Certain Shrimp and Shrimp Products* (1998) WTO Doc. DS58/AB/R (1998) (*US-Shrimp I*), at para. 141.

nature of the measure and its relation to the end. If so, panels may reject a measure that appears too broad for the given end.[20]

If the BTA fits under either Article XX(b) or (g), it still must accord with the opening words or Chapeau of Article XX. The Chapeau states that the measure must not be applied in a manner which is "a means of arbitrary discrimination between countries where the same conditions prevail or a disguised restriction on international trade". The Appellate Body has read a few key conditions into the Chapeau.[21] First, the measure must be flexible with respect to how other countries design their own environmental measures. Second, the country taking the measure (putting in place the BTA in this case) must have followed due process in putting its measure in place, including administrative processes such as notice and comment and appeal procedures. Third, the country implementing the measure must have negotiated in good faith with countries subject to the measure (in this case, with countries whose goods would be subject to the BTA on entering the United States).

Under the Chapeau, then, there are constraints on both the substance of the rule and how the rule is created. The constraints on the substance may be the hardest in the case of BTAs. For example, the BTAs under proposed U.S. legislation are based on the levels of greenhouse gases emitted in the production of the good. In order to ensure any BTA is applied "flexibly", it will be necessary to have an accurate picture of how goods are produced and emissions treated in other countries. This information is obviously difficult and costly to obtain. It is in theory possible to use different types of measures for emissions in other countries but they may fall under WTO rules. For example, using average emissions for the industrial sector in particular countries may be too broad, as exporters who emit less than average may argue they are discriminated against.

The WTO rules on BTAs therefore are similar to constitutional rules in the sense of limiting the content and form of measures governments can take. These rules appear to favour carbon taxes rather than emissions trading given uncertainties such as whether the emissions trading permits would fall within the definition of "charges". Further, even if a particular BTA is found to fall under Article XX(b) or (g), it must survive scrutiny under the chapeau. While compliance with some of the procedural requirements of the Chapeau may be possible, it may be very difficult and expensive to survive its substantive requirements.

---

[20]To meet Article XX(g), the measure must also relate to the "conservation of exhaustible natural resources" which the Appellate Body has read broadly in the past by relating the term to the subject matter of international environmental agreements (Appellate Body, *U.S. – Shrimp I*, 1998). In addition, the measure must also be "made effective in conjunction with restrictions on domestic production or consumption". The Appellate Body has set a low threshold for this provision, requiring only "even-handedness" in restrictions between domestic and foreign producers. See Appellate Body, *U.S. – Reformulated Gasoline*, 20-22.

[21]Appellate Body, *U.S. – Shrimp I* and Appellate Body, *United States – Import Prohibition of Certain Shrimp and Shrimp Products – Recourse to Article 21.5 of the DSU by Malaysia*, WT/DS58/AB/R, 22 October 2001, at para. 134.

WHO IS OBLIGATED?

In addition to the similarities in terms of constraints on the type and form of instruments, there are interesting parallels between constitutional law and WTO rules in terms of who is covered by the rules. The Canadian Constitution applies to government bodies. It sets out the powers and responsibilities of the federal and provincial governments and has made municipal governments creatures of provincial governments. Moreover, the *Canadian Charter of Rights and Freedoms* applies to government actions and decisions. Individuals can challenge government action and, if successful, there are potentially effective remedies against the government.

Similarly, WTO rules apply to governments. The WTO members are all governments, and any challenges under WTO agreements are brought by and against member governments and not individuals or companies. However, there are a few important differences. First, not all governments are covered directly by WTO agreements, which means that there is a difference in application to the federal as opposed to provincial governments. Second, the WTO agreements do not in practice apply to all governments equally because of how the rules are enforced. Moreover, only member governments can bring a WTO complaint. Individuals and companies must act through their government if they are concerned about a WTO violation by another WTO member. These issues will be discussed in turn.

Federal-State Relations

One issue in terms of the application of WTO rules to carbon pricing policies (including BTAs) is how these rules apply if the measure is put in place not by the federal government but by a province or a group of provinces. The difficulty for the WTO is that federal governments sign onto and take obligations under GATT rather than states or provinces. In order to overcome this difficulty, the GATT has a federal state clause that requires each member to "take such reasonable measures as may be available to it to ensure observance of the provisions of this Agreement by the regional and local governments within its territories" (GATT, Article XXIV(12)). These provisions have been interpreted strictly to impose a positive obligation on members such that there is little room for federal states to argue they are not responsible.[22]

For example, Ontario has decided to create an emissions trading scheme. Suppose the program covers the steel industry and that Ontario decides to impose a BTA on steel imports in order to overcome the competitive effects on its steel producers. Other WTO members could challenge these BTAs. If a panel finds that BTAs are not an "internal tax or other charge" or the BTA otherwise violates GATT provisions (and is not saved by Article XX), the Canadian federal government has a positive obligation to take "such reasonable measures"

[22]See *EC – Selected Customs Matters*, (2006) WTO Doc. WT/DS315/R (Panel Report, 16 June 2006) and Trebilcock and Howse (2005).

available to it to have Ontario bring the BTA into compliance with GATT (including possibly removing it altogether).

If a panel finds a regional or local government of a member has violated GATT and the member has not been able to secure compliance, the remedial provisions of the WTO apply against the federal government. Under these provisions, the complaining member can seek compensation from the federal government and, if compensation cannot be agreed on, the complaining member can seek permission to impose countermeasures against the country taking the measure. These countermeasures take the form of increased tariffs by the complaining member against imports from the country at a level up to the harm caused by the measure.[23]

In our example, the implication is that the BTA imposed by Ontario could lead to countermeasures against Canada as a whole. There is therefore a political dynamic that may become important. Ontario is taking a measure to protect its own industry. However, because the measure is non-compliant with WTO rules, countermeasures may be imposed on all provinces. Ontario is creating an externality by obtaining the benefits of the BTA in the form of political and possibly economic returns and shifting at least part of the cost onto other provinces. This dynamic may make for difficult federal-provincial negotiations. The other interesting point to note is that these effects are only for economic law (that is, WTO obligations) because of the agreements the federal government signed. It is not true of the federal climate change obligations, at least as current climate agreements are written, as they do not contain similarly enforceable federal-state provisions.

Not All Governments Equally

The Canadian Constitution applies to all governments equally. While there may be some divergence in how citizens in different provinces can access legal aid or the courts, in large measure all governments are exposed to challenge if they take unconstitutional measures. The WTO, on the other hand, in practice does not apply to all members equally. In large part, the differences in application arise because there is no central government body which enforces the commitments made under the agreements. Instead the agreements are self-enforcing in the sense that members agree to be bound by the agreements and can withdraw if they wish.

As noted previously, if a member puts in place a BTA that is found to not comply with the WTO agreements, another member may bring a complaint before a WTO panel. If the complaining party succeeds, the member taking the measure is to remove the measure. If it fails to do so, the members can attempt to agree on compensation, but if no agreement is reached the complaining party may seek permission to impose countermeasures against the member taking the

[23]The main remedial provisions for the WTO are found in the Dispute Settlement Understanding, Article 22.

measure. These countermeasures take the form of increased tariffs by the complaining party.

These remedial provisions potentially lead to differing abilities and willingness of members to use the WTO dispute settlement system. First, as the countermeasures take the form of increased tariffs, the country opposing the member imposing the countermeasures must have sufficient trade with the violating country to actually induce action. It is difficult to have any economic effect from the sanction if the country imposing the countermeasure cannot harm the exporting industry of the non-complying member. As a result, large importing countries such as the United States and the European Union have much greater ability to use these countermeasures to induce compliance by others.[24] This difficulty in finding leverage for some members has led to calls to revamp the remedies system to allow, for example, monetary sanctions against non-compliant members or allow successful complaining countries to auction off to other members the right to take countermeasures (Schwartz and Sykes 2002; Trachtman 2007; Bronckers and van den Broek 2005; Green and Trebilcock 2007).

Second, as the countermeasures take the form of increased tariffs, they in effect harm the complaining party at the same time they impose costs on the non-compliant party. Countries which increase tariffs in general face a welfare loss as the tariff aids domestic competing industry but this benefit is outweighed by the harm to consumers and others from the higher prices (Krugman and Obstfeld 2005). There may, of course, be political benefits from the increased tariffs, particularly where those harmed domestically by the increased tariffs (such as consumers) face collective action problems. However, any country which wishes to take action under the WTO must be willing to consider harming itself to take action against the non-compliant member. Developing countries may be unwilling to bear such costs of applying WTO remedies.

Third, the remedies under the WTO dispute settlement system are prospective. If a country is found to have violated a WTO commitment and it has neither removed the measure nor agreed on compensation with the complaining member, the complaining members may seek permission to impose countermeasures. In ordinary civil litigation, any remedies for breach of contract would in most cases at least equal the harm that the breaching party has caused to the complaining party. The remedies may be such as to put the complaining party in the position it would have been but for the breach. However, in the case of the WTO, the countermeasures can only relate to the level of harm at the time that the member was found to be in violation of its WTO commitments – that is,

[24]There are, of course, other aspects of the WTO dispute settlement system that can either increase or decrease the probability of compliance with the WTO agreements or bringing WTO complaints. For example, reputation may play a significant role in members' decisions. Developed countries may be hesitant to impose severe countermeasures against developing countries to the extent that they face a reputational cost (either at home or abroad) from harming the developing countries chances of economic growth. Developing countries for their part may not bring complaints against developed countries if they fear that the developed country will retaliate in other ways such as through reduced aid (Trachtman 2007).

it does not cover any of the harm that the member caused from the time it implemented the measure up to the finding of non-compliance.

There are a number of reasons given for the prospective nature of WTO remedies (such as that the members are presumed to be acting in good faith such that any non-compliance is merely a good faith disagreement over an uncertain agreement). However, the result is that countries do not face any immediate costs (beyond reputational costs) for putting in place a non-compliant measure. The costs only start once another member has complained and a panel has found non-compliance. This prospective nature of WTO remedies reduces the incentives against non-compliance with the agreement, and may in some circumstances encourage countries to take short-term measures, which can be removed in the event of a finding of non-compliance.[25]

The above features of the WTO dispute settlement system are relevant to the discussion of carbon pricing and BTAs. Take, for example, the U.S. bills that propose imposing border measures in conjunction with an emissions trading program (see footnote 1). The U.S. provisions may be compliant with WTO agreements, although there is considerable uncertainty as noted above. However, to the extent they are not (either unconsciously or consciously if the United States is attempting to gain leverage in the climate change debate), other members face considerable costs in taking action and may not even be able to change the U.S. system after a successful challenge.

As Barrett (2007) has noted, any action taken to induce other countries to act where there is free-riding must be both severe and credible. The action must be severe enough to cause the non-participating country to act and must be credible enough that the non-participating country will believe that the action will be taken. The same concern arises in the case of one member attempting to address non-compliance by another, such as in the case of BTAs that do not comply with WTO commitments. Developing countries or smaller WTO members may, for example, not be able to impose sufficiently severe countermeasures against the United States to induce it to change because their trade with the United States is too small. Further, even if it could impose significant countermeasures, the country may be unwilling to bear the costs of large sanctions to its own economy – that is, the threat of action is not credible.

Further, even significant countermeasures may not be enough to induce some countries to change where the issue is particularly politically salient. For example, in the dispute over EU banning of imports of meat grown using hormones, the EU ban was found to be non-compliant with its WTO obligations and Canada and the United States imposed significant countermeasures against the European Union.[26] Because of political pressure at home, the European Union did not change its measure but instead sought to build further evidence

---

[25]See, for example, Green and Trebilcock (2007) discussing how the prospective nature of WTO remedies allows countries to use one-time export subsidies to gain control of a market without facing direct countermeasures.

[26]Appellate Body Report, *EC – Measures Concerning Meat and Meat Products*, WT/DS26/AB/R, WT/DS/48/AB/R, 13 February 1998.

that the ban was justified.[27] The climate change context may be similar. If the U.S. public, for example, feels sufficiently strongly about either climate change or the unfairness of the United States taking action on climate change while other countries appear not to be, the U.S. government may not respond to countermeasures by removing non-compliant BTA provisions.

As a result, while the WTO agreements do impose limits on the types of BTAs members may put in place, it is not clear that all countries face the same incentives to respect these limits. Some larger countries may be willing to put in place the measures to ensure that their industry remains competitive and wait until challenged or even later to remove the measures (if at all). For this reason, smaller countries such as Canada must try not to be on the list of countries whose importers face these import measures rather than rely on an imperfect WTO enforcement system. On the other hand, given that Canada is a small economy, it is difficult to purposely hold in place measures that do not comply with WTO commitments.

INSTITUTIONS AND TRADE-OFFS

One final feature of the Canadian constitutional debate that has an interesting overlap with the WTO is that there is concern about who is making the decisions about what government action is valid. In the constitutional context, there is an ongoing debate in Canada about the benefits of provincial versus federal governments making particular policy decisions. The value of competitive federalism and experimentation as opposed to national action is widely debated, including in the environmental area. There is a further debate about the role the courts should play in defining social policy. Are the judges too "active" in making policy choices or are they merely exercising a necessary role that was given to them under the Constitution?

These same debates are mirrored in the WTO context, particularly in areas of social risk such as climate change. Is it better for member governments to make decisions on the appropriateness of particular climate policies, should the members as a whole attempt to agree on which measures are appropriate, and to what extent can and should WTO panels or the Appellate Body make these determinations? Which institution is chosen – domestic governments, WTO members as a whole, panels – determines who decides what is efficient or fair.

In the case of environmental measures, WTO agreements are incomplete, using standards rather than detailed rules to set limits of permissible behaviour. The extent to which domestic governments have space to decide on policy is not clear and can in part be set by panel or Appellate Body decisions. The choice of which institution is best suited to determining whether a climate measure is appropriate in large part depends on two key features. First, who has the relevant information and expertise? Climate change policy will depend on information about the potential harm for individual countries and for the globe of climate

---

[27] Appellate Body Report, *Canada – Continued Suspension of Obligations in the EC – Hormones Dispute*, WT/DS321/AB/R, 16 October 2008.

change, the costs and benefits of particular climate policies and the extent to which individuals in the particular country care about climate change. If a country puts in place a climate measure such as a BTA and another member challenges it, the response of the country implementing the measure will be that it complies with WTO commitments and, even if not in compliance with substantive WTO commitments, that it falls within the Article XX exceptions. To the extent the member government has the best information on these factors and cannot provide this information to a panel reliably or at a sufficiently low cost, it may be best to provide the domestic member more scope to make the policy choice. If, on the other hand, a panel can obtain relatively good information about the policy and is not prone to errors in interpreting the information, they may be in a position to defer less to the domestic decisions.[28]

The other factor that is important in determining the appropriate climate change policy is the extent to which the choice of appropriate policy depends on the values of the party making the choice. As noted earlier, the decision of when and how to address climate change rests not only on strict costs and benefits but also on ethical choices. These ethical choices relate, for example, to how much we care (or should care) about the well-being of future generations or of people living in other countries. Allowing member countries to themselves make decisions about climate policies allows policies to reflect the values of that particular country (depending of course on the political system). It therefore provides scope for countries to decide, based on their preferences, the well-being of others. Of course, this option leaves open the possibility that countries prefer to impose costs on others of climate change and have their climate policies reflect this through more protectionism. Allowing panels or the Appellate Body to decide if a climate policy is appropriate, on the other hand, raises fears that it is the choices of the panel or Appellate Body members that count – that is, panels members will decide on these issues not on whether the citizens of the regulating country care about the impacts of climate change, but on whether they themselves care.[29] Decisions by panel or AB members will be of particular concern if they tend to have particular preferences because, for example, they tend to have trade rather than environmental experience.

A further option for determining the appropriateness of climate policies is a multilateral agreement specifically on permissible climate measures that impact trade. Such an agreement could take the form of a set of detailed rules about when countries can take particular measures and the form these measures could take. The agreement could, for example, explicitly set out when BTAs can be used, whether they can be used to cover emissions trading, and whether they can take into account PPMs in other countries. The difficulty, of course, is that it is costly, if not impossible, to obtain the information and design the agreement to

[28]Guzman (2004), for example, argues that domestic governments have better information about their citizens' preferences concerning health risks and therefore the WTO should defer to judgments in these areas or risk large error costs.

[29]There is a large and growing literature examining whether judges vote in particular cases in line with their own personal policy preferences (for example, for discussion of these models in the U.S. context see Segal and Spaeth, 2002).

take into account the various ways in which climate measures may be validly used. Climate policies are evolving rapidly and it is difficult to know *ex ante* which forms of policies will be best. A multilateral agreement may either freeze innovation or steer it in a suboptimal direction. As importantly, however, member countries may not be able to reach agreement on these issues. WTO members are having difficulty reaching agreement on core economic issues in the Doha Round, let alone attempting to define acceptable climate policy.

There are, therefore, interesting parallels between some of the debates over Canadian federalism and the WTO. Both the Canadian Constitution and the WTO place limits on the form of measures particular governments can take. They both raise issues of institutional competence and the role of tribunals in assessing policy decisions. They both for the most part use standards rather than detailed rules to set requirements. The similarities should not, however, be taken too far.[30] The WTO agreements are sets of commitments by independent countries which depend on the willingness of countries to participate. The main form of enforcement under WTO agreements is not available equally to all members depending as it does on countermeasures to attempt to induce compliance. Border measures provide a useful and timely example of how the WTO interacts with domestic policy decisions – one that will likely play out in the not-too-distant future.

REFERENCES

Barrett, S. 2007. *Why Cooperate? The Incentive to Supply Global Public Goods*. Oxford: Oxford University Press.

Bodansky, D. Forthcoming. "Is There an International Environmental Constitution?" *Indiana Journal of Global Legal Studies*.

Bronckers, M. and N. van den Broek. 2005. "Financial Compensation in the WTO: Improving the Remedies of WTO Dispute Settlement", *Journal of International Economic Law* 8(1): 101.

Charnovitz, S. 2007. "The WTO's Environmental Progress", *Journal of International Economic Law* 10(3): 685.

Epps, T. and A. Green. 2008. "Is There a Role for Trade Sanctions in Addressing Climate Change", *UC Davis Journal of International Law* 15(1): 1.

Green, A. and M. Trebilcock. 2007. "Enforcing WTO Obligations: What Can We Learn From Export Subsidies?" *Journal of International Economic Law* 10(3): 653.

Guzman, A. 2004. "Food Fears: Health and Safety at the WTO", *Virginia Journal of International Law* 45: 1.

Hogg, P. 2008. *A Question of Parliamentary Power: Criminal Law and the Control of Greenhouse Gas Emissions.* C.D. Howe Backgrounder (August).

Horn, H., G. Maggi, and R.W. Staiger. 2006. "Trade Agreements as Endogenously Incomplete Contracts". NBER Working Paper 12745 (December).

Howse, R. 2000. "Democracy, Science and Free Trade: Risk Regulation on Trial at the World Trade Organization", *Michigan Law Review* 98: 2329.

[30]Trachtman (2006) warns that focusing on the constitutional nature of the WTO can lead to a "false rigidity".

Hsu, S.-L. and R. Elliot. 2009. "Greenhouse Gas Regulation in Canada: Constitutional and Policy Dimensions", *McGill Law Journal* 50 (forthcoming).

Jackson, J.H. 2004. "International Law Status of WTO Dispute Settlement Reports: Obligations to Comply or Option to 'Buy Out'?" *The American Journal of International Law* 98(1): 109.

Krugman, P. and M. Obstfeld. 2005. *International Economics: Theory and Policy*. 7th ed. Boston: Addison Wesley.

Leiberman-Warner. 2008. *Climate Security Act of 2008*.

Nordhaus, W. 2007. "The Stern Review on the Economics of Climate Change", *Journal of Economic Literature* 45(3): 686.

Panel Report, Argentina. 2000. *Measures Affecting the Export of Bovine Hides and the Import of Finished Leather*. WT/DS155/R (19 December).

Panel Report, U.S. 1987. *Taxes on Petroleum and Certain Imported Substances*. B.I.S.D. (34th Supp.) 136 (17 June).

Pauwelyn, J. 2007. *US Federal Climate Policy and Competitiveness Concerns: The Limits and Options of International Trade Law*. Working Paper 07-02. Nicholas Institute for Environmental Policy Solutions, Duke University (April).

Regan, D. 2007. "The Meaning of 'Necessary' in GATT Article XX and GATS Article XIV: The Myth of Cost Benefit Balancing", *World Trade Review* 6(3): 347.

Schwartz, W. and A.O. Sykes. 2002. "The Economic Structure of Renegotiation and Dispute Resolution in the World Trade Organization", *Journal of Legal Studies* 31: 179.

Segal, J. and H. Spaeth. 2002. *The Supreme Court and the Attitudinal Model Revisited*. Cambridge: Cambridge University Press.

Stern, N. 2007. *The Economics of Climate Change: The Stern Review*. Cambridge: Cambridge University Press.

Sykes, A.O. 2002. "Domestic Regulation, Sovereignty and Scientific Evidentiary Requirements: A Pessimistic View", *Chicago Journal of International Law* 3(2): 353.

— 2003. "The Least Restrictive Means", *University of Chicago Law Review* 70(1): 403.

Trachtman, J.P. 2006. "The Constitutions of the WTO", *European Journal of International Law* 17(3): 623.

— 2007. "The WTO Cathedral", *Stanford Journal of International Law* 43(1): 127.

Trebilcock, M. and S. Giri. 2005. "The National Treatment Principle in International Trade Law", in E.K. Choi and J.C. Hartigan (eds.), *Handbook of International Trade, Volume II*. Oxford: Blackwell.

Trebilcock, M. and R. Howse. 2005. *The Regulation of International Trade*. 3rd ed. London: Routledge.

Waxman-Markey. 2009. *American Clean Energy and Security Act of 2009*.

Weitzman, M. 2007. "The Stern Review of the Economics of Climate Change", *Journal of Economic Literature* 45(3): 703.

VI

The Political Economy of Climate Change

12

The Political Economy of Carbon Pricing in North America

Bryne Purchase

INTRODUCTION

Climate change represents the ultimate public policy problem. It is the result of an externality in decentralized decision making. But it is on a global scale and deeply embedded in the nature of our society – over 80 percent of the world's primary energy comes from fossil fuels. A ton of carbon dioxide emitted anywhere on earth has the same impact on the future global atmospheric temperature.

Accordingly, mainstream economic policy advice is to price carbon emissions into all decision making and implement this price on a global scale. Nothing could be more straightforward. Indeed, the fundamental economic policy advice has remained essentially unaltered since AC Pigou in 1920.

Of course, Pigouvian taxes were not the only way to deal with an externality. Ronald Coase showed that another was to create property rights. Clearly, ownership of the earth's atmosphere is not a viable alternative. But in 1968 John Dales of the University of Toronto demonstrated that a property right in the form of a tradable emission allowance could produce the same efficient result as a tax (Dales 1968). The modern terminology is "cap and trade". Again, this policy instrument is well known and understood.

With the same sector coverage and the same target emission quantity, both carbon taxes and cap and trade imply the same carbon price. But, as will be outlined below, a carbon tax is an administratively more efficient policy instrument. Yet in the political marketplace, these policy instruments are not considered equal by practicing politicians. Indeed, cap and trade has, so far, emerged as the revealed preference of federal politicians in both Canada and the United States *and* in European politics.

What is perhaps even more startling for would-be economic policy advisors is the clear political preference for "command and control" initiatives such as President Obama's new fleet fuel efficiency standards or the widespread use of

I would like to acknowledge the assistance of my colleague, Tom Carpenter, at the Queen's Institute for Energy and Environmental Policy.

"renewable portfolio standards" in electricity generation or the requirement for carbon capture capability in new coal-fired generating stations.[1] Or, a little closer to home, there is Ontario's decision to exit from the use of coal altogether to generate electricity. This clear political preference is despite the fact that economists can demonstrate that all of these initiatives are both less efficient *and* less effective than carbon pricing in reducing emissions.

This essay reflects on this conundrum in the context of where public policy in the United States and Canada might be headed on this global policy problem. The central argument is that "politics" cannot be taken out of the choice of policy instrument. More fundamentally, it is the structure of the political marketplace that determines instrument choice. And, in that regard, the future of carbon pricing in North America is still highly uncertain.

THE TECHNICAL SUPERIORITY OF CARBON TAXES

Box 1 illustrates the benefits of an excise tax applied upstream on the carbon content of a fossil fuel. Box 2, on the other hand, illustrates what to expect under a cap-and-trade regime. The weight of the observations is that carbon taxes are the superior instrument by which to achieve a given carbon emission reduction target.

Box 1: Advantages of a Carbon Tax
(applied as an upstream excise tax on the carbon content of fossil fuels)

- Applies to whole economy, not just specific sectors.
- Administratively simple (few taxable entities), using existing tax machinery.
- Minimizes evasion because government has a vested interest (revenue loss) in preventing cheating.
- Provides price/cost certainty to emitters and to technology inventors, although there is still some "political" risk.
- Can be phased in gradually, and with revenues recycled to minimize macroeconomic impacts *and* impacts on the poor.
- No international revenue flows. Countries keep their own tax revenue.
- Easier to monitor international compliance with a "net carbon tax" in each country.
- Similar to GST and applies only to Canadian consumption, not exports.

[1]The new American Clean Energy and Security Act, HR 2454, also known as the Waxman-Markey bill, which recently the U.S. House of Representatives has all of these instruments in it.

Box 2: What to Expect With Cap and Trade

- **Highly Political Market:** A new property right created by governments will create a scramble by various constituencies to secure those rights free of charge. There will be a tendency to issue too many such permits (including domestic and international offset opportunities).

- **Limited Coverage:** While all sectors and emitters are intended to be covered, the program will struggle to get beyond the initially targeted large emitters.

- **Emission Price Volatility:** Carbon prices will be extremely volatile, reflecting the underlying volatility in the demand for energy and the perfectly inelastic supply of emission rights. Large scale, long term capital investments, typical of the energy industry, are less likely with this type of volatility. Price floors and ceilings could be implemented to reduce the volatility and move cap and trade closer to a carbon tax.

- **Very High Administrative Costs:** There will be a very large number of emitters whose pre- and post-trade emissions need to be subject to audit. In addition, the management and trading of the emission allowances will create a vast new, largely private, bureaucracy.

- **Political Resistance to International Extension of the Market:** While gains from trade and market liquidity are increased by extending the emissions market internationally, doing so also increases the potential for large scale international revenue flows. The U.S. congress is unlikely to agree with this. Also this approach demands a high level of trust in the willingness and ability of foreign governments to monitor and enforce the regulations.

GORDIAN KNOT OF CANADIAN POLITICS

Setting aside for a moment the matter of carbon taxes versus cap and trade, it should be noted that Canadian federal politics are simply not conducive to national leadership on the issue of carbon pricing. This political reality is notwithstanding the existence of federal constitutional authority to implement a national program.

There are some obvious reasons for this political incapacity. The provinces own their natural resources and, for some, these are a source of both significant provincial economic development and substantial revenues to provincial treasuries. A federal tax on the carbon content of fossil fuels, with the clear intent to diminish the consumption of such fuels,[2] is certain to be opposed, notwithstanding the potential benefits to the world.

---

[2]Carbon capture and sequestration appears to be a potentially viable technical and commercial option for the production of electricity from coal. This could prove life-saving to the Alberta and Saskatchewan economies, and explains the intense interest of both the federal and provincial governments in proving this technology.

Even with the proceeds of the national tax (or the emission allowances) given to the provinces, the future must be considered highly uncertain, if not bleak, for those provinces producing fossil fuels. This is particularly the case for Alberta's oil sands given their already comparatively high cost (and hence vulnerable competitive position in a presumably shrinking future global oil market) *and* their heavy carbon footprint.[3]

Federal politics are complicated also by the fact that energy intensive resource extraction and processing (forestry, pulp and paper, mining and smelting, fishing and agriculture) are concentrated in northern, coastal and rural ridings. Then there are urban and rural low income groups who spend a disproportionate share of their income on energy.

Ontario politics on this issue are not much easier than Alberta politics. The bankruptcy in the United States of General Motors and Chrysler points to the problem of carbon pricing and its potential impact on the competitive position of these struggling companies. The new vehicle market in the United States and Canada is already dramatically shrunken, and highly likely to remain so for some time to come.[4] A carbon price, when it is finally implemented on the transportation sector, can only compound the problems. Notwithstanding the Ontario government's great public fanfare around its decision to exit coal-fired electricity generation by 2014, it never endorsed a carbon tax – a far more pervasive and efficient policy to reduce carbon emissions.

Moreover, suburban ridings would suffer under any aggressive carbon pricing regime. This comes not from simply an increase in the cost of commuting, but also from the capital loss suffered by suburban properties. Few politicians would welcome the opportunity to explain why that is necessary to save humanity.

The three mainstream national political parties also have serious "legacy" constraints on their ability to lead aggressively on this issue. The Conservatives have their power base in the most "at risk" part of the country. The NDP still must appeal to what is left of unions in heavy industry and, of course, to the urban and rural poor. The Liberals have the heritage of the National Energy Policy and Western alienation. And all parties hope to grow in Ontario, a province already undergoing profound economic dislocation.

Curiously, it was the National Energy Program and the political reaction to that policy that led to the Canada-U.S. Free Trade Agreement, with its Energy

---

[3]That stock markets do not appear to be excessively concerned suggests that announced target reductions of greenhouse gas emissions in Canada and the United States by 65 percent and 80 percent respectively by 2050 are very heavily discounted by investors.

[4]The market going forward may be on the order of 12 to 15 million units compared to the 17 to 20 million units of the recent past. The new smaller market is a reflection of the prior excess consumption by U.S. consumers related in turn to excess mortgage lending and mortgage interest tax deductibility. Carbon pricing will both shrink the market further and shift it to smaller, more fuel efficient vehicles. But the "Detroit Three" have never been able to make money in the small car market, largely because of labour costs. Even the Japanese have struggled to find profitability in that market (see DesRosiers 2009).

Chapter, subsequently confirmed under NAFTA. A North American market in natural gas as well as oil[5] has emerged. As a result, a national energy policy no longer makes any sense compared to a North American energy policy.

Accordingly, the same is true for a policy on carbon pricing. But aside from the logic imposed by an integrated North American marketplace, the fact that Canada will now sit back and let the United States lead on this policy issue may have been the only way through the torturous and highly risky Canadian political scene. Now all federal politicians can say, "The devil made us do it!"

And the United States will gladly oblige Canada in that regard. Continued market access to the United States will require a "commensurate" carbon pricing regime in Canada. Now Canada need only wait on what emerges from the legislative processes of the United States. Unfortunately, while cap and trade is the preferred instrument, the details are by no means clear. And even after a definitive piece of U.S. legislation emerges, the long lead times and the inevitable differential start dates for different sectors will provide ample opportunity for change.

THE POLITICAL PREFERENCE FOR CAP AND TRADE

In the United States, no presidential hopeful in the primaries for either party espoused a carbon tax. Indeed, both Senator McCain and then-Senator Obama opted for cap and trade in last year's presidential campaign.

In Canada, both the Conservatives and the New Democratic Party attacked the carbon tax proposal of the Liberals in the last federal election. The NDP espoused cap and trade and the Conservatives offered an even more convoluted form of carbon regulation for large emitters. Mr. Dion's defeat, along with the defeat of Elizabeth May of the Green Party, might well be taken as proof that the carbon tax is a political non-starter, at least at the federal level in Canada.[6]

Carbon taxes versus cap and trade appear to be the political equivalent of Beta versus VHS in the video technology wars of the past. Why is this so? A plausible answer is that a tax is a transparent instrument of public policy. Voters believe they understand what it means to them. And even with revenues recycled as tax reductions, few seem to believe that they will not be net payers.

---

[5]Oil is really a world market now. Natural gas is a North American market and awaits the further development of international trade in liquefied natural gas. Coal is largely a North American market.

[6]There are, of course, those who will argue that the success of the provincial Liberals in British Columbia demonstrates that the carbon tax can be a politically viable instrument of public policy. But I would note that British Columbia already has a relatively low carbon footprint, thanks to the prevalence of hydro-electric power. This is similar, of course, to Quebec. Moreover, British Columbia has no compelling producer interest in the transportation sector or, as yet, in the production of fossil fuels (although this could change).

Cap and trade, on the other hand, is not transparent. It would be the rare voter who accurately could describe how it works. Moreover, cap and trade appears initially as a regulation/tax on large polluting businesses. Most people take that fact as an indication of its incidence. And few politicians would take the opportunity to disabuse them in that regard.

Also, cap and trade typically has a two-stage implementation process, with households (that is, voters) not fully affected initially, and perhaps not until years later. This has additional appeal to political representatives whose own time horizons are short in any case. Even successful prime ministers are rarely around for more than two terms in office and, of course, presidents of the United States are constitutionally restricted to two terms. It is highly unlikely that any of today's leaders will be around to take responsibility for the full impact on voters.

Rent-seeking activity will thrive under cap and trade. The fact that cap and trade requires a new private army of auditors, lawyers and market experts also creates a powerful professional constituency in its favour. And while there are great plans, in principle, to auction the emission allowances, they are most likely to be distributed for political benefit only.[7]

Some large emitters are certain to be given generous allowances. Indeed, regions are likely to be benefitted. For example, in Canada, once a national cap is chosen, it is highly plausible that each province would be allocated emission allowances based on their historic emissions profile, clearly benefitting Alberta. And, as I will argue below, regional politics in the United States is almost certain to play a role in the allocation of allowances in that country.

WHITHER THE U.S. CONGRESS?

We know that Canada will follow the U.S. lead on carbon pricing. But we have no idea, as yet, where that will take us. Our only definitive clue is that it will be a form of cap and trade. There are many details yet to emerge from the legislative process. Box 3 illustrates some of the questions.

There are a number of bills in the U.S. Congress, plus the president's budget proposals, which imply answers to all these questions. But if one were to speculate on outcomes from the U.S. political process, it might be equally instructive instead to focus on two crucial sectors of the U.S. economy and one key region. The sectors are electricity production and transportation, and the region is constituted by several Great Lakes states.

---

[7]President Obama's budget proposals call for auctioning of the allowances in his version of cap and trade, but the political process is far from complete. The new American Clean Energy and Security Act, HR 2454, also known as the Waxman-Markey bill, which recently narrowly passed the House of Representatives, provides for only 15 percent of the emission allowances to be auctioned initially. The rest are to be allocated to a wide variety of sectors free of charge. Some sectors, such as utilities, are explicitly mandated to ameliorate the impact on consumers. The net effect of such a direction is highly likely to blunt the purpose of the initiative. The Senate has yet to pass legislation.

Box 3: Cap and Trade Uncertainties

- Setting the Cap: how stringent and what timing?
- What coverage (what GHGs, sectors and size of emitters)?
- What will the new emission rights look like – annual, bankable?
- How will emission rights be allocated/auctioned?
- How will extreme price variability be dealt with?
- Will offsets be allowed? Will Kyoto CDM be allowed?
- Will credit for early action be allowed?
- What protocols/institutions will surround the effective reporting, monitoring and enforcement?

Coal and oil are highly specialized in their end use. Ninety-one percent of U.S. coal use – the most carbon emitting fossil fuel – goes into the production of electricity. Seventy percent of oil – the second most carbon intensive fossil fuel – goes into the transportation sector where it accounts for 96 percent of all transportation fuels. In short, effectively dealing with carbon requires a vast transformation in the production of both electricity and transportation services.

The United States produces roughly 50 percent of its electricity from coal-fired generating stations. But, as one might expect, not all states are created equal in this regard. Box 4 illustrates coal-fired electricity generation and the percentage of households also heating with electricity in six key Great Lakes states.

Clearly, the impact of carbon pricing in these states will be tremendous. But these same states are also home to the "Detroit Three" North American auto makers and their nearby "just-in-time" parts suppliers. The United Auto Workers play a key electoral role in these states.

Can President Obama restructure the vulnerable automobile companies in such a way as to be able to cope with the further impact of carbon pricing when it is implemented in the transportation sector? He clearly appears ready to try, at great public expense.[8] But can he also survive the impact of markedly higher electricity prices and heating costs in these states, not just for businesses, but also for households?

---

[8]The current bailout numbers are estimated at roughly $50 billion for the U.S. government and $10 billion for the Canadian federal and Ontario governments (going along for the ride).

Box 4: Coal Dependency

Percent of Coal-Fired Generation in Great Lakes States*

- Indiana 95.8
- Ohio 85.1
- Wisconsin 66.4
- Michigan 65.5
- Pennsylvania 52.1
- Illinois 48.5

Percent of households heating with electricity*

- Indiana 22
- Ohio 18
- Pennsylvania 17
- Illinois 12
- Wisconsin 11
- Michigan 7

*Source: U.S. Energy Information Administration.

Figure 1 clearly demonstrates that he and his fellow Democrats have an incentive to try, if they are to retain the White House for two terms. It should be remembered that in the 1970s, both Presidents Ford and Carter argued in favour of world prices for oil to American consumers, along with recycling of windfall petrodollars. This was sound economic policy. But neither won a second term. The U.S. federal gas tax has not increased since 1993.

The six states shown in Box 4 represent 100 Electoral College votes. All went Democratic in the 2008 election and were clearly instrumental to President Obama's electoral success; equally they will be important in his 2012 bid for re-election. How will they be treated in the legislative process? So far, President Obama has orchestrated, along with the Congress, a massive public subsidy to rescue two failing auto companies. What more will be done to alleviate the impact on these states, in the context of a national cap-and-trade scheme, remains to be seen.

GOING GLOBAL

While the politics of Canada and the United States will clearly be difficult, it is a harmonized global regulation that is ultimately required. There can be no free riders, or at least no consequential free riders. The non-OECD developing world must somehow be included. Otherwise there will be an even more rapid deindustrialization of the West, for no climate benefit.

Figure 1: Electoral College Vote Distribution

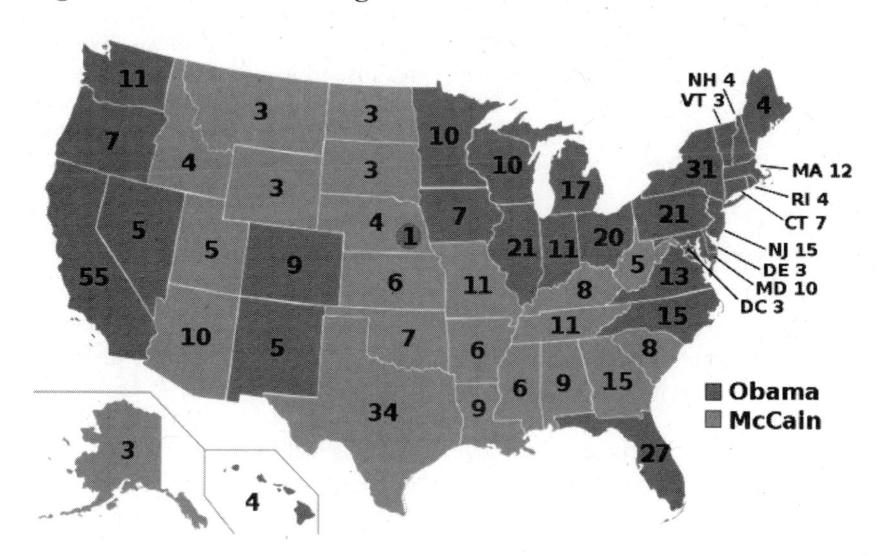

In this regard, some observers estimate that China's total emissions already exceed those of the United States. And it is total emissions, not per capita emissions, which matter to global warming. Together, the United States and China account for roughly 40 percent of global emissions. They must be co-leaders in any world-wide initiative.

But over 70 percent of China's primary energy comes from coal, which is its only secure fuel source. Will it make economic or political sense for China to bear the costs of sharply reduced economic growth in a country of 1.3 billion people and huge economic disparity? And how exactly does one acknowledge the dramatic inequality of income per capita between U.S.-Canada and China, or the fact that current post-industrial revolution greenhouse gas concentrations in the earth's atmosphere are largely attributable to the West?

The political and economic complexities of India and a host of other developing nations are no less daunting. Asia Pacific countries already use *three times* the coal used in North and South America combined!

To further complicate matters, the international distribution of costs and benefits are not likely to be aligned, and they are shrouded in scientific uncertainty and further obscured by debates in economics and philosophy about the appropriate discount rate covering potentially more than a hundred years (Nordhaus 2008).

And if cap and trade is the model to be internationalized, are legislators prepared to see large international revenue flows? And how will we develop trust in foreign regulatory regimes? All of these questions must be answered if the world is to deal with this issue in a pre-emptive fashion and by international agreement.

Perhaps most problematic for economists going forward is the distinct possibility of carbon-based tariffs if the United States does manage to move aggressively on carbon pricing. It is highly plausible politically, especially if the current protectionist political climate prevails. But no one, certainly no economist, should take comfort in that. It would be an administrative nightmare, a high risk to global commerce, and perhaps even, most importantly, a potential threat to global peace.

CONCLUSIONS

What then can one conclude is the future of carbon pricing in North America? Perhaps the only certain thing is that no detail is certain. Carbon taxes are unambiguously technically superior, but perhaps politically inferior as national policy. Unless, of course, not enough thought has gone into the political marketing of this instrument. But better marketing will not disguise its impact. And in that regard, even cap and trade has its greatest political appeal as an abstract concept, and not necessarily when the reality of higher energy prices hits home.

Other far less efficient policy instruments have a higher stealth component (such as product efficiency standards and renewable portfolio standards) and, as a result, more appeal to practicing politicians. They also allow playing with electoral timing. Despite their economic inefficiency, they may be far more prevalent in the future than most economists would advise.

In Canada's case, a potentially compelling U.S. initiative will take us forward on carbon pricing through cap and trade. That is appealing as a way through the tangle of Canadian regional politics, but it leaves open the question of how serious that initiative will really turn out to be. U.S. politics may be no less tortuous or, in the end, any more able to move aggressively forward. And when it comes to a global initiative, the future is even more uncertain.

So what begins as an astonishingly straightforward market failure – an externality in decision making – and with a simple, but elegant, and well understood solution available, proceeds with great uncertainty and deep complexity. Such is the nature of political economy.

REFERENCES

Carpenter, T. 2007. "The Limits of Emissions Trading". Working Paper, QUIEEP. At http://www.queensu.ca/qieep/files/publications/workingPapers/TheLimitsOf EmissionsTrading.pdf.

Cooper, R.N. 2006. *Alternatives to Kyoto: The Case for a Carbon Tax.*. Harvard University. At http://www.economics.harvard.edu/faculty/cooper/papers/Kyoto_ct.pdf.

Dales, J.H. 1968. *Pollution, Property and Prices: An Essay in Policy Making and Economics*. Toronto: University of Toronto Press. Reprinted with a new foreword by WE. Oates. Northampton MA: Edward Elgar Publishing Inc. 2002.

DesRosiers, D. 2009. Observations, *DesRosiers Automotive Reports* 23(7) 15April.

Hanson, C. and J.R. Hendricks Jr. 2006. *Taxing Carbon to Finance Tax Reform*. Issue Brief. Duke Energy and World Resources Institute (March).

Nordhaus, W. 2008. *A Question of Balance*. New Haven CT: Yale University Press.

Shapiro, R.J. 2007. *Addressing the Risks of Climate Change: The Environmental Effectiveness and Economic Efficiency of Emissions Caps and Tradable Permits*. American Consumer Institute. At http://www.theamericanconsumer.org/Shapiro.pdf.

13

Key Questions for a Canadian Cap-and-Trade System

Matthew Bramley

INTRODUCTION

Experts agree that the centerpiece of any national climate change plan should be a policy that "puts a price on emissions" broadly in the economy. The two main policies that can do this are a carbon tax and a cap-and-trade system.

The Government of Canada has committed to announce some form of cap-and-trade system for greenhouse gas emissions before the end of 2009. This paper aims to identify the key questions that should be asked to evaluate any Canada-wide cap-and-trade proposal.

A cap-and-trade system puts a mandatory cap on some portion of national emissions, and allows firms to buy and sell rights to emit within the cap. (The rights to emit may be called "allowances", "permits" or "credits".) This has the effect of putting a price on emissions because firms have to buy extra emission rights if they want to emit more. The market price of emission rights is commonly referred to as the "carbon price".

The carbon price will become a factor in all decisions affecting emissions taken by firms and households. The higher the price, the stronger the incentive to switch from high-emission options to low-emission options. The first fundamental question that should therefore be asked of any cap-and-trade system is: *is the carbon price likely to be high enough to adequately reduce Canada's emissions?*

It may seem odd to ask what the carbon price will be, rather than simply what level the emissions cap will be set at. But cap-and-trade systems commonly include mechanisms (elaborated later) that allow real, new reductions in domestic emissions to be replaced by reductions whose reality is dubious, reductions that already happened or may only happen in the future, or foreign reductions. The expected carbon price is therefore a surer measure of effectiveness than the level of the cap, particularly when we are concerned about cutting Canada's own emissions.

The second fundamental question is: *who will receive money when someone pays the carbon price – or lose money when someone avoids paying that price?*

This is critical because the total amount of money involved could be very large – in the tens of billions of dollars annually.

Each of these two questions raises important related design issues for a cap-and-trade system, such as the use of *"offsets"* or *"intensity targets"*. Below we examine each of these two fundamental questions, and three key related issues for each.

We assume in this paper that there will be a federal cap-and-trade system. It is important to note, however, that provincial governments too have the power to implement cap and trade, and several have already taken steps in that direction. We will not comment further here on provincial action except to say that a federal cap-and-trade system should make accommodation for well-designed provincial policies that put a price on emissions, such as British Columbia's carbon tax; but if necessary a federal system should override ineffective provincial policies, such as Alberta's greenhouse gas regulations (as argued below; cf. page 241).

WILL THE CARBON PRICE BE HIGH ENOUGH TO TRANSFORM OUR ENERGY SYSTEM?

To avoid the worst impacts of climate change, greenhouse gas emissions need to be reduced far below current levels as quickly as possible. The national science academies of the "G8+5" countries say that "limiting global warming to 2°C" – a goal now endorsed by those countries' governments (*Declaration* 2009) and by numerous leading climate scientists (University of New South Wales 2007) – "would require a very rapid worldwide implementation of all currently available low carbon technologies" (National Academies 2009). Since most greenhouse gas emissions come from burning fossil fuels for energy, that means an urgent transformation of our energy system.

How high would the carbon price need to be to achieve such a transformation?

- The National Round Table on the Environment and the Economy has found that Canada would need a carbon price of *$100 per tonne*[1] by 2020 to meet the federal government's current target to reduce national emissions to 20 percent below the 2006 level by 2020 (NRTEE 2009).
- McKinsey and Company (2009, 8-14) estimate that all worldwide opportunities to cut emissions at a cost of up to €100 per tonne[2] *($158 per tonne*) would need to be implemented starting in 2010 to have a good chance of limiting average global warming to 2°C. This suggests that we need a global carbon price of at least that level.
- Research commissioned by the Pembina Institute and the David Suzuki Foundation shows that Canada needs a carbon price starting at $50 per

[1]In this paper, tonnes refer to carbon dioxide equivalent emissions.

[2]In 2005 Euros.

tonne in 2010 and reaching *$200 per tonne* in 2020 to cut national emissions to 25 percent below the 1990 level by 2020. This target is at the least stringent end of the range of what industrialized countries need to do for the world to have a chance of staying within the 2°C limit, according to the UN Intergovernmental Panel on Climate Change (Pembina Institute and David Suzuki Foundation 2008).

Timing is important. We need a significant carbon price as quickly as we can manage, even if only to meet the government's relatively modest emissions target for 2020. And because a simple cap-and-trade system can be designed and brought into effect more quickly than a complicated one, simplicity is important.

The carbon price cannot, of course, be predicted with certainty in a cap-and-trade system because it depends on the actual costs of achieving the level of the cap, which are not known precisely in advance. But economic modeling can provide an indication of the expected carbon price.

In particular, the prices above are much higher than those currently expected in the United States. The Waxman-Markey cap-and-trade bill, passed by the House of Representatives in June 2009, is a big and important step forward from past inaction. But it is projected to generate a carbon price of just US$16 per tonne by 2020 (U.S. Environmental Protection Agency 2009). Accordingly, even though it includes major energy efficiency provisions as well as cap and trade, the bill would make only limited progress towards transforming the U.S. energy system by that year (*ibid.*).

This matters because it is often asserted that Canada's carbon price must stay close to that of our largest trading partner in order to maintain the competitiveness of Canadian industry. The Pembina Institute believes this assertion is wrong, for three reasons:

- The competitiveness impacts of varying national carbon prices tend to be exaggerated. The World Trade Organization notes that "studies to date find generally that the cost of compliance with an emission trading scheme is a relatively minor component of a firm's overall costs" (Tamiotti *et al.* 2009, xviii). Accordingly, a recent C.D. Howe Institute study found that "the overall competitiveness and leakage impacts associated with climate change policy in Canada are likely to be small", even in a scenario where Canada has a carbon price of $115 per tonne by 2020 and our trading partners have none (Bataille, Dachis, and Rivers 2009).
- A cap-and-trade system can incorporate measures to protect industry sectors expected to suffer substantial impacts on international competitiveness (elaborated later).
- To prevent the extreme consequences of global warming with little time available to act, the world desperately needs leaders: countries willing to show they can do what needs to be done without waiting for the slow-to-convince. A country as well endowed as Canada has a clear moral responsibility to do its fair share even when some of our peers have not yet started to do so. Polling suggests that Canadians agree (Pembina Institute 2008).

The Trouble with Offsets

The reason why the projected U.S. carbon price is so low is that the proposed U.S. cap-and-trade system relies massively on "offsets". Offsets are credits granted for reductions in emissions (or removals of carbon dioxide from the atmosphere) from projects outside the cap. Common examples include tillage practices that store more carbon in agricultural soils, and hydroelectric projects in developing countries. Firms are allowed to replace emission reductions achieved under the cap by offsets.

Offsets present two major risks. The first is that large volumes of offsets will flood the market and depress the carbon price to a point where it will become ineffective. This is what is currently expected to happen in the United States.

The second risk comes from the fact that it is very difficult to avoid awarding offset credits for emission reductions that would have happened anyway, even in the absence of offsets. This is due partly to technical challenges, and partly to the fact that a lax offset system is in the interests of most buyers and sellers, who lobby accordingly. Some researchers who have examined the world's largest existing offset system, the Kyoto Protocol's Clean Development Mechanism (CDM), have concluded that "only a fraction of CDM projects actually reduce emissions" (Victor 2009, 13). Under the Government of Canada's draft rules for offsets (June 2009), no attempt would be made to determine whether an offset project would have happened in the absence of offsets, and a significant volume of credits would be granted to projects that have already been implemented.[3]

The result is that emissions will be considerably higher in a system with offsets compared to a system without offsets. This represents a form of fraud: the claim will be made that emissions have been reduced to the level specified by the cap, but a significant proportion of the emission reductions will not be real, new reductions.

The Pembina Institute has come to believe that these two risks created by offsets are so serious that offsets should not be included in a Canadian cap-and-trade system. Instead, industry's desire to contain costs could better be satisfied through a straightforward ceiling on the carbon price – as long as the ceiling is set at a high enough level (David Suzuki Foundation, Pembina Institute, and WWF-Canada 2009, 9). To secure emission reductions above and beyond those achieved with the cap-and-trade system, the government could still purchase offsets from sectors like agriculture or forestry, or from developing countries, as long as the offsets met strict standards.

Linking a Canadian cap-and-trade system to another country's cap-and-trade system could present the same two risks. Importing too many cheap foreign emission rights would both depress the domestic carbon price and create greater demand for offsets if they exist in the foreign system. Linking a

[3]See the Pembina Institute submission to Environment Canada, to be published in August of 2009. It will be available at http://climate.pembina.org.

Canadian cap-and-trade system to a foreign system with similar stringency makes sense, but linking to a weaker foreign system does not.

Oil Sands: The Elephant in the Room

Oil sands account for close to half (44 percent) of the projected increase in total Canadian emissions between 2006 and 2020 in a "business-as-usual" scenario, and virtually all (95 percent) of the projected increase in industrial emissions (Government of Canada 2008b, 42).[4] How oil sands are treated in a Canadian cap-and-trade system will therefore have a large bearing on the overall effectiveness and fairness of the system.

Most obviously, if the cap level is set so as to accommodate large growth in emissions from oil sands, then it will not likely be compatible with Canada reducing its emissions overall. On the other hand, if the cap is set at a level consistent with significant reductions in Canada's overall emissions, and if the carbon price is not kept artificially low by offsets or other means, then the carbon price will rise to a level reflecting the cost of large-scale carbon capture and storage in the oil sands. This is an expensive technology, costing as much as $100 or more per tonne of emissions reduced (Alberta Carbon Capture and Storage Development Council 2008, 22).

In other words, the rapid expansion of oil sands production and the high cost of reducing the associated emissions are responsible for driving up Canada's "marginal cost of abatement" of greenhouse gas emissions, which translates into the need for a high carbon price to reduce them.

The Pembina Institute believes that it is unfair for the oil sands sector to create a significantly higher carbon price and consequent costs for all other sectors. To prevent this, we believe that the use of carbon capture and storage, or a technology achieving equivalent emissions levels, should be mandatory for all new oil sands operations.[5] New oil sands operations without carbon capture should be viewed as unacceptable in the same way that new coal-fired electricity generation without carbon capture is now widely seen as unacceptable in light of what we know about climate change.

The oil sands industry may argue that it cannot afford to pay a carbon price of $100 per tonne or more, or the cost of carbon capture and storage. However, even if $100 were paid on each and every tonne emitted, this would be equivalent to only about $6 per barrel of oil produced from a state-of-the-art new oil sands operation.[6] A few dollars is a small proportion of the likely future gap between the world price of oil and production costs – the profit margin or

---

[4]We include electricity and heat generation in "industrial emissions".

[5]The Pembina Institute intends at a later time to specify more fully the types and vintages of the operations to which this would apply, as well as the required emission levels.

[6]For example, Shell's current Alberta oil sands operation generates about 65 kg of greenhouse gas emissions per barrel of bitumen, which is equivalent to less than 65 kg of emissions per barrel of synthetic crude oil (Shell Canada Limited 2007, 30).

"rent" shared between the Alberta government (via royalties) and the oil companies. There appears therefore to be no good reason to grant free emission rights or other shares of the carbon value to oil sands firms in a Canadian cap-and-trade system.

The Importance of a Broad-as-Practical Cap-and-Trade System

Canadian discussions about cap and trade often assume that the cap would only cover heavy industry (including electricity generation), which accounts for about half of Canada's emissions. However, there is a consensus among experts that a carbon price should, to the extent possible, cover the whole economy. For example, the National Round Table on the Environment and the Economy, the official advisory body to the Minister of the Environment, has repeatedly emphasized the importance of an economy-wide carbon price (NRTEE 2007, 51).

It is often believed that it would be impractical to cap the numerous small emitters – small businesses, buildings and vehicles – that make up most of the other half of Canada's greenhouse gas emissions. But this is not true. These emissions mostly come from burning fossil fuels, and the amount of emissions is directly proportional to the amount of fuel burned. They can therefore be capped by regulating fuel wholesalers, who would pass on the carbon price to consumers through the price of fuel. This is the approach being pursued in the United States, where the Waxman-Markey bill passed by the House of Representatives (H.R. 2454, July 6, 2009) would cap about 66 percent of U.S. emissions starting in 2012, rising to about 85 percent by 2016.

If the Government of Canada chooses to rely solely on cap and trade to put a price on emissions, then there are three key reasons why a Canadian cap-and-trade system needs to be similarly broad:

- We need to transform our whole energy system, not just heavy industry's half of it. It is true that households and small businesses face barriers that prevent them from responding efficiently to a price signal, such as a lack of information or access to financing. A carbon price must therefore be complemented by policies like efficiency regulations for buildings and vehicles, financial incentives for building retrofits, and investments in public transit. But a significant carbon price will create a real incentive to cut emissions. For example, $100 per tonne is equivalent to 24 cents per litre of gasoline.
- The space left in the atmosphere to dump greenhouse gases is a scarce resource that belongs to everyone. An emitter who is allowed to use that valuable resource free of charge is therefore being subsidized by everyone else. It is unfair for whole sectors of the economy to be subsidized in this way.
- Canadians need to have confidence that their federal government is on track to meet the national emissions targets that it has committed to. Canada is much more likely to meet its targets with a cap on 85 percent of our

emissions that with a cap on just 50 percent of emissions. If the government pursues a narrow cap-and-trade system, it will need to provide a very convincing explanation of how it will meet its emissions target for 2020 using policies targeting the other half of our emissions.

WILL THE VALUE OF CARBON BE DISTRIBUTED RATIONALLY AND FAIRLY?

In a standard cap-and-trade system, firms have to hold a government-issued allowance for every tonne they emit, and the cap is set by the total number of allowances. If Canada had a cap-and-trade system covering most of its emissions, roughly 600 million allowances would be issued each year. If the carbon price reached $100 per tonne, the allowances would have a total annual "carbon value" of $60 billion, or about $2,000 for every Canadian.

As noted above, the dwindling space left in the atmosphere for greenhouse gases is a resource that belongs to everyone. So the value of that resource – the carbon value – also belongs to everyone. This means that the carbon value should be distributed among citizens, firms and governments in the best interests of society as a whole. Since we could be talking about tens of billions of dollars every year, the way this is done is of the utmost importance.

Governments can distribute the carbon value in two forms – by handing out allowances free of charge, or by auctioning off allowances and handing out the proceeds in dollars. People tend to think of these two options quite differently, but they are financially equivalent, because allowances can be converted into dollars – on a carbon exchange or through a broker – at any time. If a firm receives carbon value in the form of free allowances, this is just as much a subsidy as if it receives carbon value in the form of dollars, as a grant or a tax break.

Satisfying the best interests of society as a whole means carbon value should be distributed to meet transparent policy objectives, not to reward those who are the best at lobbying or to seek narrow political advantage. *Any cap-and-trade proposal should be accompanied by a clear statement and justification of the uses to which the carbon value will be put.*

The Pembina Institute believes that the carbon value should be used for the following priority purposes:[7]

---

[7]Another commonly proposed use for carbon value, not included here, is to compensate firms that have not yet recouped their investment in high-emitting operations that will suffer a large reduction in profitability under a cap and trade system. However, the need to reduce greenhouse gas emissions has been clearly understood and enshrined in international law for 15 years now. Our view, therefore, is that any investor in new industrial operations since the mid-1990s should reasonably have anticipated the imposition of a carbon price within a few years, and does not merit compensation. Older operations are likely to have already paid off most or all of their initial investment.

- To protect specific industry sectors that would otherwise be expected to suffer substantial "carbon leakage" – a transfer of production to foreign competitors with similar emissions levels, which would be counterproductive because it would not reduce emissions. The potential for carbon leakage is, however, often exaggerated. Protection should be targeted at sectors that are independently shown to have a high likelihood of suffering a substantial impact.
- To protect low-income Canadians. Just as it is widely agreed that tax changes should not result in an increase in the cost of essential goods and services for those on low incomes, so any increase in energy prices resulting from a cap-and-trade system should be compensated for the same people.
- To ensure regional balance. Carbon value should be distributed so as to prevent excessive net financial flows from one region of Canada to another as a result of the cap-and-trade system. Excessive financial flows are unlikely to be seen as fair.
- To ensure adequate public spending on greenhouse gas reductions. An effective federal climate plan will need to include substantial public investment in areas where infrastructure is publicly owned (e.g., transit, electricity grids), where it is difficult to regulate (e.g., building retrofits), or where the carbon price may not initially be high enough to produce needed results (e.g., renewable electricity).
- To help developing countries combat climate change. There is a strong legal, moral and pragmatic case for rich countries to provide substantial financial support to assist with emission reductions in emerging economies and to help the most vulnerable cope with the impacts of climate change (Demerse 2009).

Carbon value that is left over after these objectives have been met could be distributed to Canadians through tax cuts (or public debt repayments) or equal per capita rebates.

It should be noted that governments can put a price on greenhouse gas emissions through alternative regulatory approaches that may further obscure the question of who will get the carbon value. Sometimes these approaches are described as "cap and trade" even though no allowances are issued. Alberta's greenhouse gas regulations are one example. They set a target for an industrial facility, at a level quite close to the facility's business-as-usual emissions, and require the owner to pay a carbon price for any tonnes emitted above that target. Because the owner pays nothing if the facility meets its target, this is analogous to a cap-and-trade system where the facility receives free allowances up to the level of its target. Most of the carbon value is therefore being distributed straight back to the industrial emitters. We will revisit Alberta's regulations below.

The Importance of Auctioning Emission Allowances

The best way to ensure transparency and accountability about who gets the carbon value is to have a cap-and-trade system in which 100 percent of the allowances are auctioned off by the government. In this case the full carbon value would be obtained as proceeds of the auction, and it could be put to legally mandated purposes that would naturally be subject to a thorough public debate – rather than be concealed in complex rules for allocating allowances or setting emissions targets.

One hundred percent auctioning, when combined with an avoidance of offsets, would have the benefit of resulting in the simplest possible cap-and-trade system. As noted earlier, we need a significant carbon price as quickly as we can manage, and a simple system could be brought into effect more quickly than a complex one. There is a risk that the complexity of allocating allowances free of charge could show up as equal complexity in the distribution of auction proceeds. But with 100 percent auctioning we expect that the transparency of the debate over the distribution of carbon value would produce a quicker and simpler outcome.

One hundred percent auctioning would also automatically reward early action – emission reductions achieved before a cap-and-trade system takes effect – because those who have already reduced their emissions would have fewer allowances to buy.

It is often suggested that to avoid an economic shock, most allowances should be allocated free of charge when a cap-and-trade system starts up. This argument is bogus, because the financial impact of the system depends only on the carbon price and the distribution of the carbon value. As long as a firm receives a given amount of carbon value, it makes no difference financially whether that value is distributed in the form of allowances or dollars.

Business associations' tendency to oppose auctioning of allowances may be due to a misunderstanding of this point. But another possible explanation is that the lack of transparency over the distribution of carbon value when allowances are handed out free of charge increases business lobbies' opportunities to secure a larger portion of that value for themselves.

It is true that many will be uncomfortable with the idea of entrusting governments with billions of dollars of extra annual revenue from the auctioning of allowances. Hence the importance of clearly specifying the uses of the revenue in legislation.

The Trouble with Intensity Targets

In the past, both Liberal and Conservative federal governments have proposed regulatory approaches that set greenhouse gas "emissions intensity" targets for all heavy industry; Alberta's greenhouse gas regulations do the same. Emissions intensity is the amount of emissions divided by the amount of production, e.g., emissions per barrel of oil. So if a firm has met an intensity target, it can emit extra emissions without penalty if it expands its production volume.

Since there is a chance that intensity targets could live on in future Canada-wide cap-and-trade proposals, we need to examine what they mean, particularly for distribution of the carbon value.

In a standard cap-and-trade system based on allowances, a firm that receives free allowances in proportion to its actual production level is effectively being given an intensity target, because if the firm expands its production, it can emit extra emissions free of charge.

Since free allowances are a financial subsidy, intensity targets are a subsidy for increased industrial production. If oil sands producers are given intensity targets, we are subsidizing an expansion of oil sands production. The subsidy reduces the effective carbon price for new production. If a firm has an intensity target set, say, 10 percent below the intensity of its new production, then it will pay only 10 percent of the carbon price for the extra emissions.

There is only one justification for using carbon value in this way to subsidize production in high-emitting industries: prevention of carbon leakage. If the carbon price would cause a transfer of production to foreign competitors with no reduction in emissions, then the only way to prevent it is indeed to reduce the effective carbon price through some form of production subsidy.

Intensity targets can, therefore, be justified only for sectors that can be independently shown to have a high likelihood of suffering substantial carbon leakage – and only as long as the targets are set at a level no more generous than needed to prevent the worst of that leakage.[8] Otherwise, intensity targets are an unfair diversion of carbon value to firms that meets no justifiable policy objective.

However, even if there is a justification for intensity targets in some sectors, the policy will be more transparent and simpler if this production subsidy is provided in dollars instead.

A more familiar objection to intensity targets is that they create uncertainty about the level of actual emissions, because if future industrial production is higher than expected, emissions will be higher too. This is an important objection if some (or all) sectors receive intensity targets and no adjustments are made to maintain a fixed overall cap on emissions. The resulting policy will provide certainty neither about the carbon price nor about the overall level of emissions.

However, use of intensity targets for certain sectors is no reason not to have a fixed overall cap. In some cap-and-trade systems allowances are set aside to be allocated free of charge to firms that build new facilities; this is equivalent to an intensity target because firms receive free allowances in response to increased production. But the overall cap level is preserved by adjusting the number of allowances available for other firms.

The Trouble with the Technology Fund Mechanism

Another feature of past federal regulatory proposals has been the option for firms to comply with emissions intensity targets by making payments at a fixed

[8]Border carbon adjustments (tariffs on the emissions associated with the production of imported goods and rebates of the carbon price paid on exported goods) are sometimes proposed as an alternative way to address carbon leakage. The Pembina Institute does not support border carbon adjustments for several reasons, notably the fact that they would likely be applied without reference to a sector's actual vulnerability to carbon leakage. For a more detailed discussion, see Demerse and Bramley (2008, 43–48).

carbon price into a "technology fund" with a mandate to invest in technologies to reduce greenhouse gas emissions. There is no certainty about when or by what amount emissions will be reduced as a result of these investments.

There are two major problems with the effectiveness of this mechanism. First, the proposed fixed carbon price has been far too low to represent an adequate direct incentive to reduce emissions. Second, like a lax offset system, the technology fund mechanism results in a form of fraud: the claim will be made that emissions have been reduced to the level specified in regulations, but a significant proportion of the "emission reductions" will, in fact, be investments in an unknown amount of future reductions occurring at an unknown date.

Technology funds also have major implications for the distribution of the carbon value, because they could hold a significant share of it. In the most recent federal proposals, the technology funds would be "at arms-length from government" (Government of Canada 2008a, 3) and include industry representatives as board members (Government of Canada 2007, 12). But carbon value belongs to society as a whole. Giving the corporate sector a seat on a body that distributes carbon value is like giving the corporate sector a seat at the cabinet table when the government decides how to spend tax revenues.

Alberta's greenhouse gas regulations allow unlimited payments into a technology fund as a compliance option. It was noted above that they distribute most of the carbon value straight back to the industrial emitters through the use of emissions intensity targets set at a level close to business-as-usual emissions. Of the remaining carbon value, most is paid into the technology fund. Since a majority of the fund's board members represent or have recently retired from heavy industry interests (Climate Change and Emissions Management Corporation 2009), distribution of this value is likely to be dominated by those interests (Pembina Institute 2009).

CONCLUSION

The world's governments will gather in December 2009 in Copenhagen to finalize the negotiation of a new global climate treaty. The Copenhagen deal will cover the critical years up to 2020 during which the industrialized world has to start achieving deep emissions cuts if the worst climate impacts are to be prevented.

Canada's Minister of the Environment, Jim Prentice, calls 2009 "truly… a pivotal year" for action on climate change, and has promised to "outline the full suite of policies that relate to all major sources of emissions… by the time we reach the international table at Copenhagen" (Environment Canada 2009). The centerpiece of those policies will be some form of Canada-wide cap-and-trade system. The proposed system will be a crucial determinant of Canada's credibility in Copenhagen – and a key test of whether the government now recognizes the scale and urgency of the threat of climate change.

Canada has a choice: show real leadership with a cap-and-trade system that puts an adequate price on emissions and distributes carbon value fairly and rationally – or muddle along with a system that fails to urgently transform our energy system and gives billions of dollars of carbon value to those who are best at lobbying. Time is short and we need to get it right now, not be forced back to the drawing board later.

Simplicity is a key feature of a strong, fair cap-and-trade system – and a recurring theme in this paper. Auctioning 100 percent of allowances and avoiding or minimizing offsets, intensity targets and the technology fund mechanism will increase the strength and clarity of the carbon price signal, speed the system's implementation and help ensure that it serves the public interest, not narrow private interests.

There is no need for Canada to imitate the weaknesses of the current U.S. approach to cap and trade, nor its complexity. And talk of the need to "balance" the environment and the economy is dangerously misleading: the projected human, ecological and financial costs of climate change far outweigh the costs of curbing it (Stern 2006). The world desperately needs leaders on climate change, and Canada is well equipped to be one. History will surely judge us harshly if we fail.

REFERENCES

Declaration of the Leaders of the Major Economies Forum on Energy and Climate.2009. At: http://www.g8italia2009.it/static/G8_Allegato/MEF_Declarationl.pdf.

Alberta Carbon Capture and Storage Development Council. 2008. *Accelerating Carbon Capture and Storage in Alberta, Interim Report*. Edmonton. At: http://www.energy.gov.ab.ca/Org/pdfs/CCSInterimRept.pdf.

Bataille, C., B. Dachis, and N. Rivers. 2009. *Pricing Greenhouse Gas Emissions: The Impact on Canada's Competitiveness*. Toronto: C.D. Howe Institute. At: http://www.cdhowe.org/pdf/commentary_280.pdf.

Climate Change and Emissions Management Corporation. 2009. "Climate Change and Emissions Management Corporation: Board Named to Manage the Use of Provincial Climate Change Funds". News release, July 13. At:: http://www.marketwire.com/press-release/Climate-Change-And-EmissionsManagement-Corporation-1016143.html.

David Suzuki Foundation, Pembina Institute, and WWF-Canada. 2009. *Comments to the Government of Ontario on the Development of a Cap-and-Trade System for Reducing Greenhouse Gas Emissions in Ontario*. Vancouver, Drayton Valley and Toronto. At:: http://climate.pembina.org/pub/1797.

Demerse, C. 2009. *Our Fair Share: Canada's Role in Supporting Global Climate Solutions*. Drayton Valley: The Pembina Institute. At:: http://climate.pembina.org/pub/1815.

Demerse, C. and M. Bramley. 2008. *Choosing Greenhouse Gas Emission Reduction Policies in Canada.* Drayton Valley: The Pembina Foundation. At: http://climate.pembina.org/pub/1720.

Environment Canada. 2009. "Notes for an Address by the Honourable Jim Prentice, P.C., Q.C., M.P., Minister of the Environment, on Canada's Climate Change Plan". Speech, June 4. At: http://www.ec.gc.ca/default.asp?lang=En&n=6F2DE1CA-1&news=400A4566DA85-4A0C-B9F4-BABE2DF555C7.

Government of Canada. 2007. *Regulatory Framework for Air Emissions*. Ottawa: Government of Canada. At: http://www.ecoaction.gc.ca/news-nouvelles/pdf/20070426-1-eng.pdf.

— 2008a. *Turning the Corner: Regulatory Framework for Industrial Greenhouse Gas Emissions*. Ottawa: Government of Canada. At: http://www.ec.gc.ca/doc/virage-corner/2008-03/pdf/571_eng.pdf.

— 2008b. *Turning the Corner: Detailed Emissions and Economic Modelling*. Ottawa: Government of Canada.

McKinsey and Company. 2009. *Pathways to a Low-Carbon Economy: Version 2 of the Global Greenhouse Gas Abatement Cost Curve*. At: http://www.mckinsey.com/clientservice/ccsi/pathways_low_carbon_economy.asp.

National Academies. 2009. *G8+5 Academies' Joint Statement: Climate Change and the Transformation of Energy Technologies for a Low Carbon Future*. Washington, DC. At: http://www.nationalacademies.org/includes/G8+5energy-climate09.pdf.

National Round Table on the Environment and the Economy (NRTEE). 2007. *Getting to 2050: Canada's Transition to a Low-Emission Future*. Ottawa. At: http://www.nrtee-trnee.com/eng/publications/getting-to-2050/Getting-to-2050low-res.pdf.

— 2009. *Achieving 2050: A Carbon Pricing Policy for Canada (Advisory Note)*. Ottawa. At: http://www.nrtee.ca/eng/publications/carbon-pricing/carbon-pricing-advisory-note/carbon-pricing-advisory-note-eng.pdf.

Pembina Institute. 2008. "Poll: Canadians Want Action on Global Warming Despite Economic Downturn". News release, December 2. At: http://climate.pembina.org/media-release/1736.

— 2009. "Polluters Exploit Alberta Government Loopholes to Increase Greenhouse Gas Pollution". News release, April 22. At: http://climate.pembina.org/media-release/1822.

Pembina Institute and David Suzuki Foundation. 2008. *Deep Reductions, Strong Growth*. Drayton Valley and Vancouver. At: http://climate.pembina.org/pub/1740.

Shell Canada Limited. 2007. *2006 Sustainable Development Report*. Calgary, AB: Shell Canada Limited. At: http://www.shell.com/static/ca-en/downloads/society_environment/sd06.pdf.

Stern, N. 2006. *Stern Review: The Economics of Climate Change*. London, UK: HM Treasury. At: http://www.hm-treasury.gov.uk/stern_review_report.htm.

Tamiotti, L. *et al.* 2009. *Trade and Climate Change*. Geneva, Switzerland: WTO Secretariat. At: http://www.wto.org/english/res_e/booksp_e/trade_climate_change_e.pdf.

University of New South Wales. 2007. *Bali Climate Declaration by Scientists*. Sydney, Australia. At:: http://www.ccrc.unsw.edu.au/news/2007/Bali.html.

U.S. Environmental Protection Agency. 2009. *EPA Analysis of the American Clean Energy and Security Act of 2009*. Washington, DC. At: http://www.epa.gov/climatechange/economics/pdfs/HR2454_Analysis.pdf.

Victor, D. 2009. *Global Warming Policy After Kyoto: Rethinking Engagement with Developing Countries*. Stanford, CA: Stanford University Program on Energy and Sustainable Development. At: http://iis-db.stanford.edu/pubs/22383/CAD_Working_Paper_82.pdf.

VII

Summing Up and a Look Ahead

14

Carbon Pricing: Policy and Politics

Peter Leslie

INTRODUCTION

All or almost all participants at the conference evidently believed that any significant reduction in emissions of greenhouse gases (GHGs) will require imposition, sooner or later, of a high "carbon price". While most – let's call them the "steep-slope" group – wanted early, shock-inducing action to increase the cost of burning fossil fuels, a few argued that emissions charges should initially be low, and should rise slowly over a period of years, mainly after 2020. This second, "gentle-slope" group urged that the focus of policy right now should be on stimulating the development of emissions-reducing technologies through grants and subsidies, and they argued that the prospect of later, steady increases in the price of GHG emissions would be enough to induce industry (both as producers and as consumers of fossil fuels) to invest in appropriate new technologies as they became available on a commercial scale. The gentle-slopers said that voters (taxpayers, personal consumers) would not tolerate sharp increases in energy prices, and that it is futile to pretend that they would. By contrast, the steep-slopers said that political leaders should exercise the necessary political will to see that prices rise quickly to a level that would bring about an early reduction in the use of fossil fuels. A high carbon price would ensure that energy itself would become less emissions-intensive, and would force the pace at which Canada and other countries would make the transition to a low-carbon economy, in which less energy would be needed per dollar of GDP. Participants from both groups had their say on the relative merits of taxation ("carbon taxes") and cap-and-trade (C&T) as alternative – or complementary? – ways of discouraging the use of fossil fuels and bringing about needed changes in energy production and consumption.

In these notes, I shall review – and comment on – what was said (and sometimes, *not* said) about various policy choices relating to carbon pricing and mitigating climate change. I shall do so under three main headings: *political feasibility*, *federal leadership and environmental federalism*, and *cross-border issues*.

Emissions-reducing policies may be compared on the basis of their ***political feasibility***. As all participants at the conference were aware, and some emphasized, climate change policy can have its hoped-for effects only after the passage of decades, whereas the electoral cycle is seldom longer than four years. One consequence is that even those politicians who are convinced of the importance of curbing emissions of greenhouse gases frequently opt for policies with hidden costs, or costs most people believe others – not they! – will have to bear. Cynically: transparency is the enemy of the good; subterfuge, its friend. Low costs are best; postponed costs are better than up-front costs; and costs few people can see or understand are still OK. But high, visible costs run up against the brick wall of political non-feasibility.

The issue of ***federal leadership and environmental federalism*** is a prominent one in Canadian debates over climate policy. The powers of Parliament and of the provincial legislatures relating to carbon taxes and the trading of emissions permits ("allocations", under C&T) are not clearly defined; jurisdictions overlap. Partly for this reason, controversy reigns over the policy roles of different governments, as well as over substantive issues: what policies, or mixture of policies, for Canada? The economic stakes here are high, with potential for very considerable interprovincial fiscal redistribution, as well as potentially heavy burdens on economic activity within the various regions. Because regional economic interests diverge as sharply as they do, and because much will depend on future U.S. decisions on climate change, the federal government's position on reducing emissions of GHGs has been hesitant and cautious. Federal leadership has been lacking. Thus the policy initiative has been seized by the provinces and territories, just as, in the United States, the states have been the innovators. A relevant factor in Canada is that, especially in Alberta, resentment over the National Energy Program of 1980 remains fresh, and historical memories increase the likelihood of future federal-provincial and interregional battles.

Cross-border issues relating to emissions-reduction policies are multiple, and cannot be comprehensively surveyed here. However, the following topics deserve special mention:

- First is the potential for "carbon leakage", if a particular jurisdiction adopts policies that pressure domestic industries and resident individuals to reduce their emissions. Several participants made the point that a ton of CO_2e released into the atmosphere has the same climate-effect, no matter what country (or who within it) does the emitting: there is no climate benefit, only a burden on the domestic economy, if GHG-emitting activities are merely shifted to another part of the planet. Not only, then, does carbon leakage – to the extent it occurs – nullify the benefits of an emissions-reducing policy; concerns over *potential* carbon leakage stand in the way of implementing policies that would impose substantial costs on industrial consumers. Minimizing carbon leakage, or convincing voters that carbon leakage is a minor problem, thus becomes an important factor in evaluating different policy approaches.

- More subtle than carbon leakage, and in general (not just at the conference) less satisfactorily discussed, are *diffusion effects*, the extent to which emissions-reducing policies within one jurisdiction may induce other jurisdictions to take initiatives of similar thrust. Achieving such diffusion effects is a goal that has been most commonly pursued through intergovernmental agreement, whether within a federation, at the regional or continental level, or globally. It would have been helpful if more attention had been paid at the conference to the question, whether certain approaches to reducing GHG emissions are more likely to bring about strong diffusion effects than others are. However, discussion was limited to the subject of moral leadership: the adoption of policies to mitigate climate change in the hope that other jurisdictions will be persuaded to follow along. It was this hope that animated the Kyoto negotiations (1997), and seems to underlie planning for the imminent (December 2009) UN Conference on Climate Change at Copenhagen. Another example of aimed-for diffusion effects is the California-led Western Climate Initiative (WCI), under which seven participating states and four Canadian provinces (British Columbia, Manitoba, Ontario, and Quebec) have said they will commit to a broad C&T scheme, partly in the hope that other states and provinces too will join, and also (quite explicitly) that the two federal governments will take over the scheme, and make it mandatory for all states and provinces in both countries.

- An approach much less reliant on moral leadership and negotiation among governments may be described as aiming to exercise *leverage* on other governments. This involves encumbering carbon-intensive imports through regulation and through border adjustments or "carbon tariffs". Exercising leverage is a "carrots and sticks" approach that relies on unilateral action, a tactic that may violate WTO rules and that is probably much more readily available to major players in the international economy than to countries such as Canada. However, it is an approach that might conceivably be adopted by Canada if it acted in concert with the United States or even with a larger grouping of rich industrial countries including the European Union. The aim would be to induce other jurisdictions to adopt policies similar to those implemented domestically. In this respect, it marks a clear departure from the moral leadership approach, which (as at Kyoto, and prospectively at Copenhagen) has aimed for voluntary cross-jurisdictional commitment to a negotiated sharing of responsibilities in the field of climate change. Adopting policies with potential for gaining leverage within the international system on climate-change issues would mark a quite dramatic departure from the Kyoto/Copenhagen approach. From a political point of view (thinking back to the subject of feasibility), a major selling point for such policies would be their effect in protecting the domestic economy from arguably unfair competition from states with lax environmental standards – a subject that figures quite prominently in the U.S. climate-change debate. Indeed, some steps have already been taken towards what critics call "environmental protectionism", both federally and at the state level, while in the European Union, some spokespersons for the European Commission have talked of implementing a carbon tariff, applying to imports from states that refuse to curb GHG emissions.
- The two approaches (moral leadership, and gaining leverage or aiming to do so) are perhaps less distinct from each other than the above comments imply. Indeed, globally, it may be that efforts to mitigate climate change are already moving into a post-Kyoto phase less reliant on fully voluntary (and faithfully implemented) commitments by sovereign states. International

negotiations may come increasingly to be driven by unilateral initiatives of major trading nations, seeking to link the global trade regime with a set of global policies to limit and reduce GHG emissions. That was not a matter on the conference agenda. However, as earlier implied, an issue that might fruitfully have been highlighted in conference discussion is that certain types of climate policy – different ways of "pricing carbon", and also command-and-control approaches to reducing GHG emissions – will have broader external ramifications than others, and in particular may ultimately form the basis of a strategy to write principles of climate policy into the WTO rule-book.

Some Recent Events

In reviewing issues that arose at the conference, and that are of importance in Canadian, North American, and global climate-change policy, it will be necessary to take into account some post-conference events. Immediately after the American presidential election, Prime Minister Harper began touting a Canada-U.S. agreement on climate change, linking a joint approach to emissions-reduction with the United States' objective of ensuring its energy security. It appears that a key Harper objective was (and remains) to support bitumen sands[1] development, threatened by Californian legislation and by the U.S. *Energy Independence and Security Act* (December 2007)[2] – and now, perhaps, by the Waxman-Markey bill that has passed the House of Representatives but will ultimately have to be reconciled with whatever bill (if any) is passed by the Senate. Waxman-Markey, a draft *American Clean Energy and Security Act*, reflects in part President Obama's commitment to a stricter climate policy, but whether the Senate will endorse that commitment remains unknown. Also unknown, of course, are whether the legislation that eventually emerges from Congress will be effectively implemented, and how new federal policy will affect the role and policies of the states. Of particular concern to Canada will be the fate of the WCI, and whether it will be, in effect, taken over by the American federal government.

Not only evolving American policy, but also the global situation, will profoundly affect Canadian decision-making on climate-change policy, including the pricing of carbon. Conceivably some consensus will emerge at the Copenhagen conference, but that would be an unexpected triumph. What is clear is that "Copenhagen" will be the site of a good deal of international jockeying-for-position, and Canada will have to decide what role to play, or what to aim for, at the conference. It is not conceivable that Canada would take a position fundamentally at variance with the American one at a conference where the

[1]Detractors, correctly noting the hard, sticky properties of these sands, say "tar sands"; pro-development interests say "oil sands". A precise, neutral term – even if it does not roll smoothly off the tongue – would be "bitumen sands". That's what they are.

[2]*Page* notes that this act "forbids the American government, its agencies, or the armed forces from purchasing high-carbon fuel products like oil sands oil".

most significant cleavage is likely to be between the rich industrial countries (especially the United States and the European Union), and a group of poorer countries, the non-OECD states, of which the largest are China, India, Indonesia, Brazil, and Russia ("BRIIC"). In this context, it is conceivable (though seemingly improbable) that Canada will aspire to play a role in developing a rich-states position, whether with a view to forging an eventual global consensus, or in preparation for unilateral action by the rich that would put pressure on the developing world to adapt and conform – definitely a high-risk strategy. The objective here would be exactly what was referred to above: to link the global trade regime with a set of global policies to limit and reduce GHG emissions.

Mr. Harper's policy pronouncements indicate that he intends to develop a Canadian strategy for the further development of Canadian energy resources, along with technological innovation and a set of controls on GHG emissions. His government proposes a combination of C&T and mandated technologies such as carbon capture and storage (CCS). A more recent development of potential importance, as regards future Canadian policy, is Mr. Ignatieff's replacement of Mr. Dion as Liberal leader. One consequence of the change of Liberal leadership is that the party's policy on climate change – and potentially, if there is a change of government, the policy of the government of Canada – may be rethought, although the Ignatieff style is far too guarded for one to be confident that new policy directions will be put forward. About the most one can say is that under a Liberal government, whatever willingness there may be to exercise a form of leadership within the federation, would inevitably reflect in part the Liberals' distinctive regional electoral base.

Alternative Approaches to Emissions Reduction

The three subjects addressed in these notes – political feasibility, federal leadership and environmental federalism, and cross-border issues – require clarity regarding alternative policy choices, a refinement of the taxes/C&T dichotomy. Looking back over the conference presentations, it seems there were five broad options either explicitly proposed (*Tom Courchene and John Allan,*[3] and *Rick Hyndman*) or simply referred to, sometimes only implicitly. Three of the five are tax options, and the other two are for C&T (a "hard" variant, and a "soft" one). Each of the five options is actually a cluster of specific schemes. A summary of the options – for greater precision, see the various papers in this volume – is as follows:

- A simple carbon tax, imposed at a flat rate on energy consumption in proportion to the emissions of greenhouse gases (measured in CO_2 equivalent, CO_2e) that result. The B.C. carbon tax is an example; a more complex variant is the Mintz/Olewiler proposal for extending the existing

[3]All uncited references in this paper are to other papers in this volume, and authors' names are italicized.

federal gasoline excise duty in Canada to all other fossil fuels;[4] their proposal may have been the inspiration for Stéfane Dion's "Green Shift" carbon tax. Under a simple carbon tax, thermal-electric producers (for example) will pay more for their hydrocarbon inputs, while "clean" electricity (hydro, nuclear, wind, solar, biomass) will be exempt, although hydrocarbon fuels consumed in constructing new installations will have been taxed. Oil and gas firms will be taxed according to the hydrocarbon fuels used up in extraction and refining; the tax will affect netbacks (a burden, for example, on the bitumen sands industry). Firms that purchase hydrocarbon fuels (e.g., cement factories, newsprint producers, owners of high-rises, airlines, etc.) will be charged according to resulting CO_2e emissions, and whenever they can will pass on those costs to their customers. Individual consumers (e.g., of gasoline, heating oil, and natural gas) will also pay the tax, though it is likely to be levied on distributors. Whatever the modalities, this is a "polluter pays" tax, which discourages consumption of high-emissions goods or services such as air travel, and (unless rebated at the border, as Mintz and Olewiler propose) imposes a burden on exporters proportional to the energy-intensity of what they produce. The aim of a simple carbon tax is to raise the cost of emissions across the board, and thus to discourage the consumption of fossil fuels within the domestic economy.

- A carbon tax imposed on industrial emitters, but only above a certain intensity standard (an important feature of the *Hyndman* proposal[5]). Under this option there would be a relatively high marginal rate on emissions but a low average rate – this combination of rates would provide a strong incentive for firms to invest in low-carbon technologies, while avoiding a heavy burden on the industry, and perhaps any burden at all on relatively carbon-efficient firms. Implementation would require, for each industry, estimating average emissions per unit of output (say, a KWH of electricity,

---

[4]Jack Mintz and Nancy Olewiler (2008): *A simple approach for bettering the environment and the economy: restructuring the federal fuel excise tax*, prepared for the Sustainable Prosperity Initiative (University of Ottawa, Institute of the Environment). This study is summarized in *Courchene/Allan*, who note that Mintz and Olewiler propose a trade-neutral form of carbon tax, with border tax adjustments (taxes on imports, rebates on exports). This feature makes their proposal rather more complicated than what I here describe as "a simple carbon tax". It is not clear to me how their proposed border tax adjustments can be implemented, other than by adopting a carbon-added tax, as *Courchene/Allan* have proposed.

[5]Hyndman proposes a simple carbon tax at a low rate – \$5/t (that is, \$5 per tonne) at first – on individual and most corporate consumers, but in the "energy-intensive, trade-exposed" (EITE) sector a tax at differential rates, nil below the set standard of energy efficiency, but considerably higher above it (Hyndman proposes \$20/t at first). The differential rates would apply to facilities in export-oriented industries, or in energy-intensive industries selling into the domestic market and exposed to import competition. Under the initial rates he proposes, firms in the EITE sector with average emissions per unit of output would pay \$3/t on total emissions. For more on the Hyndman proposal, see note 11.

or a ton of cement or newsprint, or a barrel of synthetic crude), and setting an efficiency standard somewhat below the average (say, at 85%). Emissions above the standard would be taxed, while a firm that kept its emissions below the standard might be entitled to a rebate or subsidy. The low average rate would protect the competitiveness of the industry internationally and would ensure that there was not much tax to pass on to consumers, thus reducing political opposition to the tax. This is definitely a gentle-slope proposal, but none the less establishes a high *marginal* tax rate and, therefore, a carbon price sufficiently high to provide an incentive for emission reductions. Implicitly: no taxes on imports, no rebates for exports (no border adjustments).

- A carbon-added tax and tariff (CATT – the *Courchene/Allan* proposal). This is a tax analogous to a value-added tax such as Canada's GST. For exported goods, the cumulative CATT paid by domestic producers is refunded at the border; for imports, the CATT is imposed on the presumed "carbon content", including shipping. As *Courchene/Allan* emphasize, a CATT would be trade-neutral.
- "Hard" C&T: for those firms, facilities, or sectors to which the scheme applies – first, a cap on total emissions (an allowance is to be required for every metric ton of CO_2e emitted); second, distribution of allowances or credits, whether free or by auction, with the possibility of purchasing additional allowances on an emissions market ("trade"); and third, severe penalties for exceeding emissions allowances, whether obtained without charge, or purchased on the market. This is the form of C&T adopted by the European Union, and proposed under the Western Climate Initiative (WCI) as well as under Waxman-Markey – all of these have or would have a combination of free allocation and auction of allowances. It should be noted that even a "hard" C&T will not necessarily curb the emission of GHGs very much, if the scheme is relatively narrow in its coverage (say, if half or more of total emissions are not regulated), if the aggregate cap is high, and/or if emitters can evade the cap through escape mechanisms such as a generous offsets scheme.[6] Also worth noting: the logics behind "hard" C&T and a carbon tax of any kind are antithetical. Under "hard" C&T, *emissions are targeted directly*, and the price of "carbon" is derivative: it emerges from the marketplace, as allocations are traded. The carbon price is subject to sharp swings, and may be influenced by speculators, certainly by prevailing moods and assumptions, and by economic conditions. By

[6]Offsets are credits, or additional allowances, granted to an emitter as a form of reward for taking some action considered likely to reduce overall emissions in the future, either in the domestic economy or abroad. (*Bramley* cites, as common examples, "tillage practices that store more carbon in agricultural soils, and hydroelectric projects in developing countries.") Of course rules or guidelines for offsets must be legislated. The effect of such provisions may be dramatic. For example, under Waxman-Markey, the total cap is to be set initially at about 4.6 billion metric tons of CO_2e, but offsets may be granted up to 2 billion metric tons annually. For an estimate of the effects of this provision, see *Bramley,* as referred to below.

contrast, under carbon taxes, *emissions are targeted only indirectly*, and with imprecision. It's the *price* of emissions that is fixed, not the quantity; the implications of that price for the level of emissions that will actually come about can only be estimated through economic modelling. Of course, should the outcome differ markedly from the policy-maker's target, the tax rate could be appropriately adjusted, although this would have the disadvantage of increasing uncertainty for the industry.

- "Soft" C&T: for firms to which the scheme applies – first, a cap on "intensities" (CO_2e per unit of output) with no overall limit; second, possibility of purchasing additional allowances on the market; and third, relatively slight penalties for exceeding emissions allowances. This is the variant of C&T already adopted by Alberta and seemingly intended by the Canadian federal government; in Alberta, firms may acquire allowances from the regulator at a price of $15, permitting them to emit one metric ton of CO_2e additional to their cap or limit (in effect, an offset); a consequence of this arrangement is that the market price of allowances can never exceed $15/t. Revenues from the sale of allowances are paid into an innovation fund, *and are analogous to a tax on marginal production (i.e., not hugely different from the Hyndman proposal, as applying to the EITE sector – see note 5, above).*

Each of these five options leaves room for substantial variation. For the three tax options, a decision must be made regarding coverage (the emissions to be subject to tax), and rates must be set, both at inception and over time. For the two C&T options, various refinements lie open to policy-makers. Relevant issues include, as *Purchase* has pointed out: how stringent the caps are to be; how wide the coverage is to be; what rules for emissions trading are to be imposed; whether allowances are to be auctioned or distributed free (and if so on what basis); what types of offsets are to be allowed; and what administrative arrangements are to be made for reporting, monitoring, and enforcement. The practical effect of the C&T options, in terms of the carbon price they will bring about, and the reductions in GHG emissions they will achieve, will depend heavily on decisions on these and other matters.

Notwithstanding the range of variation within each of the options, the differences among them are great enough to permit comparisons as regard political feasibility, federal leadership and environmental federalism, and cross-border issues. Such comparisons are essential to any evaluation of climate-change policies at the present juncture: a time when the United States is, for the first time, coming seriously to grips with climate change, when there is prospect of sharply increasing emissions-levels in developing countries (the BRIIC and others), and when the world is groping towards a climate-change regime analogous to – and perhaps linked with – the international trade regime. These issues are highlighted by the imminent Copenhagen conference, but are unlikely to be resolved there. All countries are, willy-nilly, in the climate-change business for the long haul. So indeed are *homo sapiens* and, in general, other species.

POLITICAL FEASIBILITY

The most convenient way to broach this subject is to refer to *Rabe's* suggestion, "The relationship between the 'economic desirability' and 'political feasibility' of climate policy options may be nearly inverse." His survey of American state-level policies reveals a distinct political preference for high-cost, command-and-control policies (regulation of products and production processes, and mandating the use of renewable energy sources in electricity production); next come C&T schemes; and finally – showing the highest level of political resistance – carbon taxes, the near-universal recommendation of economists on grounds of efficiency. Similarly, *Purchase*: "Carbon taxes versus cap and trade appear to be the political equivalent of Beta versus VHS in the video technology wars of the past [where the inferior technology won out, over the one that got an early start]." The two authors explain this illogicality by noting the public's resistance to any form of taxes or charges collected directly from consumers, and widespread skepticism that even with rebates to low-income individuals, "few seem to believe they will not be net payers" (*Purchase*). By contrast, regulation is seen as affecting producers, not consumers; and C&T (which is based on regulation) is simply not understood, but presumed to be a "market solution" that leaves consumers as bystanders, and ultimately as beneficiaries of the search for efficiencies in production processes. The costs, if any, will be years down the road. For *Rabe* and *Purchase*, the less transparent the policies, the easier they are for politicians to embrace.

A Steep-Slope Position

The papers by *Bramley* and *Nic Rivers* shift the discussion away from policy instruments and towards a single variable: carbon price. For these authors, it really does not matter whether a desired carbon price is brought about through C&T or through carbon taxes. The reason for this is clearly laid out by *Bramley*, who reasons that C&T can be designed in such a way as to result in a carbon price at more or less any chosen level. Of course economic modelling will be necessary to estimate, for any given C&T scheme, what the market price of emissions allowances (permits or credits) will turn out to be: low supply, together with strong enforcement, results in a high price. This, for *Bramley* (and also for *Rivers*) is good, because (*Bramley*):

> The carbon price will become a factor in all decisions affecting emissions taken by firms and households. The higher the price, the stronger the incentive to switch from high-emission options to low-emission options. The first fundamental question that should therefore be asked of any cap-and-trade system is: *is the carbon price likely to be high enough to adequately reduce Canada's emissions?*

What price is "high enough"? *Bramley, Chris Green, Rivers,* and *Hyndman* all cite studies indicating that the required level of emissions charges is upwards of \$100/t CO_2e by 2020, if Canada is to meet federally-declared targets for emissions reductions (by 2020, 2.6 percent below the 1990 level, or 20 percent below the 2006 level). The $100 minimum – equivalent to 24 cents per litre on gasoline, which compares with the current federal excise tax of 10 cents per litre – represents a scientific consensus. Thus Canada's National Round Table on the Environment and the Economy (NRTEE) estimates that a price in the range of $100 to $150 will be required by 2020. By comparison, the Pembina Institute and the David Suzuki Foundation (cited by *Bramley*) propose a figure of $50 in 2010, rising to $200 by 2020. Moreover, according to *Rivers*, some provincial targets for emissions reductions would seemingly require a CO_2e price substantially upwards of $200/t. He adds, "The carbon price that is estimated here to be required to meet commitments made by the provincial and federal governments would ... dwarf the carbon pricing policies already adopted in Europe, the current leader in climate change policy." Even more dramatically, *Hyndman* and *Chris Green* both note that an emissions charge of $115/t CO_2e – as envisioned by the NRTEE – represents a tax on coal of $329/t, compared to current market coal prices in the range of $15 - $110/t. It appears, then, that the necessary tax on coal, if emission reduction targets are to be met (assuming the accuracy of the figures cited above), would increase its price by a factor of at least 3 – or even 30 or more. One must candidly admit, that increases anywhere in that range are unthinkable, given that some provinces rely mainly on coal for the generation of electricity.[7]

No serious public debate on climate policy can ignore either the political or the economic implications of figures such as these. Unfortunately the issue was not joined at the conference, although two quite dramatically different positions did emerge from individual presentations. *Bramley* and (a little less categorically) *Rivers*[8] articulated what I earlier called the steep-slope position, which appeared to be shared by most participants at the conference; by contrast, *Chris Green*, *Purchase*, and *Hyndman* argued strongly for some kind of gentle-slope alternative.

Bramley presented a straightforwardly normative paper that set out criteria by which any Canada-wide C&T scheme should be evaluated. He argued for what I have called here a "hard" variant of C&T, one that would "transform our energy system". "Time is short", he affirmed, "and we need to get it right now,

---

[7]International figures on the use of coal for electricity generation are dramatic, and illustrate the difficulty of achieving substantial reductions in emissions on a global scale. Thus, *Purchase* reminds us, worldwide coal-fired generation is very high: about 50 percent in the United States (and up to 96 percent in individual states), and 70 percent in China.

[8]Rivers highlighted the inconsistency between targets set or commitments made by Canadian governments, and the carbon-pricing policies adopted or promised. But he also made the point that the cost of meeting commitments depends heavily on how fast emissions are to fall. He adds: "It is certainly valid to debate whether such dramatic targets should be met.... Meeting the targets will require especially stringent policies because of the short time available, and could impose relatively high costs on the Canadian economy. Environmental policies should only be pursued to the point where, at the margin, the cost of the policy matches the benefit (in terms of improvement in environmental outcome) due to policy implementation."

not be forced back to the drawing board later". Not only, he said, is a high emissions price needed to curb consumption of fossil fuels, it also is essential on grounds of fairness, since:

> The space left in the atmosphere to dump greenhouse gases is a scarce resource that belongs to everyone. An emitter who is allowed to use that valuable resource free of charge is therefore being subsidized by everyone else. It is unfair for whole sectors of the economy to be subsidized in this way.

Reflecting such arguments, *Bramley* proposed an absolute (not intensities-based) cap, broad coverage, auctioning 100 percent of allowances, effective penalties for exceeding allowances (the paper argues that "escape mechanisms" such as making investments in a technology fund, as Alberta rules allow, do not qualify), and avoiding or minimizing offsets. For him, the details of any C&T scheme were crucial, because an apparently rigorous C&T scheme may be much more lax than first appears. For example, under the Waxman-Markey bill, which incorporates very loose rules for offsets, the estimated cost of emissions allowances is only \$16/t CO_2e in 2020. The real (as opposed to the nominal) cap on emissions is correspondingly high, and the market-set price of allowances is expected, in consequence, to be low. Taking Waxman-Markey as his reference point for U.S. policy, Bramley urged that Canada should adopt "an adequate price on emissions", and affirmed, "There is no need for Canada to imitate the weaknesses of the current U.S. approach to cap and trade…." He added: "And talk of the need to 'balance' the environment and the economy is dangerously misleading: the projected human, ecological and financial costs of climate change far outweigh the costs of curbing it."

Political Feasibility and Technological Innovation

The papers by *Chris Green*, *Purchase*, and *Hyndman* were based on an entirely different logic. For them, there is no point arguing about what sort of policy might be, under different political circumstances, desirable. On the contrary, all insisted that no policy can be considered if the voters will not tolerate it; the essence of *Hyndman's* position is that "… much needed action has been stalled by the advocacy of policies more costly than the public will support. Meanwhile, time continues to pass as governments debate how to commit themselves to action…." Whereas *Bramley* argued for a carbon price that will drive the transformation of the Canadian energy system over a short time frame, these authors all dismissed a carbon price of \$100/t or more as politically infeasible. The case was put most forcefully by *Purchase*, who argued that under a carbon price high enough to seriously diminish consumption of fossil fuels, "the future must be considered highly uncertain, if not bleak, for those provinces producing [them]…. This is particularly the case for Alberta's oil sands." He pointed also to negative consequences for other parts of the resource extraction and processing sector, "concentrated in northern, coastal, and rural ridings", to how the auto industry's problems (and thus those of south-central Ontario, "already undergoing profound economic dislocation"), and to how "… suburban ridings would suffer under any aggressive carbon pricing regime" as commuting costs rise and property values decline. He dryly commented, "Few politicians would welcome the opportunity to explain why that is necessary to save humanity"; and he added the further observation, "The Conservatives have their power base in the most 'at risk' part[s] of the country." In fact, each of the other federal parties too would have its reasons to fear the political consequences of sharp increases in energy prices.

Thus, for *Chris Green, Hyndman,* and *Purchase*, the mitigation of climate change ought not to be based on setting emissions targets and attempting to enforce them through C&T. Of the three, *Purchase* was the least specific about what might be done; he commented that Canada's access to U.S. markets will be dependent on the federal government's following a carbon pricing regime "commensurate" with that eventually adopted by Congress (for more on this, see the section on "cross-border issues", below). He implied that the threat of American import restrictions will change the electoral calculus in Canada, inducing governments to be no less strict in emissions-reduction than the United States, while conversely, Canadian governments will be unwilling or politically unable to burden domestic firms more heavily than their competitors to the south.

For *Chris Green*, and for *Hyndman*, a high carbon price – at least during the next few years – is not even intrinsically desirable. Both proposed a low-yield carbon tax, the revenues being dedicated to supporting the development of low-emissions industrial processes and energy-saving products. Indeed, for both, the centerpiece of climate policy must be technological development, a strategy that will facilitate (and indeed is a precondition for) an eventual transition to a low-carbon economy. This is so in Canada; it is so globally. The main supporting argument is that, as *Chris Green* put it, a high carbon price, "however desirable [in the long term], is not sufficient to stabilize climate … without new, scalable, and breakthrough technologies". He asserted that, contrary to what the conventional wisdom – as voiced, for example, by the International Panel on Climate Change (IPCC) – maintains, the required technologies are not already "on the shelf", or merely requiring commercialization. On the contrary, he argued, it will take many years to develop the necessary technologies. Moreover, relying on market incentives to induce private firms to do the necessary research is a misguided approach, because the prospects of a firm's achieving good returns on its investment are poor. Technological break-throughs are far from being assured, and may be appropriable only in part (in which case free riding is inevitable). These are arguments for heavy public-sector involvement, especially when the technology problem is a global one.

To such arguments *Hyndman* added that actually achieving the emissions-reduction targets that governments in Canada have set – that is, squeezing the consumption of fossil fuels under existing technologies – could only result in a significant drop in levels of output (GDP). Not only would voters rebel against such a policy, it would be retrograde because of the economic pain it would entail – and for negligible benefit in terms of climate change, given Canada's size.

The Regional Politics of Climate Change

A strategy that gives top priority to technological innovation and infrastructure development implies a go-easy, gentle-slope approach to reducing emissions over the short term (say, to 2020), while trying to make up later for the slow start. Of course, for politicians, this is an attractive strategy because no one now in office will have to answer either for failure, or for such economic pain as the strategy may entail. Should one be cynical about this? Perhaps not, because as *Rivers* too acknowledges, the costs entailed by policies to bring about a high carbon price cannot be disregarded. Such costs cannot usefully be expressed merely as a percentage of GDP; what matters more is the forced pace of economic restructuring, necessarily affecting certain regions more than others, and the burden imposed not just on taxpayers and consumers in the aggregate, but especially on low-income families, rural dwellers and other specific groups.

No wonder, one might reasonably conclude, the politics of climate change are regional, and that they drive electoral wedges between groups according to income level, occupation, social values or ideology, and urban-suburban-rural residency! Thus it is banal – but still important – to make the point that the politics of reconciling divergent interests and preferences relating to emissions-reduction are tremendously complex. What is especially important as regards the subject of *environmental federalism*, is that the salience of the regional dimension in controversies over climate policy creates a minefield for federal politicians. The challenges that stand in the way of coordinating the policies of different governments, and the obstacles that confront any Canadian federal government that may seek to exercise leadership in this area, may well be insurmountable.

FEDERAL LEADERSHIP AND ENVIRONMENTAL FEDERALISM

As *Page* notes, jurisdictional issues have historically played a key role in environmental policy. The federal government cannot, simply by entering into international commitments (such as Kyoto), extend the range of its powers, and has sometimes resorted to criminal penalties – the criminal law being one of the enumerated and exclusive federal powers – to achieve regulatory objectives. On the other hand, the provinces have extensive regulatory powers over most types of contract ("property and civil rights"); they gained substantial powers over resource management under the Constitution Act, 1982 (the "resources" clause); and they are owners of resources located on crown lands. Thus, as regards the environment, both federal and provincial powers are relevant, and they overlap substantially.

None the less, it seems clear that the Canadian federal government (more formally, the Government of Canada, GOC) has constitutional authority to enter into international commitments to reduce GHG emissions, and that Parliament is similarly empowered to enact measures that will raise the price of carbon, the intent being to meet such commitments. This was the conclusion of *Chalifour* as regards taxing powers, and of *Elgie* as regards emissions trading (implementation of C&T). Both affirmed that some caution would be necessary in framing the necessary legislation, but they held that so long as due attention was paid to the words through which legislative powers are conferred under the constitution, the powers of Parliament are adequate to the task of setting a carbon price. Although the Chalifour and Elgie papers did not review constitutional provisions relating to the setting of product standards, infrastructure development, or support for technological innovation, two things about these subjects seem clear: first, that federal powers in these areas are extensive, and second, that they are not exclusive (provinces may also enact legislation in these same areas, including legislation authorizing the expenditure of monies). Finally, given WTO rules, which *Andrew Green* treated as analogous to constitutional limitations on legislative powers, the extent of the GOC's capacity to implement a scheme of border adjustments is somewhat murky; however, one might add, there is no doubt at all that, to the extent the relevant powers inhere in any Canadian government, those powers are federal rather than provincial.

It appears, then – and no one at the conference challenged what *Chalifour, Elgie,* and *Andrew Green* said on these matters – that the GOC is constitutionally equipped to exercise a leadership role in climate policy, including as regards the pricing of carbon. While relatively few participants explicitly referred to "federal leadership", I highlight the term here because I think the supposed desirability of federal leadership was a major underlying theme of the conference. In any case, most participants appeared to have opinions about the role appropriate to "Ottawa", and to support the idea of a larger federal role in climate policy than has been evident so far. Up to now, as the papers by *Harrison* and *Courchene/Allan* emphasize, such leadership as has occurred on carbon pricing has been provincial. Some of the conference participants expressed discomfort about this, because as long as the provinces are left with primary responsibility, some clearly will choose to do little or nothing. In that situation, those who would like Canada to take a strong stand on climate change have wanted the GOC to take the lead, establishing a "floor" on emissions-reduction that all provinces will have to adhere to, though some may exceed. In general, it has been argued that if Canada is to play its part internationally in reducing GHG emissions – or indeed to take any other approach to mitigating climate change – Ottawa cannot simply stand on the sidelines. I do not think that at the conference there were any dissenters from this rather limited proposition, but equally, it was obvious that consensus simply did not exist on two key questions. These are: first, is the federal government *politically* (as opposed to constitutionally) equipped to play a leadership role, and second, assuming it has at least some political capacity to act, what should it be attempting to do? Taking these two questions together, there seemed to emerge three broad normative positions:

- "just do it!"
- follow the U.S. lead (meaning: until Congress passes legislation on carbon-pricing, no policy of significance will emerge in Canada, at least not from Ottawa)

- exercise leadership, broadening the agenda from "pricing carbon" to encompass a new "environmental federalism".

"Just Do It!"

This was explicitly the *Bramley* position, seemingly that of *Rivers*, and perhaps that of *Harrison.* All three took note of, or implicitly endorsed, the consensus (or strong-majority) view of scientists, that climate change is occurring at a rate that exceeds earlier worst-case scenarios, that GHGs are a major contributor, and that the consequences of spiraling-higher emissions will probably be disastrous for many species, including human beings. A corollary is that even if catastrophe can be avoided (which many doubt), the costs of acting now to reduce GHG emissions are much smaller than seeking to adapt or compensate later on. *Bramley* explicitly made this point. One might add (not at all originally) that even if the scientific consensus is unconvincing to some, the magnitude of potential or predicted change – on a par with earlier geological/evolutionary epochal changes, ones that have occurred over millions of years – is such that observance of the precautionary principle becomes a moral absolute. If it should turn out (long after we are all dead) that acting to mitigate climate change was futile or unnecessary, then all that will have been lost is a few percentage points of GDP. But conversely, if it should turn out that the scientists (in the main) got it right, but nothing was done, the consequences will be immeasurable, even inconceivable. Prudence demands that one pay attention not only to the cost of acting, but to the costs incurred by refusing, in this early twenty-first century, to face a few inconvenient truths. From such a viewpoint, the only thing that really matters is to transform the global economy, and perhaps human society, in ways that respect the limited carrying capacity of the planet and its atmosphere. So the message to politicians cannot be other than, "Just do it!" Do locally, or nationally, what can be done, and work towards extending domestic accomplishments globally.

Follow the U.S. Lead

This is a "waiting for Waxman-Markey" stance. It is not so much a normative position, as an empirical judgement that until Congress passes legislation on climate change, political considerations will force the GOC to do essentially nothing. This thesis was developed most fully and most forcefully by *Purchase*, and reflects his analysis (reported in the previous section) of the political obstacles facing any Canadian federal government on the climate-change issue. His conclusion, after reviewing the electoral pitfalls awaiting any federal government that chose to act decisively on emissions-reduction, was, "Canadian federal politics are simply not conducive to national leadership on the issue of carbon pricing."

He implies exactly the same conclusion for the United States, noting how heavily the Democrats – and, before them, the Republicans – depend on electoral college votes from swing states in which most electricity production is coal-fired, and thus would be vulnerable to any action that substantially increases the price of coal. The astuteness of this analysis has been borne out by the legislative history of Waxman-Markey, an as-yet unfinished story. *Rabe*, more inclined to see in the policies of the recent Bush administration an ideological aversion to environmental activism, simply records the unwillingness, pre-Obama, to do anything of significance on climate change. But that has not been so, as he demonstrates and as *Harrison* also notes, at the state level. Indeed, referring to "a state-centric American climate policy", *Rabe* offered a detailed and informative review of initiatives and policies at the state level, and summarized his findings by noting, "those policies that tend to maintain the strongest base of support from policy analysts appear to have the greatest difficulty of being adopted by state legislators and governors".

Courchene/Allan, likewise, highlighted the relative passivity or inaction at the federal level, in both countries, and took note of the leadership role played by some of the states/provinces. In Canada, the leaders have been British Columbia (with its carbon tax), Ontario, Quebec, and Manitoba. All four, which together (as *Courchene/Allan* have emphasized) make up more than half the Canadian population, have indicated their intention to participate in the California-led Western Climate Initiative. However, these provinces' hope that Ottawa will thereby be goaded into taking a more active role in emissions-reduction faces implacable opposition from oil-producing provinces: Alberta, Saskatchewan, and Newfoundland and Labrador. It may well be also that the Maritime provinces and the three territories, all of which are heavily dependent on thermal generation of electricity (as also are Alberta and Saskatchewan), will continue to refuse to go along with any policy that would significantly raise the price of coal and other fossil fuels. Quite simply, the "green" provinces – the WCI adherents – are the producers of hydro-electricity, although the case of Ontario is distinctive and somewhat puzzling. As *Purchase* has emphasized, Ontario relies heavily on thermal generation (nuclear, coal, gas) for its electricity supplies, and its industrial economy is vulnerable to any increase in the cost of energy. For him, then, the divergence of regional interests as regards the pricing of carbon is conclusive: Ottawa will not move and cannot move, unless and until the United States does.

Courchene/Allan say more or less the same thing, but less emphatically. They seemingly believe that ambitious programs of emissions-reduction, if adequately supported by the public, will eclipse the weaker position taken by other governments in the system. Thus, they comment that a 2008 agreement between Ontario and Quebec "committed these provinces to an effective C&T system based on Kyoto's 1990 baseline, thereby effectively relegating the weak Alberta and Ottawa C&T systems to the sidelines". They seem here to be formulating an inverted Gresham's law, according to which strong climate-change policies drive out weak ones. The logic is not explained. I would have thought it equally plausible to argue that as long as the "soft" versions of C&T remain the essence of federal and of Alberta policy, the weakness of those schemes will threaten to undercut the "green" provinces' more ambitious intentions. The general question here, which unfortunately was not addressed at the conference, is to what extent it is possible for one jurisdiction, or a group of them, to move forward when others refuse to do so? At what point do the voters of the more strongly committed jurisdictions revolt against the taxes or raised costs (because of C&T) that they have to pay, while other jurisdictions sit on the sidelines? Of course, these questions cannot be answered merely in the abstract, but a review of relevant considerations – perhaps by analogy with Mancur Olson's *The logic of collective action* (1965) – would be well worth undertaking.

It may be that this – or more precisely a "logic of intergovernmental action" on the environmental front – is what *Harrison* has been working towards. She asks, under what conditions different federations act to facilitate or to deter the adoption of policies to reduce GHG emissions. Her cases are the United States, Canada, and the European Union. She concludes that institutions do play a role – positive in the case of the European Union and the United States, negative in the case of Canada – but that they do so in conjunction with other factors including public opinion, and the regional distribution of the costs of implementing emissions-reducing policies. In the European Union, multi-level governance has contributed to policy effectiveness; in the United States, some of the states have been able to exercise leadership in spite of the federal government's past indifference or hostility; while in Canada, federalism has, at least until late 2006, produced only deadlock. Why is it, then, that in the absence of federal leadership several U.S. states have moved forward, while until recently nothing comparable has emerged on the provincial side in Canada? For Harrison, "arguably the most important" factor may be the regional distribution of the costs of GHG emissions: "In both the European Union and the United States, the largest and wealthiest states are 'green and keen', while the opposite has been the case in Canada." I find that generalization a little shaky, preferring instead (as earlier suggested) to see which provinces are producers of fossil fuels, and which ones are most richly endowed with hydro-electricity resources. The general point, however, is well taken: regional politics reflects regional economic differences. But Harrison says more than this: her emphasis is on institutions (federal structures) and how they interact with other factors. I agree with that approach, but a corollary is this: it becomes impossible to appraise the importance of institutions in determining outcomes. Likewise, the importance of public opinion and underlying values in a society is difficult to assess. When Harrison concludes, "... if we [in Canada] are to assume our responsibility to address a global problem to which we have contributed far more than our share, greater resolve will be needed from both the electorate and our political leaders", the discussion has shifted back to where it began: the reliability of the science, prudence and moral imperatives and, on the part of decision-makers, political will. Perhaps it is necessary, in Canada, simply to resign ourselves to waiting for Waxman-Markey, on the assumption that, as regards climate change, only a dramatic shift in U.S. policy will change the regional dynamic in Canada.

Exercising Leadership: Environmental Federalism

We are not done, though, with the subject of federal leadership. For starters: leadership on climate change is not the same thing as imposing a policy on provincial governments. *Page*, for one, is emphatic about this; he is strongly critical of the federal government's action at Kyoto in first consulting the provinces, and then, under pressure from Al Gore, breaking the agreement he had made with them. The more general point is that leadership is about persuasion, negotiation, and consensus-building, directed toward coordinating the actions of Canadian governments in support of emissions-reduction. The provinces cannot be marginalized simply because the task is important. A federal government committed to leadership will work to bring them onside, partly because the politics demand it, and partly because the provinces have powers that they can use either to complement and support federal objectives, or to undermine and frustrate them. Maybe leadership is impossible anyway, given regional differences, but there were hints at the conference, in papers as widely different as those of *Bramley* and *Courchene/Allan*, that it may be possible to move forward on climate change through a set of strategically astute moves. Or, at least, that the obstacles can be made a little less formidable.

The aim of a strategic policy-maker must be, in the first instance, to put together a package that will move forward while pacifying opponents, if they cannot simply be marginalized. *Bramley* suggests this when he argues for the auctioning of allowances under C&T, and using the revenues in a priority sequence, the first three items being: to protect industry sectors vulnerable to carbon leakage, to protect low-income Canadians, and to ensure regional balance. This means, in effect, that emissions-reduction must be embedded in a larger framework of what *Courchene/Allan* called "environmental federalism".

The term is not defined, but in their hands evokes the standardized procedures of bureaucratic cooperation and political negotiation that have characterized – with obvious lapses and exceptions – fiscal federalism in Canada. They implicitly appeal for an equivalent on the environmental front, to the meetings of finance ministers and their officials on the fiscal arrangements, from the mid-1950s on. By comparison, they say, "The structures and processes of environmental federalism are in a state of disarray." Substantively, they argue against returning revenues from carbon pricing to the provinces and territories in which they are generated, and propose instead (along with *Bramley*) using such revenues for a variety of purposes including, most prominently, offsetting the impact of carbon-pricing on low-income Canadians. Beyond this, they also note that the sums of money involved are potentially so huge that it will be necessary to redesign the equalization program. It remains to be seen whether broadening the environmental agenda, such that it embraces key aspects of fiscal federalism, would envenom federal-provincial relations or, conversely, would smooth the way for a major federal initiative in matters of climate change. What does seem beyond doubt is that it will be necessary, in order to explore the potential for federal leadership on emissions-reduction, to follow the *Courchene/Allan* lead, and to start exploring in a comprehensive way potential linkages among different aspects of what they call "environmental federalism". The machinery may count for something, but the substantive issues are what matter most of all. It must be remembered that consultation and negotiation produce little result, if governments are fundamentally at loggerheads because regional interests, and perhaps the interests of the governments themselves, are too far apart.

I return to the subject of federal leadership, with three observations.

- First, if the United States does succeed in moving forward on climate change in a dramatic way, and in Canada the federal government makes itself the enforcer of U.S. policy vis-à-vis the provinces, that is not (though *Courchene/Allan* seem to believe otherwise[9]) exercising leadership. It is just acknowledging the *realpolitik* of continental economic relations.
- Second, there was no discussion at all at the conference on whether estimates of the cost of emissions reductions are at all reliable, except for an assertion by *Chris Green* that estimates are, in general, too low, and *Bramley's* counter-assertion (citing Nicholas Stern, and a C.D. Howe Institute report, one of the co-authors being Nic Rivers) that the costs are small. This is a matter that requires far more comprehensive attention, both on a global scale and within Canada; it will be necessary to take into account both the short-term employment effects, and longer-term effects from technological change. What is particularly striking is that no one at the conference raised the question whether there are potential economic *gains* from making the transition to a low-carbon economy. This seems odd, since it is widely accepted that the Second World War definitively put an end to the Great Depression, and war is the epitome of wasteful (though often absolutely necessary) expenditure. An essential feature of leadership consists in finding new ways of putting things together ("thinking outside the box"), but the effort to do so, as regards the costs/benefits of change, was not made at the conference – nor, so far as I know, is it being made among policy-makers and/or within political parties.
- A final observation: exercising leadership is not, or should not be, confined to Canada; it is not just an aspect of federal-provincial relations. Rather, leadership is, at least potentially, about the role that Canada plays internationally, or will do in the post-Kyoto era. That observation leads us to our final subject, cross-border issues and the search for new directions in climate policy.

CROSS-BORDER ISSUES

My main point in this part of the paper is that choices among possible options for emissions-reduction may significantly affect – and be affected by – the policies of other jurisdictions, beyond any possible moral leadership effect. There are two sets of cross-border issues. The first has to do with political feasibility: is it possible to move forward in some way, when or if other

---

[9]They write: "… the prospect of Canada buying into a U.S. cap-and-trade system after it has been wrung through the geo-economic and political rent-seeking and rent-keeping machine called Congress would be quite another matter [than devising, independently of the United States, a Canadian C&T regime]. The upside in all of this is that Ottawa, by committing itself to link up with the U.S. C&T, will have finally jettisoned its criticized role as a Kyoto signatory but non-implementer, and will have asserted the leadership role on the environmental file that its earlier inaction had defaulted to the provinces."

jurisdictions lag behind? (The greater the discrepancy among jurisdictions in carbon pricing, the greater the potential for carbon leakage – and, correspondingly, the greater the political obstacles to taking effective action.) The focus on political feasibility draws attention to how other countries' policies affect domestic debates; in Canada, the impact of U.S. policy moves is obvious and important. The second set of cross-border issues has to do with the leverage that "leader" countries (acting individually, or together) may gain in relation to laggards, somehow inducing them to act more effectively on climate change. Leverage-effects may be gained through bargaining or simply from the way that market forces play out. Also, they are potentially of importance at several different levels: within a given federation, regionally or continentally, or globally. The question here is how domestic decisions and policies are likely to affect, or can be made to affect, the neighbours (bearing in mind that, in climate change, *all* countries are neighbours). Global relations are particularly important. But since Canada, by itself, can have little impact on the global dynamics of climate change, what is most at issue is whether Canada is a passive or an active player within an informal coalition of wealthy countries (mainly, a North Atlantic group).

U.S. Policy and the Canadian Politics of Climate Change

A recurring theme at the conference, as earlier parts of these notes attest, was that changing U.S. policies will profoundly affect what Canada chooses to do in the area of climate change. If Waxman-Markey, or some version of it, passes the Congress, the United States will have in place a carbon-pricing regime supported by border adjustments (tariffs or surcharges) that will burden Canadian exports of fossil fuels, unless Canada obtains an exemption. If Congress fails to adopt that bill, or a new bill of similar thrust, it is still possible that regulatory measures will be resorted to under the authority of the President, with comparable or even more import-restrictive effect.

Participants at the conference – the point was made especially by *Purchase* and *Courchene/Allan* – were well aware than any such policies from the U.S. side would dramatically change the dynamics of political debate in Canada over carbon pricing. As *Purchase* said, Canadian federal politicians might well be forced to follow the American lead, but would be able to say, facing their critics (especially in Alberta), "The devil made us do it!" The most obvious course of action for Canada would be to implement, nationally, a C&T scheme similar in design to the one put into place by the United States. In that case the provinces – more because of American action than because of initiatives by Ottawa – would be brought under a policy umbrella "made in the USA" but applying in practice equally to Canada. Whether Ottawa would be eager, wanting to move in that direction anyway, or reluctant (being rolled on by the American elephant), would be irrelevant. There was no debate at the conference on such matters, but that was presumably because everyone was perfectly well aware that Canadian policy must be "commensurate" (*Purchase*) with U.S. policy.

Is a Global Agreement Necessary?

Of greater import, and unfortunately not grappled with at the conference, was this question: whether certain options for pricing carbon stand to contribute more strongly to gaining leverage over states, such as the BRIIC group, who will not join in on any global emissions-reducing set of policies unless pressured to do so – and perhaps not even then. The issue here is whether broader diffusion effects can be achieved through linkage between the global trade regime and a putative carbon-pricing regime not necessarily embraced by all.

The *Courchene/Allan* proposal for a carbon-added tax and tariff (CATT) implicitly raises this question, although the authors shied away from it. And, perhaps, since they did not highlight this issue – to me, an important implication of their proposal – the conference did not address it. So let me try.

The Kyoto/Copenhagen approach to mitigating climate change wagers all on getting a near-global agreement in which each signatory commits to a specific target, and on the supposition that such commitments can, in one way or another, be enforced. There must be penalties for non-compliance; those penalties must be applied to defaulters; and they must be more severe than the economic cost of living up to one's promises. But enforcement is the Achilles heel of any global agreement. Even if all major emitters do participate in a new global agreement, and are willing in principle to take the necessary steps to ensure enforcement – both of which are hard to imagine – they must first have revisited several of Kyoto's major provisions, or added on new elements or principles, answering questions such as the following:

- Are Kyoto principles for estimating each country's emissions to be incorporated into a new agreement? Under Kyoto, the *principle of origin* applies (emissions, at every step of a production/distribution chain, are counted), but it is arguable that a *principle of destination* (in which emissions are attributed to the final consumer) would be fairer, as *Courchene/Allan* and also (implicitly) *Page*, have argued. Acceptance of the principle of destination favours resource-producers, but would greatly disadvantage the European Union, of which *Page* remarked: "the production emissions attributable to the E.U. consumption [of fossil fuels] were [under Kyoto] largely off-loaded into the totals for Russia and OPEC".
- How, and by whom, are emissions to be monitored?
- Enforcement: what mechanisms might be devised to ensure that countries making commitments will actually live up to them?
- In setting targets, should account be taken of population growth, of growth in output (GDP) per capita, and/or the supply of energy or energy-intensive raw materials to other signatories? (One way of doing some of these things would be to adopt the destination principle.)
- What offsets, such as carbon sinks, are to be permitted?
- Are developing countries to be expected to achieve reductions comparable (either on a per capita or an aggregate basis) to those the wealthy industrial nations will commit to?

- What financial support, or support in the form of technology transfers, will the wealthy countries offer to developing countries, as an inducement to reducing their emissions, and on what basis will such assistance be allocated among the recipient states?

The most telling thing about this rather daunting list is that all the issues raised here come up under C&T, but – except for those relating to financial support and, perhaps, monitoring to preclude carbon-tax evasion[10] – are absent under a carbon-tax approach to emissions reduction. Not that the Kyoto Protocol prescribed how individual countries would go about meeting their targets, or that a Copenhagen treaty would be likely to do so, but the thinking behind such an international negotiating agenda is all C&T. Under this approach, setting the targets comes first and the carbon price is derivative. If, by contrast, it is the carbon price that is directly set, as is the case with the taxation approach, emissions reductions are implied and aimed for, but not actually negotiated. With a carbon tax, it becomes unnecessary to argue about absolute *versus* intensities caps, because there are no caps. Making allowance for increases (or decreases) in population, or in GDP, is irrelevant; these items disappear from the negotiating agenda. The same goes for offsets, and for the question whether developing countries ought to be treated differently from industrialized ones. Development assistance, whether in the form of grants or of technology transfers, remains an important subject, but such aid need not be linked in any explicit way to reaching an international agreement on climate change. In fact, *making progress on the climate change issue would not even require a global agreement.*

If there is no global treaty on climate change, either each jurisdiction will act as it chooses, or a group of countries ("leaders") will adopt a common or comparable set of policies, either C&T, or harmonized carbon taxes – not necessarily setting standard rates, but doing things like avoiding double taxation. A corollary to such action would be that they would also act to burden imports from the "laggards". The very largest jurisdictions – the European Union, the United States – would be able to adopt their own policies, one feature of which would likely be border adjustments applying to goods from those countries deemed to have lax emissions-reducing policies. Any such measures would

---

[10]Within a regional or other non-global agreement, a condition for exemption from a carbon tax imposed at the border would necessarily be that a comparable tax had already been imposed in the exporting country. Exemptions would be granted only on the basis of careful monitoring, which is why one might expect a regional or developed-countries' agreement to be put into place. Imports from countries not participating in such an agreement would be taxed at a level that assumed the exporting countries had no carbon tax. The issues that would arise in this situation are similar to those arising when a country imposes a countervailing duty – the absence (or low level) of a carbon tax in an exporting country would be considered, by the importing country, to be a subsidy. The analogy, of course, underscores the risks to the international trading regime, if a group of climate-change leaders – or great economic powers individually: entities such as the United States or the European Union – attempt, through import levies, to lever the laggards into taking more vigorous steps towards climate-change mitigation.

certainly be challenged under the WTO, leading to years of litigation and potentially to serious disruption of the international trade-and-investment regime, along with mounting tensions between the developed and the developing countries. It seems likely that global uncertainties and conflict would drive the leaders on climate change towards a non-global agreement, to smooth trade-and-investment relations among themselves.

Diffusion Effects: Gaining Leverage in the International System

How strong an impetus would there be towards the formation of a leaders' club? That would depend to a large extent on what approach various jurisdictions in this group take to reducing emissions. Here, my earlier list of options for mitigation of climate change becomes pertinent.

The weakest approach, in terms of driving an international but non-global agreement, would be a simple carbon tax imposed at a low level, or, as per the *Hyndman* proposal for the EITE sector,[11] a tax with relatively high rates at the margin but a very low average rate. In both cases, the low rate or low average rate would avoid burdening domestic industry (export or import-competing sectors), making border adjustments largely unnecessary. The low carbon price, or at least the gentle-slope feature of these options, would not reduce consumption very much, nor would it drive international coordination. A corollary would be that no jurisdiction could afford (on pain of damage to its economy) to get very far out in front of others, in terms of its rates of carbon tax.

The strongest policy approach, in terms of achieving diffusion effects and enabling climate change leaders to exercise leverage within the international system, would be a fully trade-neutral approach. To realize this is one of the main objectives of the *Courchene/Allan* proposal for a carbon-added tax and tariff (CATT). The fact that the CATT would be imposed on imported goods at the border would ensure that domestic producers would not, on account of the carbon tax, be undercut by import competition; moreover, removal of the tax on exported goods would ensure that domestic industry would not operate at a disability in foreign markets. A similar effect would be aimed for under the

---

[11]The essence of the Hyndman proposal is that it envisions two tax systems, one applying to the domestic consumption of goods produced in non-export, non-import-competing industries, and one applying to the EITE sector (emission-intensive industries either export-oriented or import-competing). Industries in the EITE sector would suffer if they were subject to high carbon taxes of general application. To avoid this, most of their carbon consumption – i.e., of fossil fuels resulting in GHG emissions – goes untaxed. But at the margin (i.e., above the set standard) they pay a relatively high carbon tax, which sets the carbon price. For goods produced outside the EITE sector, the tax ultimately falls upon the consumer, who for this reason has at least some incentive to reduce consumption. For goods produced in the EITE sector, the high marginal rate (the high carbon price) encourages development of carbon-saving technologies and provides incentives for investments that will use such technologies. Whether or not firms could pass on this high carbon price will depend on competitive conditions. Where price is determined internationally, the firms, not their customers, are likely to pay the tax.

proposal of Mintz and Olewiler for what would otherwise be a simple carbon tax, imposed on final consumers.

Trade-neutrality would allow for much higher domestic carbon prices, out of line with the policies of some of that country's trading partners. The high domestic price, together with accompanying border adjustments, would have three important consequences:

- it would induce individual consumers of fossil fuels in high-price countries to search out the most carbon-efficient goods, and to induce firms to invest in carbon-saving production techniques, without resulting in carbon leakage
- it would encourage some sort of trade deal among the high-price group, a sort of mutual recognition strategy that would exempt other countries in the group from border adjustments, thus smoothing trade and investment relations among themselves – but at the price of fragmenting the international trade-and-investment regime, and (correspondingly) raising international tensions
- it would put pressure on the "laggards", to the extent they seek out export opportunities in the high-price group, to move towards a higher carbon price – in other words, it would create a diffusion effect, as emissions-reducing countries, especially if acting as a coordinated group, gain leverage in the international system.

These possible consequences of a CATT are not discussed in the *Courchene/Allan* paper, and the authors may disavow any intention to bring about the sorts of diffusion effects referred to here. They have said nothing about the climate change leaders gaining leverage in the international system, nor about their forming a particular group within the WTO, either undermining or transforming it. These are clearly important issues, deserving attention in the concluding section of these notes.

CONCLUSION: THINK GLOBALLY, ACT REGIONALLY

As I write, the extent of regional conflict in the Canadian debate over climate change and approaches to emissions reduction has burst into full public view. On 22 October the federal minister of the environment, Jim Prentice, was reported to have "been consulting with provinces on a plan that would impose a cap of industrial emissions, but allow Alberta's energy-intensive, emissions-heavy oil sands to continue expanding".[12] A week later the Pembina Institute and the David Suzuki Foundation released estimates of the cost of meeting the 2020 federal target for reducing emissions by 20 percent relative to 2006, or, as the Kyoto Protocol envisions, by 25 percent relative to 1990. The findings of

[12] *Globe and Mail*, 22 October 2009, "Ottawa dashes hope for climate treaty in Copenhagen".

that study, which relied on modelling by M.K. Jaccard and Associates,[13] are largely foreshadowed in the paper by *Bramley* in this volume, and reflect assumptions about the carbon price that would be needed to meet those targets. However, two important things are new.

- First, it is estimated that federal revenues from carbon taxes or from auctioning of allowances would be a minimum of $40 billion in 2020, about the same as current federal revenue from the personal income tax; the money would then, Pembina/Suzuki recommends, be redistributed in various ways, including through the reduction of income tax rates. But it is hard to imagine any government surviving an election, after introducing a policy that would swell federal revenues by so large an amount, no matter what is done with the money.
- Second, the study envisions that almost the entire cost of the program, in terms of GDP growth foregone, would be borne by the economies of the three most westerly provinces (and presumably also Newfoundland and Labrador, although the relevant figures were submerged in a total for the four Atlantic provinces).[14] There could scarcely be a more dramatic illustration of former Alberta premier Peter Lougheed's warning, cited in the papers by *Page* and by *Courchene/Allan*, about potential strains on Canadian unity if the federal government were to force a slowdown in the production of fossil fuels. Mr. Lougheed evoked a scenario under which, for environmental reasons, a go-slow policy for the development of the bitumen sands were imposed on Alberta; his prediction was that, were this to happen, regional conflict would rise to a level that would make the National Energy Program of 1980 an exercise in consensus-building. No wonder the Pembina/Suzuki report was denounced in Alberta and Saskatchewan, and was dismissed by Mr. Prentice as "irresponsible". He added that Canada's – that is, his government's – targets can be met with a much lower carbon price than the report envisions, that Canadian policy must be aligned with that of the United States, and that the costs must be acceptable in all regions.

[13]The study was funded by the TD Bank Financial Group, and is summarized in a *Special Report* of TD Economics, 29 October 2009, printed as an appendix to this volume.

[14]To meet the federal government's target of a 20 percent reduction in emissions in 2020, relative to 2006, Canada's GDP would be 1.5 percent lower than it would otherwise be (but would still rise, relative to current levels of output). However, Alberta's output – while still allowing for higher per capita incomes than today – would be 8.5 percent lower than under "business as usual" assumptions, with lesser declines in British Columbia and Saskatchewan. Canada, and especially the western provinces, would be more adversely affected if Canada were to take the measures necessary to meet Kyoto commitments. Most of those who have reacted to these figures have evidently assumed them to be accurate and reliable, but there has been outrage that such a scenario could be contemplated.

The contrast between what appears to be the federal government's preferred regional allocation of the costs of emissions reduction, and the recommendations of the Pembina/Suzuki study, would appear to demonstrate the near-impossibility of developing a Canadian emissions policy along Kyoto/Copenhagen lines. Whether the targets are those of the Harper government, or those proposed by environmentalists, a substantially higher carbon price would be required, and that would impose terrible strains on the country. Under any imaginable allocation of costs across industries and regions, the prospect of gaining an interprovincial consensus would appear to be negligible – unless, as is possible, the regional dynamic changes dramatically as a result of U.S. policy.

Shift, now, to the global scene. On emissions reductions, the parallels with Canadian debates are eerily strong, though with the important difference, that there can be no appeal to a world government to knock heads together and "just do it!" It is not simply that there are sharply divergent views among governments within the international system. More than that: the divergences arise for the same reasons as they do in Canada, though at the global level they have a starkness and a scale that far exceeds our own regional or interprovincial differences. Specifically, countries with growing populations and with ambitions for economic growth refuse to suppress or abandon those ambitions. This is understandable. Their governments could scarcely survive the civil disorder – never mind the mere threat of electoral defeat – that would be sure to break out if the goal of obtaining a better life were to be seen as unreachable. "A better life": not a self-indulgent consumerism, but, for many, a footpath out of destitution, malnourishment, ill-health. One should recall that the origins of revolt seem as a rule to lie not in unrelieved poverty, but in increasing precariousness of existence (as is threatened, for example, by drought and other climate-related disasters) and dashed hopes.

At Copenhagen, and in the aftermath of the conference, there will be much discussion of an "unfair" and "disproportionate" distribution of the costs of mitigating climate change: expecting others to do the heavy lifting. This is exactly what occurs, as well, in Canadian domestic politics. It is an embarrassing comparison, given our enormous wealth, but if anything is clear about the political dynamics of climate change, it is that a sense of proportion – of sacrifice, or willingness to forego gain, given existing levels of wealth and comfort – is utterly lacking. In such circumstances, consensual solutions to problems are elusive. It is so within Canada, and it is so globally. Specifically, at the global level, the wealthy countries seem unwilling to commit to significant reductions in GHGs as long as major emitters among the developing countries refuse to do likewise; but conversely, the developing countries evoke the principle of "ability to pay" (and remind the wealthy that it is they who are responsible for current levels of GHGs), insisting that they should have their turn. Attempts to negotiate towards an equitable outcome through technology transfers and development assistance are an essential element in mitigating climate change, but a global consensus on *absolute and enforceable* emissions reductions is difficult to imagine. One of the problems here is the absence of credible enforcement mechanisms: for this reason alone, a re-run of Kyoto would be unpromising.

There is a lesson here for Canada, as for other wealthy countries. Arguably, it would be desirable to focus our domestic debate over climate change on the need for change at the global level – as well as nationally, or continentally – and to shape our domestic policies accordingly, *but without placing all eggs in a very fragile Kyoto/Copenhagen basket.* The advocates of a strong climate-change policy have done exactly that: they have assumed the necessity of reaching a global agreement, and they want Canada to do its part in ensuring that such an agreement will commit its signatories to sharp reductions in emissions. They reason that in this process, Canada will gain credibility and perhaps influence *only* if it commits to sharp reductions in GHGs and adopts policies commensurate with its commitments.

A contrary position – to which I personally subscribe – would be this: that it is time to remove our "Kyoto blinkers" – to abandon the assumption that the only effective approach to mitigating climate change is to reach a global consensus on emissions reductions, based on ability to pay and thus on a negotiated sharing of the economic pain. A better strategy for Canada and for other wealthy developed countries would be to turn the searchlight on the diffusion effects likely to flow from domestic policy innovations. The policies of great economic powers such as the United States and the European Union may have, automatically as it were, significant diffusion effects even in the absence of regional agreements, but for countries of Canada's size it will be necessary either to just tag along behind regional leaders, or to seek to gain influence within a regional grouping (North America, Europe) or among OECD countries generally. The environmentalist slogan, "Think globally, act locally" should probably be revised: Think globally, act regionally. A shift in public debates over approaches to mitigating climate change, focusing on global outcomes of policies adopted nationally and regionally, or within an OECD grouping, would be desirable.

The papers in this volume, taken together, point to the need to raise carbon prices in the home economy, in order to discourage the consumption of carbon-intensive goods. There was considerable disagreement among authors on the subject of *how fast* to raise prices, but – as noted at the beginning of this article – all were agreed on the general proposition that the price of carbon must go up. If, as a result of such increases, wide disparities emerge internationally, border measures will probably be required in order to curb both carbon leakage and the economic costs (in terms of output foregone) of emissions-reduction. Such policies, while not *directly* doing anything to reduce emissions in the developing world – thus being open to the charge that they will do little to mitigate climate change – would *indirectly* do precisely that: they would induce, through the workings of the market, the BRIIC and other developing countries to reduce their GHG emissions.

For the countries of the North Atlantic or generally for the OECD, merely striving for moral leadership in global negotiations is too weak an approach. On the other hand, to adopt policies that disrupted the world trading system or otherwise entrenched or widened global disparities in economic well-being, would be neither morally defensible nor conducive to mitigation of climate change. Moreover, international tensions would rise sharply. Thus, on multiple grounds, policies that developing countries would experience as aggressive or punitive must be ruled out. Indeed, the wealthy and economically developed countries, individually and as a group, must not only bring developing countries onside in mitigating climate change, but must support them in improving health, nutrition, housing, living standards, and economic security for their people. Only if they are able to do so, will they come on board for reducing GHG emissions.

It is not at all obvious how to reconcile so many objectives potentially in conflict with each other. There is thus plenty of room for very real leadership, perhaps seemingly altruistic but in the longer term self-interested, in helping the developing countries make the transition to a lower-carbon economy. One way of doing this is through technology transfers and assistance.

In general, in regard to climate change, the focus of public debates has been on individual jurisdictions and what they can do, rather than on the dynamics of a world system which is political and cultural as well as economic. The papers in this volume illustrate the common tendency to discuss policy alternatives and options on a country-by-country basis. The greatest shortcoming of these papers, taking all of them together and seeing them in relation to each other, is insufficient attention to the international system, and to the question of how different policy approaches to emissions reduction will play out, or can be made to play out, beyond national borders. Surely it is necessary to base policy prescriptions for individual jurisdictions on their supposed or likely consequences *globally*, not just in the sense of making a marginal contribution to reducing GHG emissions, but in the sense of nudging others along as well, through a combination of pressure, encouragement, and help. In this, the lead role will fall to the wealthy developed nations, which have the task of working together to re-shape global action in relation to climate change.

Underlying that task will be an intellectual challenge, as well as a political one, of major dimensions, to consider what a global approach may imply in terms of economic well-being. Climate change and its mitigation is only in part a technical or scientific issue; it involves also cultural attitudes, or basic social values – a challenge to the conventional assumption that GDP per capita is the primary measure of economic performance. It will be necessary to consider also things like the distribution of incomes (measures of equality and inequality), economic security, and enjoyment of non-priced goods ranging from health to self-esteem to an unsullied environment. Above all, there must be recognition of a basic fact, that steadily-rising levels of resource consumption will inevitably result in the degradation of nature and the over-burdening of the planet's atmosphere, developments that are utterly destructive of well-being by any measure, and for all.

15

Epilogue…Lessons from Copenhagen for Canadian Climate Policy

Nancy Olewiler

THE COPENHAGEN ACCORD

Copenhagen was the stage for this decade's last international climate conference under the auspices of the United Nations' Conference of Parties (COP). It was a climate play that has been performed in previous runs in places such as Rio, Kyoto, and Bali. The actors – delegates from each country, politicians, environmental groups, the media – were well studied for their roles. There was hope by many that that this time there would be a different ending, with all countries agreeing to specific and binding targets, but it was the same old story – the talk resulted in an accord consisting largely of promises rather than legally binding targets and specific policies. The drama occurred in the final act when it appeared that no accord of any kind would be reached. China's prime minister, Wen Jiabao, stormed out of the conference leaving a low-ranking protocol officer in charge after Barack Obama's speech to the conference chastised China's intransigence. Lumumba Di-Aping from Sudan said the proposed accord would be a suicide pact for Africa, compared the accord to the Holocaust, and urged all African countries to reject the proposal. The climax came when the accord was salvaged from defeat in the middle of the night by British Secretary of State for Climate and Energy, Ed Miliband. The basic problem is that countries in the rapidly developing world – China, India, Brazil will not cede sovereignty over their use of fossil fuels to an international entity. They demand the right to pursue their own path of economic growth without constraints, as did the countries that enjoyed the use of these fuels for the past century. No amount of political cajoling and threats would alter their stance. The Copenhagen Accord allows the United Nations to say that some progress was made, albeit little of any substance.

The Copenhagen Accord has five main components.[1]

[1]United Nations Framework on Climate Change, Conference of the Parties, Draft Decision/CP15, Copenhagen Accord, December 18, 2009. Accessed at http://unfccc.int/

- Annex I countries (e.g., the European Union, United States, Canada, Japan) will commit to emission reduction targets by January 31, 2010. As all of these countries except the United States ratified the Kyoto Protocol and set targets already, this provision is unlikely to lead to major increases in targets from these countries. In any event, as evidenced by Canada's failure to come close to its Kyoto targets, there is no binding enforcement mechanism so targets may continue to be "hot air".
- Non-Annex I countries (e.g., China, India, Brazil, the African countries) will implement mitigation strategies and report on their emission levels every two years to the Conference of Parties.
- The incentive to report accurately and to reduce emissions is found in the third main provision of the accord – the establishment of a fund to help developing nations by supporting mitigation, adaptation, technology development and transfer, as well as provisions to reduce deforestation and forest degradation. There is a target to raise $30 billion for this fund from the wealthier nations over the period 2010 to 2012 and to increase that amount to $100 billion a year by 2020. Multiple entities may administer these funds, but the United Nations intends to have a significant amount managed by its creation – the Copenhagen Green Climate Fund. Non-Annex I countries wishing to access the fund will be subject to international verification of their mitigation, adaptation, and technology actions. This is the one component of the Accord that may be beneficial, if these large sums of money can be raised and administered effectively.

There is historical precedent to the new fund. The Global Environmental Facility (GEF) was established in 1991 to fund the incremental costs incurred by developing countries to help them participate in international environment treaties. The GEF covers a wide spectrum of environmental activities and in 2001 under the Marrakesh Accord (COP 7) the GEF's authority was expanded to assist implementation of the Kyoto Protocol and United Nations Framework Convention on Climate Change (UNFCCC) activities. Two funds directly connected to climate change were created: the Special Climate Change Fund (SCCF) and the Least Developed Countries Fund (LDCF).[2] The SCCF's mandate sounds very similar to the provisions in the Copenhagen Accord, namely, to finance projects related to capacity building, adaptation, technology transfer, climate change mitigation, and economic diversification for countries highly dependent on fossil fuels. As of May 2009, 13 countries including Canada had contributed a total of $100.5 million. Canada's contribution totalled $13.5 million

resource/docs/2009/cop15/eng/l07.pdf. The parties also agree to keep working toward a fuller agreement at the next meeting in late 2010 in Mexico.

[2]See Global Environmental Facility (May 26, 2009) *Status Report on the Least Developed Countries Fund (LDCF) and the Special Climate Change Fund (SCCF)* for information about the source and use of these funds. Accessed at http://www.gefweb.org/uploadedFiles/Documents/LDCFSCCF_Council_Documents/LDCFSCCF6_June_2009/LDCF.SCCF.6.Inf.2.Status_Report(1).pdf. All amounts are in U.S. dollars unless specified otherwise.

(Canadian dollars), of which $11 million was allocated to adaptation activities. Cumulative outlays for climate projects under the SCCF to May 2009 were $71 million. The LDCF is to support special work projects in developing countries. Cumulative funding of projects to May 2009 was $87.4 million, while total contributions from 19 countries amounted to $135.4 million. Canada's contribution was $10 million (Canadian). If it took eight years to raise a combined total of $236 million, one might be sceptical that the call for a new fund will raise $30 billion in three years, yet alone $100 billion per year by 2020.

- Signatories agree that efforts must be made to stabilize GHG emissions so as to keep global temperatures from rising more than 2°C. The Accord notes that deep cuts in emissions will be necessary to reach this goal and that all countries should be encouraged to reduce emissions as quickly as possible given their state of economic development.
- Finally, there is a call to quickly develop mechanisms to provide incentives to reduce deforestation and forest degradation, including REDD activities.[3] I will have more to say about payments to enhance forest practices below.

Canada received its share of international attention in Copenhagen, chastised by environmental groups for its intransigence on implementing meaningful cuts to GHGs and "winning" the "Colossal Fossil – Fossil of the Year" award. In making the award, Canada was recognized for "bringing a totally unacceptable position into Copenhagen and refusing to strengthen it one bit. Canada's 2020 target is among the worst in the industrialized world, and leaked cabinet documents revealed that the governments is contemplating a cap-and-trade plan so weak that it would put even that target out of reach.... Canada's performance here in Copenhagen builds on two years of delay, obstruction and total inaction. This government thinks there's a choice between environment and economy, and for them, tar sands beats climate every time."[4] Prime Minister Harper and Minister of Environment Jim Prentice both attended the meeting; Harper's visit was brief with no substantive policy statements made.

LESSONS FROM COPENHAGEN

The failure at Copenhagen to move the world to binding and enforceable targets to reduce GHG emissions mirrors Canadian policy at the national level. Canada,

---

[3]REDD stands for "reducing emissions from deforestation and degradation", a UN-backed program to provide incentives to developing countries to protect forest ecosystems and improve forest management. The United Nation estimates that loss of forests contributes to 20 percent of the overall GHGs entering the atmosphere annually. REDD programs can be controversial because of the moral hazard problems – forest owners/users can exhort payments for not cutting the forest even if they had no intention of doing so. Verification of actions to reduce forest degradation is another challenge.

[4]Ben Wikler of Avaaz.org, accessed at http://www.fossil-of-the-day.org/.

as a federal state, represents a microcosm that parallels the challenges of reaching agreements when dealing with multiple political jurisdictions with diverse interests, dependence on fossil fuels, and economic conditions. Many commentators had low expectations of Copenhagen for good reason.[5] Chances of getting the developing world to agree to incur the economic costs of reducing their dependence on cheap fossil fuels (cheap relative to non-GHG-emitting fuels) were slim to none. All the best efforts of countries committed to reducing their GHGs, environmental groups and scientists emphasizing the importance of taking action now could not produce an agreement with any teeth. The first COP was in 1995, Copenhagen was COP 15, and future Conferences are scheduled until 2012 for COPs 16 through 18. The world continues to talk about climate change, but does little to act, much as Canada has done over the past twenty years.

Other authors in this volume have discussed Canada's dismal performance on GHGs at the federal level and the challenges the country faces in addressing climate change; I will not repeat all of those. I offer four lessons from Copenhagen and in the next section I use these lessons, plus the wisdom of the other authors in this volume to offer policy proposals for Canada.

Lession #1 : Establishment of binding targets is a very challenging strategy whether you are the world or Canada. Reasons for this are many: At what level should the target be set? How will the targets be allocated among jurisdictions? What policy instruments will achieve the target? How are policies to be enforced? Canada, like other countries, has reams of environmental legislation and administrative practices that look great on paper, but do not deliver the promised outcomes. There are far too many examples to discuss here; one must serve. The Canadian Environmental Protection Act (CEPA) was promulgated first in 1988 and amended in 1999. CEPA 1988 covers such things as setting emission standards for toxic compounds, hazardous wastes, ozone-depleting compounds, sulphur in gasoline, ocean dumping, pulp and paper effluent and more. There are successes from the act – the elimination of ozone-depleting compounds and dioxins and furans from pulp and paper, restrictions on sulphur content in fuels; the rest is a work in progress. Environment Canada has been struggling for over twenty years to address the regulation of emissions of toxic compounds. A key reason is that it is scientifically challenging to determine the "safe" level of emissions. GHGs are no exception: what is the appropriate target for each emitter, for each province, for the country. Enforcement is costly. Under CEPA 1988, in the 2000-2001 fiscal year, a total of 605 inspections were done, 8 prosecutions, 6 convictions.[6] Under CEPA 1999 in the same fiscal year, 2642 inspections were done, 3 prosecutions, and 1 conviction obtained.

---

[5]For example, see McColl 2009-2010.

[6]See *Canadian Environmental Protection Act (1988 and 1999), National Enforcement Statistics, 2000 and 2001.* Accessed at: http://www.ec.gc.ca/alef-ewe/5C63F879-0E5A-4B2D-9C41-312FC3F888A9/cepa_natl_2000_2001_e.pdf.

Lesson #2: A debate framed as the economy versus the environment is a false dichotomy, one that proponents of little to no action on environmental issues use repeatedly. It doesn't matter whether the economy is flourishing or floundering, the message remains the same – it is too costly to [fill in the blanks – reduce GHGs, reduce air contaminants, clean up toxic waste sites, etc.]. Copenhagen sank because countries argued that GHG targets and policies would reduce economic growth rates and that was unacceptable to them. My view is that the only hope is to continue to emphasize that a healthy environment is a necessary condition for a strong economy, and to employ policy mechanisms that work to enhance both the economy and the environment and minimize adverse impacts on GDP. These mechanisms do exist, as discussed in many papers in this volume.

A corollary to this lesson is that rich countries (and Canada is one) need to demonstrate to poorer ones that it is possible to take unilateral action on GHG emissions and not suffer significant adverse consequences.

Lesson #3: Regional inequality is a huge barrier to agreement on climate change policy. Copenhagen marked one turning point: the inclusion of the United States (and Australia) in the fold of rich countries who supported meaningful action on GHG emissions. But the barriers between rich and poor are simply too great to permit an accord. Until more is done to improve the economic well being of the poorer nations, any sort of meaningful agreement will be difficult. The climate funds may help, but they require the richer countries to ante up and to have the funds distributed in a way that generates real reductions in emissions. Policy mechanisms – whether they are regulations such as renewable portfolio standards, or pricing instruments such as carbon taxes and/or a cap-and-trade systems – need to recognize the differential impacts they will have on regions/countries and do something about these impacts.

Lesson #4: A combination of policies is needed to tackle climate change. The Copenhagen Accord, like many of its predecessors, has a multi-pronged approach: mitigation, adaptation, R&D, technology transfer. Canada's climate agenda also includes these components. What is needed is demonstration of concrete and measurable actions on all these fronts.

A CLIMATE POLICY STRATEGY FOR CANADA

Canada can and should take concrete action to reduce its GHG intensity and emission levels. The current policy of our federal government appears to be "wait for the Americans to act". The lessons I take from Copenhagen, combined with the papers in this volume, can be translated into a policy framework that calls for immediate actions, rather than our stance of wait and see. The framework involves five components: dump a specific emission reduction target as a policy goal for 2020; rather than a cap-and-trade system, implement a carbon tax at a low rate with a schedule of regular increases (as in the B.C. carbon tax); recycle a significant portion of the revenue raised under the carbon tax in the form of reductions in individual and corporate income taxes; use the rest of the tax revenue to create a technology fund to reduce carbon intensity, sequester GHGs, support cold fusion, and whatever might reduce our dependence on fossil fuels; and implement a REDD program in Canada targeted initially at First Nations.

Policy #1: Dump the 2020 Target

Return to Rick Hyndman's graph that shows demand for emission reductions (the willingness to pay to reduce GHG emissions) and the marginal costs of reducing emissions. The problem Hyndman identifies is illustrated clearly in the figure: the 2020 target cannot be achieved unless: (1) people and the federal government dramatically increase their willingness to pay for carbon emission reductions (there is a demand curve that intersects the cost curve in the area of the uppermost ball); (2) the marginal cost of reducing GHG emissions falls dramatically so that the shaded cost curve intersects one of the lower balls; (3) both occur so as to reach one of the lower balls; or (4) we reduce the 2020 target, e.g., move it to the left to coincide with the striped ball.

In the current economic and political climate, the only strategy that is compatible with reality is to give up the facade that Canada will reach its 2020 target (see the papers on the costs of doing so under current technology). The 2020 target should be replaced with a schedule of achievable outcomes given current prices and technology. This means that we need carbon prices; on to policy #2.

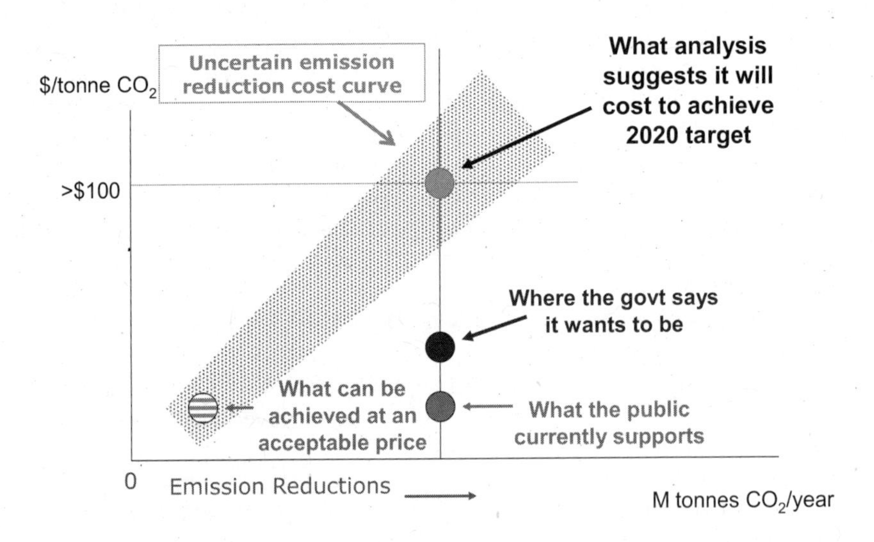

Policy #2: Immediately Implement a Federal Carbon Tax at a Rate that Starts at a Modest Level, Say $10 Per Tonne, and Rises Predictably Over Time

As part of the next federal budget, the government should implement a carbon tax set, for purposes of exposition, at $10 per tonne CO_2e.[7] I suggest as broad a base as is technically feasible (e.g., the B.C. base plus emissions from oil and gas extraction)[8]. Let's suppose, as does Rick Hyndman, that the tax will cover 80 percent of total emissions. Announce at the onset that the price will rise each year by $5 per tonne until say, the year 2015. That will make the price $35 per tonne of carbon in 2015. Using Hyndman's chart, that tax will achieve a reduction in carbon emissions at about the level of the striped ball initially, but each year thereafter total emissions should fall from their previous levels. Covered sources of GHGs will cut emissions up to the point where their marginal costs of emission reduction equals the tax rate. The downside of a tax compared to cap and trade is that we won't know exactly how far emissions will fall each year until after annual data is collected. The federal government can use this data (that is already being collected) to report each year on emission levels. If GHGs are not falling at a rate deemed fast enough, the government can announce that tax rates post-2015 will rise more rapidly.

Why no cap and trade? There are many reasons.[9] First, I believe a cap-and-trade policy will fail to contain GHG emissions until well into the future. Studies done by the National Round Table on the Environment and Economy (2009, 2007) show that the sooner a jurisdiction implements a carbon-pricing policy, the cheaper emission reduction will be because, in the absence of policy, emissions will continue to grow, making it that much more costly to achieve a given level of reduction in the future (unless technological breakthroughs emerge). But without mechanisms to help fund research into GHG reduction/control technologies (as Chris Green argues), Hyndman's emission-reduction cost curve won't shift down sufficiently rapidly over time, making it necessary to have very high carbon prices to get to any sort of targeted reduction level. Cap-and-trade systems are massively complex, and complex systems take a long time to develop. The Western Climate Initiative has been at work for a number of years to design its system that is scheduled for implementation in 2012. Whatever bill the U.S. Congress ultimately passes (if they do) won't likely be implemented until 2015. The carbon tax proposed would be already hard at work to reduce emissions long before 2015. A carbon tax can be implemented very quickly as the administrative apparatus is already in place; governments know how to levy and collect excise-type taxes. For example, B.C.'s carbon tax was announced in the February 2008 budget and came into effect on July 1st of that year.

---

[7]By way of comparison, the current B.C. rate is $15 per tonne and rises to $30 per tonne by 2012.

[8]For details on the B.C. carbon tax see http://www.sbr.gov.bc.ca/individuals/Consumer_Taxes/Carbon_Tax/carbon_tax.htm.

[9]Rick Hyndman and others have already mentioned a number of these.

Secondly, cap-and-trade systems require the establishment of the caps. A federal cap will have to be apportioned provincially unless the system is one that auctions every permit. But if that were the case, we'd basically have a tax system, but one where the price (the tax rate) is uncertain and determined by the market. No functioning cap-and-trade system auctions all its permits (e.g., New England's Regional Greenhouse Gas Initiative, the E.U.'s Emission Trading System), neither does current and proposed legislation (e.g., the Western Climate Initiative, Waxman-Markey bill). Lesson #3 from Copenhagen says this apportionment will be very challenging: think Alberta and Saskatchewan versus Quebec, British Columbia, Manitoba and Ontario, and then there are the Atlantic provinces. Rent-seeking behaviour will be the name of the game, as industries jockey to get as big an initial allowance of permits as possible.

Third, cap-and-trade systems will create a new market for the financial sector that portends to generate substantial rents to that sector.[10] Every dollar that goes to financial intermediaries is a dollar that would not be available to help mitigate the impact of the climate policy on low income people, or to invest in new technologies to reduce our carbon dependence. Unless the government auctions the permits and collects the revenues, it will be difficult to address regional and income inequality across the country. Under the WCI design proposals, there is to be a minimum auction of 10 percent of each jurisdiction's total allowances in the initial period (2012 to 2015), rising to 25 percent by 2020. Under a tax, every dollar that is raised can be recycled back to the economy in ways that governments will be held accountable for by their electorate.

Policy #3 and #4: Recycle a Significant Proportion of the Carbon-Tax Revenue to Individuals and Businesses by Reducing Income Taxes; Put the Balance into a Fund to Stimulate Technological "Fixes" to Our Carbon Problems

This proposal combines lessons 2, 3 and 4 from Copenhagen and supports some of the other papers in this volume. Many economists have called for the return of carbon-tax revenues to the economy in the form of tax cuts, particularly taxes that distort positive activities such as saving and investment. Using the carbon-tax revenue to cut income taxes will enhance the efficiency of the economy without creating an incentive to increase carbon emissions while also providing

---

[10]Estimates of the total value of a national U.S. cap-and-trade market range from $130 to $500 billion per year, if all the permits were auctioned annually. See Paltsev *et al.* (2007). Using these estimates, if say, 25 percent of the total number of permits are traded each year using financial intermediaries who set a commission at 2.5 percent of the value of the trade, rents to the financial sector could range from $500 million to $3 billion annually.

additional after-tax income to offset the costs of the carbon tax.[11] A key issue that was mangled by politicians and the media in all the rhetoric over carbon pricing during the last federal election was that any form of pricing (either taxes or cap and trade) that is effective in reducing GHG emissions will have an impact on the prices of carbon-intensive goods. The extent of that impact is, however, uncertain because it depends on the nature of the markets (the shape of supply and demand curves) for carbon-intensive goods, and how prices are determined (world markets, domestic markets, degree of competition and so on). Rather than try to guess these impacts before the tax policy is implemented and engage in all sorts of compensation measures to specific groups, fiscal neutrality over a major share of the carbon tax revenues should prevail.

The need to shift down the marginal cost curve for reducing carbon emissions calls for some share of the carbon tax revenues to go into a technology fund. While details of the exact distributions between recycling revenues in tax cuts versus the technology fund could be the subject of a federal/provincial first minister's meeting, suppose for sake of argument, that the revenues from three-quarters of the carbon tax rate each year ($7.50 initially, or an estimated $4.3 billion in the first year) are used to reduce income taxes and increase GST rebates to low-income individuals. As noted, this lessens the tradeoff between the economy and environment, helps address the public's aversion to taxes of any sort (admittedly a challenge for politicians), and puts the decision of how best to use the additional after-tax income in the hands of individuals and businesses rather than the federal government. It is a policy that if brought in by the Harper government and communicated effectively to the public would be difficult for the opposition parties to reject and fight an election on.[12]

The remaining 25 percent of the tax revenues would go to the technology fund, or about $1.5 billion initially (using Rick Hyndman's estimated revenue from a $5 per tonne carbon tax). $1.5 billion is a lot of money for one country to use to help fund investment in new technologies. Compare that to the $236 million in the GEF for climate assistance. See Chris Green's paper in this volume for how to spend this money and why it is necessary.

Policy #5: Implement a Canadian REDD Program Starting with Territorial Lands of Canada's First Nations

This proposal follows directly from the Copenhagen Accord's goals and lessons #3 and #4. Canada's First Nations have heritage rights to a significant

[11]As in the case of British Columbia, the federal government could also increase the GST credit for low-income people to help soften any impact of the carbon tax on prices of fuels and goods.

[12]The NDP in British Columbia will focus on issues other than the carbon tax in the next election. It learned from the May 2009 provincial election that opposition to the carbon tax not only incurred the wrath of environmental groups, but did not incite enough of the population to vote against the Liberals. The HST may be another matter!

proportion of the country's forests. A number of First Nations in British Columbia are actively engaged in looking for ways to manage the forests that enhance productivity while sequestering carbon and reducing GHG emissions. Funding for these initiatives is being sought from voluntary markets, but the current prices are low because there are few binding offset policies in place in North America. Canada could serve as a leader in REDD and, at the same time, combine climate policy with environmental stewardship and economic policy to support the creation of sustainable jobs for First Nations communities. Funding could come from a realignment of Indian and Northern Affairs Canada and Natural Resources Canada and/or part of the technology fund noted above.

CONCLUDING COMMENTS

These policy proposals focus on the areas that are not current federal policy and not likely to be popular with the current government. Other federal, provincial, and municipal policies such as promoting energy efficiency, public transportation, land-use strategies such as compact communities, climate-adaptation initiatives, and so on are all necessary as well, and much less controversial. But we need more than these to reduce our emissions and to show Canada's commitment to substantial reductions in GHG emissions in coming years. Addressing Canada's dependence on fossil-fuel extraction and use is a challenging issue and lesson #4 from Copenhagen states the obvious – we need a suite of policy initiatives. First and foremost, we need to price carbon; the existing policies have not led to reductions in GHGs. I've offered reasons why I believe a tax is a superior instrument to cap and trade. I do not believe Canada must follow exactly the United States. Our Canadian provinces are implementing their own policies, as are U.S. states, the European Union and other countries. The longer our federal government waits to act, the more difficult it will be to harmonize climate policies across the country. Differential policies can create undesirable leakage of economic activity from one region to another. The federal policy proposed here would create an incentive for the provinces to harmonize their carbon policies with the federal government. We do not have to wait until the United States acts. They've set a policy direction that supports pricing carbon and a carbon tax will likely be easier than a made-in-Canada cap-and-trade regime to show comparability when border adjustments are imposed. A growing number of industrial leaders and not just economists are pointing out the relative merits of taxes over cap and trade. Finally, acting swiftly and decisively on carbon pricing might just earn our federal elected leaders an award at the next Climate Conference of the Parties that they could be proud of.

REFERENCES

McColl, V. 2009-2010. "The Politics of a Global Climate Change Deal", *Policy Options* December 2009/January 2010: 56-60.

National Round Table on the Environment and Economy. 2007. *Getting to 2050: Canada's Transition to a Low-emission Future.*

— 2009. *Achieving 2050: A Carbon Pricing Policy for Canada.*

Paltsev, S. *et al.* 2007. "Assessment of U.S. Cap-and-Trade Proposals", MIT Joint Program on Science and Policy of Global Climate Change. Report No. 146. Cambridge, MA: MIT.

Appendix 1

Executive Summary

Adopted from

Achieving 2050: A Carbon Pricing Policy for Canada
A 2009 Report from the National Round Table on the Environment and the Economy

In 2009, Canada finds itself facing both new and familiar climate policy challenges. The past several years have seen the emergence of federal and provincial plans to arrest and ultimately reduce greenhouse gas emissions (GHGs) in Canada. A variety of policy instruments have been ventured – from carbon taxes to trading regimes to technology funds to regulations. A deeper understanding by many Canadian interests of the likely scale of the problem and solutions to it is taking root.

Yet, the collective result has been perhaps less than anticipated. Carbon emissions remain on a rising path; Canadian businesses and consumers confront the prospect of a fragmented patchwork of federal, provincial, territorial, and regional carbon pricing policies sprouting across the country and continent; and now we are dealing with the onset of a global economic recession more complicated and profound than we have experienced in decades.

But with these challenges come opportunities. A new administration in the United States has committed to significant climate policy action domestically and internationally. A growing international consensus to develop a post-2012 framework implicating all emitters is emerging. And, economic recession will ultimately give way to renewed economic growth, giving Canada the opportunity to position itself now for a truly sustainability-oriented recovery based in part on an effective, unified national carbon pricing policy.

The movement toward a low-carbon world is inevitable. But our place in it is not. Like our economy as a whole, Canada's long-term competitiveness in a low-carbon future will not be served by inter-jurisdictional carbon competition here at home or by allowing protectionist carbon barriers to be raised at our expense abroad. The link between the two is obvious. Engagement internationally needs to be reinforced by harmonized action nationally. Canada's national environmental and economic interests jointly demand such an approach.

The National Round Table on the Environment and the Economy believes now is the time to press forward on the design of the right climate policy for Canada and Canadians. A year of research and consideration has reinforced our view that it is urgent to act decisively, even in the face of current economic turbulence and evolving climate science. Now is exactly the time to seize the opportunity before us – of preparing for a sustainable economic recovery and actively engaging the United States and our other major trading partners. Now is the time to lay the groundwork for a truly effective long-term climate policy framework through a nationally collaborative approach to a unified carbon pricing policy in Canada and an internationally harmonized approach in North America.

This report recommends a unified carbon pricing policy for Canada – a policy aimed at meeting one clear objective: *the greatest amount of carbon emission reductions, at the least economic cost.* Following more than a year of research and consultation, our report sets out what we believe is the most effective, realistic, and achievable carbon pricing policy for current and anticipated Canadian circumstances.

The scale of transformation to the Canadian energy system to meet the federal government's 2020 (20% below 2006 levels) and 2050 (65% below 2006 levels) emission reduction targets should not be underestimated. Greenhouse gases are so widely embedded in the energy we use that to significantly reduce emissions will have wide-ranging economic and social implications. Our collective challenge now is to transition the emerging fragmentation of current carbon pricing policies to a unified policy framework across all emissions nationally. The negative consequence of not doing this, and maintaining this fragmentation of differentiated carbon prices across emissions and across jurisdictions, will be significantly higher economic costs, intensified environmental impacts, entrenched barriers that will make it harder to act in the future, and the real risk of not being able to meet Canadian emission reduction targets.

A CARBON PRICING POLICY FOR CANADA

The carbon pricing policy proposed in this report has two main goals. *First*, it seeks to achieve the Government of Canada's medium- and long-term greenhouse gas emission reduction targets at least cost. *Second*, it seeks to minimize adverse impacts of achieving these targets on regions, sectors, and consumers.

A nationally integrated carbon pricing policy is required to meet these goals based on four main elements. At the core is an *economy-wide cap-and-trade system* to price carbon and provide real market incentives for firms and households in Canada to change their technology choices and behaviour in order to reduce emissions. *Complementary regulations and technology policies* are then needed to improve the cost-effectiveness of the cap-and-trade system by broadening coverage across all key emission sources, while supporting targeted technology development and deployment. Participation in *international emissions markets* through trading and credit purchases will help reduce economic costs at home by allowing Canadian firms and consumers access to credible reductions internationally. Finally, a climate *governance and implementation strategy* is needed to establish new, collaborative institutions and coordinating processes to implement and adapt the carbon pricing policy over time, making sure it sends a clear and certain price signal to industry and consumers, while remaining responsive to new information and situations.

These are our conclusions:

- An economy-wide carbon price signal is the most effective way to achieve the Government of Canada's medium- and long-term emission reduction targets and reduce cumulative emissions released into the atmosphere.
- That price signal should take the form of an economy-wide cap-and-trade system that unifies carbon prices across all jurisdictions and emissions and prepares us for international linkages with our major trading partners.
- An effective carbon pricing policy needs to find a balance between certainty and adaptability – it should be certain enough to transmit a clear, long-term price signal to the economy upon commencement to encourage technology and change behaviour, yet adaptable to changing circumstances and future learning.
- There is a cost to delay in the form of higher carbon prices later to meet targets, and a cost to maintaining Canada's current fragmented approach to carbon pricing policies in the form of reduced GDP and higher carbon prices over time.
- Canada's economy will continue to grow under this policy – it is forecast to be twice as large in 2050 than today – but this will be smaller than if no carbon pricing policy were adopted.
- New federal/provincial/territorial governance mechanisms and processes should be put in place to achieve a harmonized Canadian carbon pricing policy.
- Technology development and deployment, along with the electrification of the energy system, is central to emission reductions and is stimulated through an economy-wide carbon price signal, as well as appropriate public investment in carbon capture and storage and renewable energy.
- Complementary regulations and technology policies in the transportation, buildings, oil and gas, and agricultural sectors are also required to ensure broad-based emissions coverage at an overall lower price, reduce total emissions, and meet government targets.

GUIDING PRINCIPLES FOR A CANADIAN CARBON PRICING POLICY

Getting started with the right national carbon pricing policy is the first, best step Canada can take to achieve its ambitious medium- and long-term greenhouse gas emission reduction targets. Our research indicates that Canada has the capacity to successfully achieve these targets while maintaining a high standard of living and continued economic well-being. But our research also shows that this transformation will require us, as a country, to take three steps:

First, we need to implement a carbon pricing policy that is both certain and adaptable. Investors and consumers will have the confidence to change their behaviour if they are certain the policy and prices are real; at the same time, the policy must be responsive to changing information and circumstances to secure our own interests.

Second, we must unify carbon policies and prices here at home. That means transitioning from the current, fragmented patchwork of federal, provincial, territorial, and regional policies to a unified or harmonized carbon pricing policy that covers all emissions in all jurisdictions.

Third, we need to link our carbon pricing policy and trading system with the world next door. Enabling international emissions trading, particularly with our largest trading partner, the United States, will help address competitiveness concerns and manage our costs.

Unify at home; link with abroad; implement with certainty and adaptability. This is the foundation for the specific carbon pricing policy guiding principles we set out below:

- *Focus on carbon prices and economic efficiency.* With Canadian targets set, an important first principle is to ensure that the policy focuses on economic efficiency so that long-term costs are minimized. This means providing a unified carbon price across emissions and jurisdictions. While adverse impacts on some segments of the economy and society can be expected, these are best dealt with through targeted income support and not through a fundamental dilution of the carbon price signal.
- *Move to uniformly apply the carbon price across all emissions.* This will make Canadian carbon policy more cost-effective by avoiding sector-specific exclusions for competitiveness or jurisdictional reasons. While there will likely be adverse and perhaps disproportionate impacts on some, the carbon pricing policy should not deliberately omit emissions as a starting point. Otherwise, overall costs will need to rise accordingly by those paying to meet the stated targets, which will be viewed as unfair and inequitable. Using revenues generated by the cap-and-trade system through the auctioning of emission permits provides flexibility within the uniform system to address specific economic or societal needs arising from the carbon pricing policy.
- *Contain costs initially and then transition the policy to deliver more certain emission reductions over time.* Uncertainties dominate climate policy, including abatement response, cost uncertainties, and most importantly the carbon prices that major competitors will be imposing on their industries. These uncertainties indicate a need for climate policy to initially contain costs as uncertainties are revealed. But with cost containment comes reduced emission reductions that must be balanced against achieving our targets. The carbon price should therefore align with the emissions reduction targets. Ultimately, there is a need to transition the initial cost containment approach to one focusing on getting the emission reductions we need through higher carbon prices over time.

- *Position Canada to participate in international policy frameworks.* Given the very high carbon prices required to attain domestic reductions sufficient to hit our long-term targets, a policy that seeks real and verifiable reductions from outside Canada to lower domestic costs makes sense. To implement this, Canada's carbon pricing policy should be designed to eventually link with major trading partner systems, particularly those of the United States.
- *Develop governance mechanisms to set policy but also to update expectations about future carbon prices.* Policy credibility over the long term is required to drive needed technology investment and behavioural change. Creating dedicated governance mechanisms that implement the carbon pricing policy in a transparent and accountable manner is central to maintaining this credibility. This requires a rules-based approach that minimizes political interventionism and future policy backsliding. Monitoring and reporting progress publicly is equally important as part of updating expectations that carbon prices or emission quantity restrictions will need to rise or fall, relative to that progress.

RECOMMENDATIONS

This report serves as a comprehensive and integrated recommendation for developing and implementing a Canadian carbon pricing policy. To reinforce the report's research, analysis, and conclusions, the NRTEE highlights the following specific recommendations for consideration:

1. Unify carbon policies and prices across emissions and jurisdictions based on three principal policy elements:
   - an economy-wide cap-and-trade system transitioned from current and planned federal, provincial, and territorial initiatives;
   - complementary regulations and technology policies in the transportation, buildings, oil and gas, and agricultural sectors; and
   - international carbon abatement opportunities that are credible, affordable, and sustainable.

2. Ensure the unified Canadian carbon pricing policy can link with current and proposed international systems and, most particularly, with a prospective trading regime likely to emerge in the United States, to ensure compatibility in pricing and action.

3. Use generated revenue from permit auctions first and foremost to invest in the required technologies and innovation needed to meet the Canadian environmental goal of reduced GHG emissions.

4. Transition the current fragmented approach to carbon pricing across jurisdictions and emissions to a unified Canadian carbon pricing regime as soon as possible and no later than 2015.

5. Establish a dedicated carbon pricing governance framework based on adaptive policy principles to develop, implement, and manage the unified carbon pricing regime over time with the following elements:
   - Federal/provincial/territorial collaboration through an ongoing forum, which would allow governments to coordinate and harmonize efforts and actions in support of the unified carbon pricing policy, and regularly consult and engage with each other to maintain progress and direction on carbon emissions pricing revenue distribution and climate policy development.
   - An expert Carbon Pricing and Revenue Authority with a regulatory mandate to collect auction revenues from emitters, set carbon pricing schedules and compliance rules, establish permit allocation rules based on principles and policy directions set by the federal government, monitor and enforce compliance, implement procedures for monitoring and reporting emissions, and ensure confidence in the long-term robustness of the policy.
   - An independent, expert advisory body to provide regular and timely advice to government on interim targets for each compliance period; on the distribution of auction revenue to meet environmental, economic, and social objectives as required; on ongoing evaluation and assessment of the carbon pricing regime; and on any proposed adjustments to the policy and pricing framework for decision makers to consider.

Appendix 2

TD Economics*
Special Report
October 29, 2009

Answers to Some Key Questions about the Costs of Combating Climate Change
A Summary of the Pembina/David Suzuki Foundation Paper

Climate change is a widely discussed policy issue that continues to rank high in public opinion polls. It is also a key international political concern, as evidenced by the forum to be held in Copenhagen, Denmark in December, in which Canada will be a participant. One of the key goals of this gathering is to answer the question of how much industrialized countries are willing to reduce their emissions of greenhouse gases.

TD has become increasingly concerned that the environment debate is largely conducted without objective analysis of the economic impacts (on a national, regional, and sectoral basis) or an appreciation of the breadth and depth of the measures that would be required to achieve the objectives.

To this end, TD helped provide funding to conduct research on what it would take to achieve the federal government's target, how much it would cost,

---

*This report is provided by TD Economics for customers of TD Bank Financial Group. It is for information purposes only and may not be appropriate for other purposes. The report does not provide material information about the business and affairs of TD Bank Financial Group and the members of TD Economics are not spokespersons for TD Bank Financial Group with respect to its business and affairs. The information contained in this report has been drawn from sources believed to be reliable, but is not guaranteed to be accurate or complete. The report contains economic analysis and views, including about future economic and financial markets performance. These are based on certain assumptions and other factors, and are subject to inherent risks and uncertainties. The actual outcome may be materially different. The Toronto-Dominion Bank and its affiliates and related entities that comprise TD Bank Financial Group are not liable for any errors or omissions in the information, analysis or views contained in this report, or for any loss or damage suffered.

and who might bear those costs. The authors of the report – the Pembina Institute and David Suzuki Foundation (DSF) – were also interested in examining the costs associated with the deeper target, supported by environmental non-government organizations (ENGOs). M.K. Jaccard and Associates Inc. was engaged to do the formative analysis using an energy economy simulation model and a macroeconomic general equilibrium model.

It is important to note upfront that TD does not endorse the Pembina/DSF report, or a particular target or set of policies related to GHG emissions. However the analysis done by M.K. Jaccard and Associates (MKJA) appears to be robust. And this report will help fill an information gap and further a productive debate on environmental policy. No doubt alternative assumptions and models could produce different results that might also be realistic. TD hopes that the release of the analysis will provoke alternative research into the economics of addressing climate change. In our opinion, an informed national debate is warranted on the policy options and the associated costs.

While the assumptions and models used shape the outcomes, TD believes that the findings provide one set of answers to some of the key questions that are at the core of the climate change policy debate.

What targets might Canada pursue?

The MKJA analysis assesses the economic impact of two different targets. First, the Government of Canada has announced a commitment to reduce greenhouse gas (GHG) emissions by 20% from the levels in 2006, which constitutes a 3% reduction from the level in 1990. Second, environmental non-government organizations (ENGOs) have argued for a more ambitious target of lowering emissions by 25% from their level in 1990 by 2020. This call is broadly consistent with the Intergovernmental Panel on Climate Change (IPCC), which argued that the industrialized countries need to reduce their GHG emissions to 25-40% below the 1990 level by 2020 if they are to make a "fair" contribution. The principle of "fair" reductions reflects the fact that developing countries were not the main contributors to the emissions in the past and their economic development should not be unfairly diminished by efforts to reduce emissions – which will be a key issue discussed at the upcoming forum in Copenhagen.

Can the targets be achieved?

The MKJA modelling suggests that either target can be met, but there is a material economic cost to each. And, the cost is naturally much deeper with the more stringent target. There is a strong regional and sectoral dimension to the costs, as they are not spread evenly across the country. There is a variety of approaches that could be taken to achieve each outcome. The MKJA analysis presents the outcomes under one set of assumptions provided by Pembina and DSF, who felt that the selected policies were the most efficient and equitable combination that achieved the targets at the least cost to individuals, businesses and society.

Can Canada pursue a more stringent emissions target than other countries?

The MKJA modelling suggests that Canada can achieve either target without other countries following suit. One of the surprising results from the modelling is that the overall economic cost is not materially higher if Canada pursues targets that are more stringent than other nations. However, the pursuit of a more aggressive Canadian target does have an impact on some of the policy actions, such as requiring the purchase of more international permits and affects the regional and industrial impact, since there is a greater burden borne by Canadian carbon-intensive industries and energy-rich provinces.

Actions Taken to Reduce Emissions Under the Government Target, Mt CO_2e (2020)

	Canada Goes Further	*OECD Acts Together*
Baseline (BUA[a]) emissions	848	848
Emissions after application of domestic policies	626	643
Domestic emissions reductions:		
Output reduction	36	21
Other GHG control	43	38
Fuel switching to nuclear	0	0
Fuel switching to renewables	22	22
Fuel switching to electricity	30	29
Fuel switching to other fuels	10	10
Carbon capture and storage (CCS)	30	32
CCS energy efficiency penalty	5	5
Energy efficiency	49	49
International permit purchases	56	73
Target (remaining emissions)=Baseline-domestic emissions reductions-permit purchases	570	570

[a]BAU=business as usual.
Source: M.K. Jaccard and Associates Inc.

Actions Taken to Reduce Emissions Under the ENGO[a] Target, Mt CO_2e (2020)

	Canada Goes Further	*OECD Acts Together*
Baseline (BUA[b]) emissions	848	848
Emissions after application of domestic policies	514	535
Domestic emissions reductions:		
Output reduction	64	36
Other GHG control	52	46
Fuel switching to nuclear	1	1
Fuel switching to renewables	33	35
Fuel switching to electricity	33	33
Fuel switching to other fuels	10	11
Carbon capture and storage (CCS)	76	84
CCS energy efficiency penalty	9	10
Energy efficiency	57	58
International permit purchases	80	101
Target (remaining emissions)=Baseline-domestic emissions reductions-permit purchases	434	434

[a]ENGO=Environmental NGO; [b]BAU=business as usual.
Source: M.K. Jaccard and Associates Inc.

Is a carbon price part of the policy solution?

Pembina/DSF, and the MKJA modelling, use a core assumption that a carbon price is applied in order to evoke a change in behaviour on the part of consumers and businesses. The purpose of the carbon price is to lower demand for high GHG emitting activities or products. For example, the application of the carbon price raises the cost of fossil fuels relative to the cost of other energy sources. The carbon dioxide and carbon dioxide equivalent emissions charge could take the form of either upstream cap-and-trade system or a carbon tax – the modelling by MKJA is agnostic between these two alternatives. For the government target, the MKJA analysis used a charge of \$40/tonne CO_2e starting in 2011 and rising to \$100/tonne CO_2e in 2020. For the ENGO target, a charge of \$50/tonne CO_2e starting immediately in 2010 was used, rising to \$200/tonne CO_2e in 2020. While these carbon prices curtail GHG emissions, they do not achieve the targets on their own.

Are regulations required on top of carbon prices?

Pembina and DSF assume the application of complementary regulations by the Federal and Provincial governments. These are deemed necessary on the grounds of efficiency (i.e., they are less costly than relying purely on carbon prices) and some of them address market failures. With one exception, the Pembina/DSF recommended set of regulations are the same regardless of which target is pursued. The regulations used in the MKJA analysis that are implemented by 2011 include:

- Elimination of non-safety related venting and flaring in the upstream oil and gas sector, with a carbon charge applied on the safety emissions.
- Increased energy efficiency for all new buildings. New commercial buildings to be built to LEED Gold standard or higher. Residential buildings to be 50% more energy efficient than current standard practices. There is an added assumption that all new buildings in British Columbia, Manitoba and Quebec are restricted to using electric heating.
- All new vehicles sold to meet the California GHG emissions standards, with these standards being gradually tightened over time. As of 2011, "white good energy efficiency standards" for all appliances to be raised to the most efficient commercially available that existed in 2008 and then improved over time.
- All landfills to be covered and the landfill gas flared or used to produce electricity and heat.

Under the more stringent ENGO target, there is one additional regulatory assumption in terms of the use of carbon capture and storage (CCS). Specifically, CCS is regulated for most emissions from new natural gas processors, new hydrogen production facilities, and new coal fired electricity plants, oil sands facilities and upgraders starting in 2016.

Are international permits required to meet the target?

The MKJA analysis finds that the use of international emission permits is required to avoid excessively high domestic carbon prices and to take advantage of lower emission reduction costs abroad. The traditional case for the use of international permits is that from a climate point of view what matters are global emissions – not the location of where the emissions are taking place. The analysis assumes that Canada buys between 56 Mt and 73 Mt CO_2e of permits in 2020 to achieve the government target – with the lower number being applicable if Canada has a more stringent target than other countries and the higher number if the OECD countries have similar policies. The reason for less permits being purchased in the case where Canada has a tougher target is a reflection of the fact that output growth by carbon-emitting industries is reduced under this scenario. To hit the ENGO target, between 80 Mt and 101 Mt CO_2e of permits are required in 2020, again depending on whether other countries are pursing similar policies to Canada or not.

Will the government reap huge tax windfalls from carbon prices?

The MKJA modelling estimates that government revenue from applying the carbon price assumptions made above to hit the government target would be at least $40 billion per year in 2020. The revenue generated from hitting the ENGO target is estimated at least $70 billion in 2020. However, the modelling shows that in order to achieve the GHG emissions reductions at the least economic cost, the carbon-related revenues should be fully recycled into the economy.

The assumptions of Pembina/DSF and applied in the MJKA modelling are that the recycled funds are used to:

- Invest in public transit, with usage increasing by 35% compared to what would otherwise occur.
- Upgrade the electricity emissions grid to allow greater use of intermittent renewable electricity generation, with the latter to reach 25% of generation in some regions.
- Provide refunds to the two most adversely affected manufacturing industries (industrial minerals and metal smelting) to maintain their output at the level recorded in 2008. More on this later.
- Fully refund individuals for the resulting higher household energy costs.
- Purchase verifiable domestic agricultural offsets.
- Purchase the needed international emissions permits.
- Once all of the above are accomplished, the remaining funds are used to lower personal income taxes to provide a boost to economic activity in order to soften the impact of the climate change policies.

It may seem odd to readers that carbon prices are applied and then that a couple of industries and all consumers receive rebates or tax reductions. The analysis shows that the combination of these actions raises the cost of high carbon-emission activities relative to low carbon-emission activities. This lowers demand for the former and raises demand for the latter, which leads to a reduction in GHG emissions.

One could argue that the policies to reduce GHG emissions are, in effect, a massive fiscal transfer that leads to a major industrial realignment. A tax (either directly or indirectly) is being applied to carbon-emission heavy activities, and then fiscal transfers are made to reduce the economic impact, which acts as a boost to low carbon-emission activities.

Does action need to be taken immediately?

The Pembina and DSF assumptions include a carbon price being applied in 2011 to reach the government target and applied in early 2010 to reach the ENGO target. All other policy actions begin in 2011, with the exception of regulations for Carbon Capture and Storage (CCS) in the ENGO target that takes effect in 2016. The modelling shows clearly that if the actions are delayed, the cost to achieve the same target will increase materially.

What is the national economic impact of reaching the targets?

Under the assumptions made by Pembina and DSF, and compared to an environment where no policy action is taken, MKJA concludes that achieving the government target reduces the level of Canadian real GDP by approximately 1.5% by 2020. Achieving ENGO target lowers real GDP by 3.2%. The cost is equivalent to a significant recession of varying magnitude depending upon the target. Unlike recessions, however, the lost economic output would not be recovered by a subsequent economic rebound.

However, it is important to stress that unlike recessions, the economic impact would be gradually felt over a decade. Under a scenario where no policy action is taken, the modellers assume that the Canadian economy would expand by 27% over the 2010 to 2020 period – or 2.42% per annum. Under the assumptions made above, the MKJA estimates that the government target can be achieved and the economy would grow by 25% (regardless of whether Canada has the same policies as other countries or more stringent ones), which is an average annual growth rate of 2.26% per year, or 0.16 percentage points less per annum than the business as usual case. MKJA finds that hitting the ENGO target would allow the economy to grow by 23% over the decade, or 2.09% per annum, and again is regardless of the policies taken by other countries.

Are some industries more impacted by the required policy actions?

The analysis by MKJA shows that economic growth continues while hitting both targets, but the carbon prices and regulations ultimately lead to a major structural change in the Canadian economy, away from heavy carbon emitting industries (like fossil fuels) and towards lower carbon emitting industries. Because the former also tend to be capital intensive businesses, there is also a shift towards more labour intensive activities – which limits the negative impact on employment (more on this below).

Change in Level of GDP in 2020 from Business as Usual (%)								
	BC	AB	SK	MB	ON	QC	ATL & RoC	Canada
GOVT OAT[a]	-2.2	-7.3	-1.2	1.9	0.6	-0.7	-0.5	-1.4
GOVT CGF[b]	-2.5	-8.5	-2.8	2.1	0.9	-0.3	-0.1	-1.5
ENGO OAT	-4.2	-11.9	-4.7	2.7	0.0	-1.3	-2.5	-3.0
ENGO CGF	-4.8	-12.1	-7.5	2.1	0.0	-1.3	-1.9	-3.2

[a]OAT=OECD acts together; [b]CGF=Canada goes further.
Source: M.K. Jaccard and Associates Inc.

Projected Cumulative Economic Growth Between 2010–2020 (%)								
	BC	AB	SK	MB	ON	QC	ATL & RoC	Canada
BAU[a]	30	57	26	20	21	15	33	27
GOVT OAT[b]	27	46	24	22	22	14	32	25
GOVT CGF[c]	27	44	22	22	22	15	33	25
ENGO OAT	24	39	20	23	21	13	30	23
ENGO CGF	24	38	16	22	21	14	30	23

[a]BAU=business as usual; [b]OAT=OECD acts together; [c]CGF=Canada goes further.
Source: M.K. Jaccard and Associates Inc.

The most adversely affected industries in terms of slower growth are petroleum refining, petroleum and natural gas extraction, and coal mining. Less affected, but still negatively impacted (particularly under the "Canada goes further scenario") are industrial minerals, freight transport, chemical products, paper manufacturing, iron and steel, and metal smelting. As one might expect, the impact is greater under the ENGO target than the government target.

However, a commitment was made by the modellers when formulating the analysis that no manufacturing industry would be allowed to experience lower output than its level in 2008. Only metal smelting had this outcome under the government target. Under the ENGO target, both the metal smelting and industrial minerals sectors failed to meet the pre-established limit. The analysis assumes these industries receive government transfers to bring output back up to the 2008 level.

There are industries that benefit from the carbon prices and the regulatory changes. For example, there is an increased demand for electricity. Ethanol and Biodiesel also experience a dramatic rise in output compared to an environment without any policy changes. The shift away from capital intensive industry and towards labour intensive industry also creates added growth in the latter.

What is the impact on employment?

The MKJA models predict that overall employment in the Canadian economy would not be reduced by achieving either target. In fact, the policies might lead to marginally higher employment. TD Economics considers this a surprising result warranting further reflection. The modelling explanation has to do with the recycling of the carbon price revenues. The loss of economic output is accompanied by a decline in wage rates, which encourages firms to hire more workers. The personal income tax cuts are so substantial that after-tax personal income rises, which induces an increase in the supply of labour. There is also a shift away from capital-intensive industry and towards labour-intensive industry, which boosts demand for workers. So, in the analysis done by MKJA, output is lower and employment is largely unchanged – which implies a weaker performance for labour productivity.

One should note that while aggregate employment is not dampened, and may actually increase slightly according to the modelling, the industrial structural change would lead to a considerable disruption to labour markets in the negatively affected sectors. Many workers in the capital-heavy GHG-emitting industries would experience job losses and they would need to be retrained and supported while moving between industries. The impact on these workers should not be dismissed just because total employment is not reduced. Moreover, pre-tax wages of workers in general are lowered by the policies, reflecting the negative impact on productivity coming from lower output growth but little impact on aggregate employment.

Will different regions be more or less impacted than others?

The MKJA modelling suggests that the structural changes at the industrial level will lead to significant regional implications. As one would expect, provinces with a greater concentration of heavy carbon emitting industries will be the most adversely affected.

Relative Change in GDP from Business as Usual in 2020

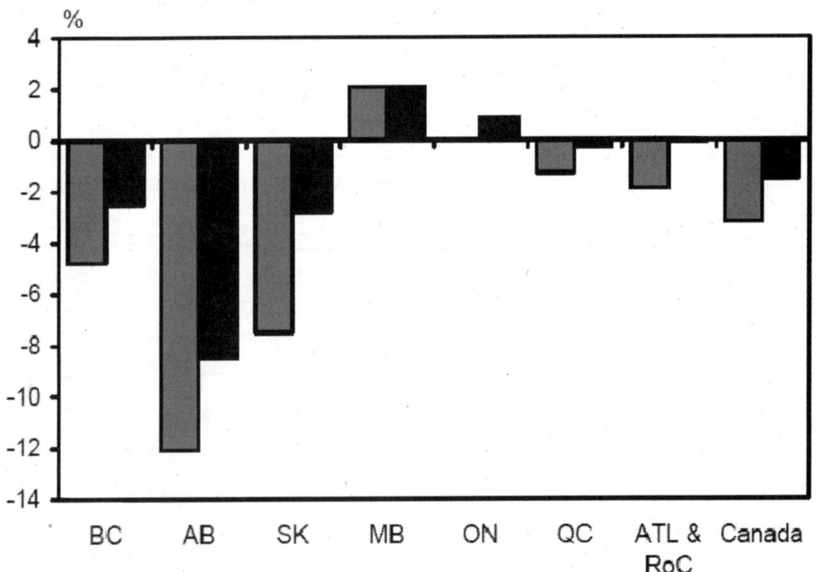

Source: M.K. Jaccard and Associates Inc.

For example, according to MKJA, hitting the government target under the "Canada goes further" scenario leads to a 1.5% decline in the national level of Canadian real GDP compared to a business as usual scenario, but output in Alberta falls 8.5%, Saskatchewan drops 2.8% and British Columbia declines 2.5%. Achieving the ENGO target leads to a greater impact. National real GDP drops 3.2%, while Alberta suffers a 12.1% decline, Saskatchewan loses 7.5% and British Columbia falls 4.8%.

Because the impact is felt over a decade, the MKJA analysis shows that the economies still grow. The average annual growth rate for Alberta, without carbon prices and carbon reduction regulations, was projected to be 4.6% between 2010 and 2020. Hitting the government target and under the "Canada goes further" assumption, Alberta growth slows to 3.7% annual pace. The ENGO target with Canada pursuing more stringent targets has Alberta growing at 3.3% annum. Saskatchewan's trend growth rates are projected to slow from 2.3% to 2.0% or 1.5% a year, while British Columbia's trend growth rate slips from 2.7% to 2.4% or 2.2% – under each scenario respectively.

Notably, under the MKJA modelling, the other provinces are significantly less affected. Indeed, Manitoba and Ontario might even see a small increase in output depending on the scenario. The reason is that these provinces have less concentration in GHG-heavy emitting industries and the reallocation of capital away from such industries leads to greater capital investment in these provinces. The modellers treated Atlantic Canada and the Territories as a group, and in aggregate, they are little affected hitting the government target and only moderately negatively impacted reaching the ENGO target, but regions in the composite with significant exposure to the energy sector would likely experience greater weakness than the average, while the others would be little impacted in terms of overall GDP.

Can't technological change reduce the cost?

Pembina and DSF only assume the use of technologies that exist today, but in some cases they assume wide use of technologies that have not yet been deployed commercially in a broad based fashion. The main example is carbon capture and storage.

It is true that new technologies can help to reduce GHG emissions. Over the next decade, however, it is not reasonable to expect that technical advances will provide a solution. The MKJA analysis shows that action would be required quickly to achieve the targets. Innovation is unlikely to provide the answer in the 2010 to 2020 time frame. Having said that, new technologies may have a significant role in achieving GHG emission objectives in the long-term, such as the 2020 to 2050 time frame. Indeed, the implementation of the rising price on carbon could prove to be a significant catalyst for the development of new carbon-reduced or carbon emissions limiting technologies.

Is this assessment reasonable?

Based on the assumptions used and the models applied, the impact assessment done by MKJA appears reasonable. The estimate on reaching the government target is broadly consistent with the National Roundtable on the Environment and the Economy (NRTEE) findings on the same subject. For example, the NRTEE analysis suggests that hitting the government target would reduce the level of real GDP by 1% to 3%, lowering the annual pace of national economic growth by 0.2% per annum into an annual range of 1.5% to 2.0% in 2020. The similarity of the analysis is not an entirely surprising outcome, since some of the same models and assumptions were used in both sets of modelling. However, a comparison between the work by MKJA and the NRTEE shows how assumptions can differ, as the NRTEE allows for a greater use of nuclear power. This simply highlights that there are different paths to reach the same outcome. The key addition to the assessment of hitting the government target by the Pembina/DSF report is to add the critical regional dimension, which was not addressed by the NRTEE. Also, the Pembina/DSF study provides a more in-depth analysis of complementary regulations and public investments, and takes into account the two years that have elapsed since the NRTEE analysis was undertaken.

The key lessons that we take away from the analysis is that the macroeconomic and regional economic impact is significant. The breadth and depth of the policy response is also remarkable. The speed at which progress must be made is also notable, and would require considerable commitment. The structural changes necessary at the industrial level are masked by the more subdued headline economic impact assessment. The fiscal transfer involved is enormous and has a significant impact on the economy.

Résumés – La tarification du carbone et le fédéralisme environnemental

---

Le combat de la politique canadienne face au changement climatique : mise en place d'une tarification carbonique
Bob Page

Ce chapitre constitue un survol historique, politique et intergouvernemental de l'expérience du Canada en matière de mesures contre les changements climatiques post-Kyoto. Il démontre notamment la complexité et l'interdépendance des politiques en ce domaine, dont l'influence se fait sentir sur pratiquement toutes les facettes de la Politique publique canadienne, et qui créent des conflits entre les paliers gouvernementaux ou les provinces productrice d'énergie fossile et leurs homologues favorables à l'accord de Kyoto. Page remarque que le Canada, à titre de pays émetteur, exportateur de ressources et à taux d'accroissement démographique très rapide, était fondamentalement désavantagé par l'accord de Kyoto : il devenait le seul pays des Amériques à avoir des obligations suite à l'entente. Il fait aussi voir le rôle crucial que jouent les sables bitumineux dans la formulation des futures politiques sur les changements climatiques, et les problèmes qui découleront fort probablement de la volonté de l'Alberta de conserver toutes les recettes provenant des énergies fossiles et de celle du Québec d'élargir la vente de crédits de carbone.

Les approches fédérales et provinciales actuelles face à l'atténuation du changement climatique : répétons-nous les erreurs du passé?
Nic Rivers

Les mesures climatiques canadiennes du passé – provenant surtout du fédéral – étaient nettement moins rigoureuses et efficaces que nécessaire pour atteindre les cibles gouvernementales de réduction d'émissions de gaz à effet de serre. Cet article évalue les politiques courantes sous l'angle des engagements à moyen terme (2020) des gouvernements fédéral et provinciaux. L'analyse, qui se base sur un modèle énergétique-économique quantitatif, suggère que les politiques actuelles seront nettement insuffisantes pour atteindre les objectifs des deux paliers de gouvernement. Cet échec est causé par le fait que les politiques en question ne sont pas exhaustives (c.-à-d. que certaines sources d'émissions ne sont actuellement soumises à aucune cible de réduction d'émissions) ni rigoureuses; il est aussi causé par la création d'instruments axé sur les forces du marché, donc trop flexibles, et par un manque de coordination, en matière de développement de politiques, entre les gouvernements des provinces et Ottawa. Si l'on tient à atteindre les cibles de 2020, il reste peu de temps pour réformer les politiques concernées.

La tarification du carbone et l'impératif technologique
Christopher Green

Le changement climatique est essentiellement une question de technologie de l'énergie. Toute politique canadienne sur les changements climatiques est illogique si le reste de la planète refuse de faire face à l'énorme défi technologique que représente la stabilisation du climat. À elle seule, la tarification du carbone ne constitue pas une manière adéquate de relever ce défi. Le Canada, de concert avec d'autres nations technologiquement avancées, devrait contribuer *directement* à la création de technologies adaptables, raisonnablement compétitives et transférables internationalement, sans quoi la stabilisation climatique demeurera impossible. Une modeste taxe ou tarification devrait être mise en place pour financer les investissements en technologie et infrastructure énergétiques. Avec le temps, cette taxe devrait augmenter lentement et continuellement, ce qui encouragerait d'un point de vue financier le déploiement de technologies propres à mesure qu'elles sont mises au point. Les articles de Rivers et Page ont décrit l'échec de l'intervention mondiale en matière de changements climatiques, ce qui soutient une approche alternative telle que celle décrite ici.

Tarification du carbone et fédéralisme
Thomas J. Courchene et John R. Allan

Cet article porte sur diverses approches alternatives à la tarification du carbone ainsi que sur leur façon d'interagir avec le fédéralisme canadien et, de manière plus générale, la gouvernance à paliers multiples. L'exposé comprend la taxe sur les émissions carboniques, le plafonnement et échange, des approches mixtes et le modèle taxe ajoutée/tarif sur le carbone semblable à la TPS ou la TVA. On examine ensuite l'interaction entre la tarification du carbone et le fédéralisme car au Canada comme aux États-Unis, ce sont les gouvernements sous-nationaux qui ont pris les devants en matière de politiques sur les changements climatiques et l'établissement des prix du carbone. L'article insiste particulièrement sur la détermination du palier apte à recueillir les recettes provenant des taxes sur le carbone ou de la mise aux enchères de permis. Le reste du chapitre est consacré à une série de défis découlant de la tarification du carbone, défis qui consistent à déterminer la marge de tolérance relative à la croissance de la population, décider si le pays exportateur ou importateur doit être tenu responsable de l'empreinte carbone générée avant l'exportation, et établir les ajustements frontaliers.

La tarification carbonique comme si la diminution des GES importait

Rick Hyndman

Les dommages liés aux émissions mondiales de gaz à effet de serre constituent l'exemple classique d'inefficacité du marché à laquelle on pourrait réagir simplement et efficacement par la mise en place, dans chaque pays émetteur important, de mesures vastes et coordonnées contre les émissions. Réglée à des niveaux qui encourageraient, pour atteindre les cibles, des choix propres en matière de production et de consommation, une simple approche de tarification du carbone entraînerait l'efficacité en termes de réductions d'émission et des investissements en nouvelles technologies propres. La réduction significative d'émissions de carbone est essentiellement un défi technologique. D'ici à ce que la volonté politique d'établissement d'une politique des émissions de carbone se matérialise et entraîne d'importants investissements privés dans les technologies propres, le Canada et les États-Unis devraient mettre en œuvre une modeste taxe initiale sur les émissions de carbone, et en canaliser les revenus vers la recherche technologique nécessaire aux réductions de l'avenir.

La gouvernance à paliers multiples et la tarification du carbone au Canada, aux États-Unis et dans l'Union européenne

Kathryn Harrison

Le fédéralisme a facilité la tarification du carbone dans certaines fédérations et l'a découragée dans d'autres. Dans l'Union européenne, l'impact de la gouvernance à multiples paliers a été largement positif. Le leadership dont ont fait preuve des états membres clés ont encouragé une dynamique horizontale d'imitation renforcée verticalement par la Commission européenne. Par conséquent, c'est l'UE qui a le mieux progressé dans l'adoption de réformes politiques sur la tarification du carbone. Aux États-Unis, où sévissait un vide politique au niveau national, le fédéralisme a facilité au niveau des États la création et la diffusion de politiques en permettant la collaboration de gouvernements d'États en matière de modèles d'échange d'émissions. En revanche, l'impact du fédéralisme sur la tarification du carbone au Canada a été essentiellement négatif puisque les gouvernements fédéral et provinciaux sont dans l'impasse depuis presque deux décennies et que, jusqu'à tout récemment, les provinces canadiennes n'ont pas aussi bien réagi d'elles-mêmes que leurs homologues américains.

La dynamique intergouvernementale de la politique américaine sur le changement climatique

Barry G. Rabe

Aux États-Unis, c'est l'innovation des États qui a entraîné les politiques de changement climatique au cours de la dernière décennie. Cependant, chaque État a son propre modèle de développement et ses habitudes particulières en ce qui a trait aux émissions, ce qui complique toute tentative fédérale de législation, mais indique aussi quelles politiques sont envisageables aux États-Unis. Tout ceci donne le ton : il faut mettre le changement climatique en contexte intergouvernemental grâce à une série d'options qui vont de la centralisation à Washington à la poursuite de la décentralisation vers les États.

L'échange de droits d'émission de carbone et la Constitution

Stewart Elgie

Cet article examine les pouvoirs constitutionnels des gouvernements fédéral et provinciaux qui permettraient de décréter des dispositions législatives de plafonnement et échange visant à réduire les émissions de gaz à effet de serre. Il est probable que le gouvernement fédéral dispose de ces pouvoirs si les dispositions sont conçues avec soin. Cette autorité découle très probablement des pouvoirs en matière de commerce et de droit criminel, bien que la disposition concernant la paix et l'ordre ainsi que le bon gouvernement et les pouvoirs de mise en place des traités puissent aussi la supporter. Dans un cas comme dans l'autre, une modeste expansion de la portée de ces pouvoirs par les tribunaux serait requise.

En ce qui a trait aux provinces, en tenant pour acquis qu'elles disposent des pouvoirs constitutionnels de réglementer les émissions de gaz à effet de serre, elles ont presque certainement le droit de contrôler les échanges de droits d'émission dans leur territoire à titre d'élément du commerce local. Cependant, leur autorité sur les échanges extraprovinciaux est incertaine. Un régime provincial d'échange de droits d'émissions aurait de meilleures chances de réussite s'il avait des objectifs économiques en plus des cibles environnementales, et s'il faisait partie d'une approche coordonnée à multiples juridictions.

L'autorité constitutionnelle de percevoir des taxes sur le carbone

Nathalie J. Chalifour

Cet article analyse la constitutionalité des taxes sur les émissions carboniques. Après avoir décrit les diverses rubriques de compétence fédérales et provinciales justifiant une tarification du carbone, l'essai examine la constitutionalité de la taxe sur les émissions carboniques de la Colombie-Britannique et de la taxation carbonique du Québec. Cet examen démontre que le modèle de la politique de tarification du carbone joue un rôle crucial dans l'évaluation de sa validité constitutionnelle. Bien qu'on s'attende à ce que les deux ordres de taxation autorisent une tarification du carbone, l'analyse montre que les mesures provinciales sont plus faciles à justifier grâce au pouvoir d'attribution des permis des provinces. Pour mettre en œuvre une tarification fédérale du carbone, cet article fait valoir que sa justification se trouverait alors dans la ramification nationale de la disposition concernant la paix et l'ordre ainsi que le bon gouvernement, renforcée par les pouvoirs en matière de commerce et de droit criminel.

La tarification du carbone, l'OMC et la Constitution canadienne
Andrew Green

Le chapitre d'Andrew Green fournit un aperçu du rôle et des pratiques de l'OMC et met l'accent sur les principes qu'elle consacrera probablement aux politiques sur les changements climatiques empiétant sur le système commercial mondial. Il est possible qu'on utilise des ajustements fiscaux frontaliers (AFF) pour surmonter les facteurs de dissuasion politiques qui freinent la mise en œuvre de mesures, et pour encourager les autres pays à agir. Cependant, ces AFF devraient se limiter à des taxes « indirectes », c.-à-d. à des taxes sur les produits et non les producteurs. Puisque la majorité des taxes sur le carbone répondent à ce critère, ou pourraient y répondre, le point de vue de Green est qu'il serait possible de mettre sur pied des AFF liés à un régime de taxation du carbone en vertu de l'OMC. Comme il le démontre, l'utilisation d'AFF conjointement à des systèmes d'échange d'émissions est beaucoup plus litigieuse et incertaine, comme l'est aussi l'interaction entre les AFF et la variation des « méthodes de production et procédés ».

La science économique de la tarification du carbone en Amérique du Nord
Bryne Purchase

Il existe depuis longtemps une solution convenue, simple et élégante au réchauffement climatique : *la tarification du carbone*. La meilleure façon de la mettre en œuvre réside dans une taxe sur le contenu carbonique des combustibles fossiles dans toutes les nations. Pourtant, dans le monde politique, cet instrument n'est toujours pas utilisé. Cet article examine l'économie politique de la tarification du carbone au Canada et aux États-Unis, ainsi que la tortueuse route politique à suivre pour en arriver à un système de plafonnement et échange.

Les clés d'un système canadien de plafonnement et échange
Matthew Bramley

Cet article identifie et examine les problèmes clés liés à la création et la mise en œuvre d'un système national de plafonnement et échange pour la réduction des émissions de gaz à effet de serre au Canada. Ces questions fondamentales comprennent : 1) la capacité de la tarification escomptée du carbone de diminuer efficacement les émissions, et 2) l'organisation des paiements. En ce qui concerne la première question, Bramley remarque que les compensations d'émissions peuvent nuire à la mise en place d'une tarification efficace, et défend la mise en place d'un système de plafonnement et échange permettant d'éviter les systèmes de compensation domestiques et étrangers. Il plaide aussi pour un système de plafonnement et échange aussi étendu et inclusif que possible. Pour ce qui est de la deuxième question, il affirme qu'il faut distribuer de manière équitable et transparente la valeur carbonique imputable à tout système de plafonnement et échange. Il fait voir que les « objectifs d'intensité » forment une caractéristique inéquitable de toute distribution de valeur carbonique et qu'ils ouvrent la porte à des émissions de carbone supplémentaires sans frais tout en alimentant l'incertitude face au niveau véritable des émissions.

Tarification du carbone : mesures et politique
Peter Leslie

À titre d'observateur de conférence, Peter Leslie fournit un aperçu détaillé des nombreux points de vue et arguments des participants, et analyse chacune des approches importantes sur le plan de la faisabilité politique, du leadership fédéral, du fédéralisme environnemental et des problèmes transfrontaliers (c.-à-d. comment la mise en œuvre de diverses initiatives de réduction d'émissions peut influencer et être influencée par les autres juridictions). Il fait valoir que les pays cherchant à prendre l'initiative au niveau international devraient porter attention à deux principales considérations, en plus de chercher à obtenir du capital moral dans les négociations mondiales : (a) comment obtenir une vaste diffusion des politiques domestiques et (b) comment influencer le système international, pour persuader les pays « retardataires » à réduire leurs émissions. Il conclut en faisant voir qu'il est maintenant temps d'abandonner l'hypothèse selon laquelle la seule manière efficace d'atténuer les changements climatiques est d'en venir à un consensus mondial et un accord international contraignant.

Épilogue : les leçons de Copenhague pour la politique climatique canadienne
Nancy Olewiler

Le dernier chapitre du volume est un épilogue (« Les leçons de Copenhague pour la politique climatique canadienne »), écrit par Nancy Olewiler après la conférence de Copenhague. Elle en examine les succès et les échecs, et explore les leçons qu'on peut tirer de la plus grande conférence sur les changements climatiques à s'être jamais tenue. Elle se concentre particulièrement sur les implications de ces leçons pour l'établissement d'une stratégie climatique au Canada, et fait valoir qu'il existe une plus grande marge de manœuvre pour une initiative canadienne, en matière de temps et de conception, que veut bien l'admettre le gouvernement quand il affirme son intention de suivre les progrès des États-Unis.

CONTRIBUTORS

John R. Allan is Associate Director, Institute of Intergovernmental Relations. He is also Vice-President Emeritus and Professor of Economics Emeritus of the University of Regina.

Matthew Bramley is director of the Pembina Institute's Climate Change Program. He is the author of numerous reports and articles on climate policy.

Nathalie J. Chalifour is an Assistant Professor in the Faculty of Law at the University of Ottawa. Her research focuses on ecological fiscal reform, social justice, forest conservation and trade and environment. She is a former Senior Advisor to the Chair of the NRTEE.

Thomas J. Courchene is the Jarislowsky-Deutsch Professor of Economic and Financial Policy and the Director of the Institute of Intergovernmental Relations in the Queen's School of Policy Studies. He is also Senior Scholar of the Institute for Research on Public Policy (Montreal).

Stewart Elgie is a Professor at the University of Ottawa, Faculty of Common Law. He is also the Associate Director of the University's Environment Institute and founder of Sustainable Prosperity.

Andrew Green is an Associate Professor at the University of Toronto Faculty of Law.

Christopher Green is Professor of Economics at McGill University. He is also a member of the Global Environment and Climate Change Centre based at McGill.

Kathryn Harrison is a Professor of Political Science, a faculty associate of the Institute for Resources, Environment and Sustainability, and Associate Dean of Arts at the University of British Columbia.

Rick Hyndman is Senior Policy Advisor, Climate Change and Air Issues for the Canadian Association of Petroleum Producers.

Peter Leslie is a Resident Fellow and former Director of the Institute of Intergovernmental Relations, and Professor Emeritus of Political Studies at Queen's University.

Nancy Olewiler is a Professor of Economics at Simon Fraser University, where she heads the Public Policy Program. She also serves as a Director for both BC Hydro and Translink.

Bob Page is the first TransAlta Professor of Environmental Management and Sustainability at the University of Calgary. He is also the Chair of the Government of Canada's National Round Table on the Environment and the Economy.

Bryne Purchase is an Adjunct Professor in the School of Policy Studies at Queen's university. He is a former Deputy Minister of Finance and Revenue and of Energy, Science and Technology in the Ontario Government.

Barry G. Rabe is a Professor of Public Policy in the University of Michigan's Ford School and also holds appointments in the School of Natural Resources and Environment and the Program in the Environment. He is a non-resident senior fellow in the Governance Studies Program at the Brookings Institution.

Nic Rivers is with M.K. Jaccard and Associates Inc. and has written/consulted on climate change policy for all levels of government, to industry, and to non-governmental organizations. He is currently pursuing a PhD at Simon Fraser University, and is a co-author of *Hot Air*.

Queen's School of Policy Studies (SPS) was established in 1987 in order to create an organizational focus to build upon Queen's long and venerable tradition as a leading contributor to Canadian public affairs and public policy. Toward this end, Queen's School of Public Administration (est. 1969) came under the SPS umbrella in 1993 so that SPS now offers a Masters in Public Administration (MPA) and a part-time Professional MPA. The School of Industrial Relations with its Masters in Industrial Relations (MIR) and its new part-time Professional MIR, along with the Institute of Industrial Relations, merged with SPS in 2003. Other policy-related institutes/centres/programs within SPS include: the Institute of Intergovernmental Relations (IIGR), the Queen's Centre for International Relations (QCIR), the Centre for the Study of Democracy (CSD), Queen's University Institute for Energy and Environmental Policy (QIEEP) and the Canadian Opinion Research Archive (CORA) as well as Defence Management Studies and the Third Sector Initiative. For more information on the publications and activities of the SPS visit www.queensu.ca/sps.

Founded in 1965, Queen's Institute of Intergovernmental Relations (IIGR) is Canada's premier university-based, multi-disciplinary institute focusing on research and debate on all aspects of federalism and intergovernmental relations, both in Canada and in federations throughout the world. Among the current and emerging policy areas addressed in recent IIGR publications are: health care, cities and multi-level governance, Senate and Supreme Court reform, Quebec-Canada relations, fiscal federalism, internal social and economic unions, open federalism and the spending power, federal-provincial fiscal relations, and climate change. The IIGR website (www.queensu.ca/iigr) not only provides information on our ongoing activities but, as well, contains pdf versions of all of the Institute's publications prior to 2005 and information relating on how to purchase the post-2004 publications.

Queen's Policy Studies
Recent Publications

The Queen's Policy Studies Series is dedicated to the exploration of major public policy issues that confront governments and society in Canada and other nations.

Manuscript submission. We are pleased to consider new book proposals and manuscripts. Preliminary enquiries are welcome. A subvention is normally required for the publication of an academic book. Please direct questions or proposals to the Publications Unit by email at spspress@queensu.ca, or visit our website at: www.queensu.ca/sps/books, or contact us by phone at (613) 533-2192.

Our books are available from good bookstores everywhere, including the Queen's University bookstore (http://www.campusbookstore.com/). McGill-Queen's University Press is the exclusive world representative and distributor of books in the series. A full catalogue and ordering information may be found on their web site (http://mqup.mcgill.ca/).

Institute of Intergovernmental Relations

The Democratic Dilemma: Reforming the Canadian Senate, Jennifer Smith (ed.), 2009
Paper ISBN 978-1-55339-190-6

Canada: The State of the Federation 2006/07: Transitions – Fiscal and Political Federalism in an Era of Change, vol. 20, John R. Allan, Thomas J. Courchene, and Christian Leuprecht (eds.), 2009 Paper ISBN 978-1-55339-189-0
Cloth ISBN 978-1-55339-191-3

Comparing Federal Systems, Third Edition, Ronald L. Watts, 2008
Paper ISBN 978-1-55339-188-3

Canada: The State of the Federation 2005: Quebec and Canada in the New Century – New Dynamics, New Opportunities, vol. 19, Michael Murphy (ed.), 2007
Paper ISBN 978-1-55339-018-3 Cloth ISBN 978-1-55339-017-6

Spheres of Governance: Comparative Studies of Cities in Multilevel Governance Systems, Harvey Lazar and Christian Leuprecht (eds.), 2007
Paper ISBN 978-1-55339-019-0 Cloth ISBN 978-1-55339-129-6

Canada: The State of the Federation 2004, vol. 18, Municipal-Federal-Provincial Relations in Canada, Robert Young and Christian Leuprecht (eds.), 2006
Paper ISBN 1-55339-015-6 Cloth ISBN 1-55339-016-4

Canadian Fiscal Arrangements: What Works, What Might Work Better, Harvey Lazar (ed.), 2005 Paper ISBN 1-55339-012-1 Cloth ISBN 1-55339-013-X

The following publications are available from the Institute of Intergovernmental Relations, Queen's University, Kingston, Ontario K7L 3N6
Tel: (613) 533-2080 / Fax: (613) 533-6868; E-mail: iigr@queensu.ca

The Role of the Policy Advisor: An Insider's Look, Nadia Verrelli (ed.), 2008
ISBN 978-1-55339-193-7

Open Federalism, Interpretations Significance, collection of essays by Keith G. Banting, Roger Gibbins, Peter M. Leslie, Alain Noël, Richard Simeon, and Robert Young, 2006
ISBN 978-1-55339-187-6

First Nations and the Canadian State: In Search of Coexistence, Alan C. Cairns, 2002 Kenneth R. MacGregor Lecturer, 2005 ISBN 1-55339-014-8

Publications prior to 2005 may be downloaded from the IIGR website: http://www.queensu.ca/iigr/pub/archive.html
The Institute's working paper series can be downloaded from our website www.iigr.ca

School of Policy Studies

Taking Stock: Research on Teaching and Learning in Higher Education,
Julia Christensen Hughes and Joy Mighty (eds.), 2010
Paper ISBN 978-1-55339-271-2 Cloth ISBN 978-1-55339-272-9

Architects and Innovators: Building the Department of Foreign Affairs and International Trade, 1909-2009/Architectes et innovateurs : le développement du ministère des Affaires étrangères et du Commerce international,de 1909 à 2009, Greg Donaghy and Kim Richard Nossal (eds.), 2009
Paper ISBN 978-1-55339-269-9 Cloth ISBN 978-1-55339-270-5

Academic Transformation: The Forces Reshaping Higher Education in Ontario,
Ian D. Clark, Greg Moran, Michael L. Skolnik, and David Trick, 2009
Paper ISBN 978-1-55339-238-5 Cloth ISBN 978-1-55339-265-1

The New Federal Policy Agenda and the Voluntary Sector: On the Cutting Edge,
Rachel Laforest (ed.), 2009 Paper ISBN 978-1-55339-132-6

The Afghanistan Challenge: Hard Realities and Strategic Choices, Hans-Georg Ehrhart and Charles Pentland (eds.), 2009 Paper ISBN 978-1-55339-241-5

Measuring What Matters in Peace Operations and Crisis Management,
Sarah Jane Meharg, 2009 Paper ISBN 978-1-55339-228-6
Cloth ISBN 978-1-55339-229-3

International Migration and the Governance of Religious Diversity, Paul Bramadat and Matthias Koenig (eds.), 2009 Paper ISBN 978-1-55339-266-8
Cloth ISBN 978-1-55339-267-5

Who Goes? Who Stays? What Matters? Accessing and Persisting in Post-Secondary Education in Canada, Ross Finnie, Richard E. Mueller, Arthur Sweetman, and Alex Usher (eds.), 2008 Paper ISBN 978-1-55339-221-7
Cloth ISBN 978-1-55339-222-4

Economic Transitions with Chinese Characteristics: Thirty Years of Reform and Opening Up, Arthur Sweetman and Jun Zhang (eds.), 2009
Paper ISBN 978-1-55339-225-5 Cloth ISBN 978-1-55339-226-2

Economic Transitions with Chinese Characteristics: Social Change During Thirty Years of Reform, Arthur Sweetman and Jun Zhang (eds.), 2009
Paper ISBN 978-1-55339-234-7 Cloth ISBN 978-1-55339-235-4

Dear Gladys: Letters from Over There, Gladys Osmond (Gilbert Penney ed.), 2009
Paper ISBN 978-1-55339-223-1

Centre for the Study of Democracy

The Authentic Voice of Canada: R.B. Bennett's Speeches in the House of Lords, 1941-1947, Christopher McCreery and Arthur Milnes (eds.), 2009
Paper ISBN 978-1-55339-275-0 Cloth ISBN 978-1-55339-276-7

Age of the Offered Hand: The Cross-Border Partnership Between President George H.W. Bush and Prime-Minister Brian Mulroney, A Documentary History, James McGrath and Arthur Milnes (eds.), 2009
Paper ISBN 978-1-55339-232-3 Cloth ISBN 978-1-55339-233-0

In Roosevelt's Bright Shadow: Presidential Addresses About Canada from Taft to Obama in Honour of FDR's 1938 Speech at Queen's University, Christopher McCreery and Arthur Milnes (eds.), 2009 Paper ISBN 978-1-55339-230-9
Cloth ISBN 978-1-55339-231-6

Politics of Purpose, 40th Anniversary Edition, The Right Honourable John N. Turner 17th Prime Minister of Canada, Elizabeth McIninch and Arthur Milnes (eds.), 2009
Paper ISBN 978-1-55339-227-9 Cloth ISBN 978-1-55339-224-8

Bridging the Divide: Religious Dialogue and Universal Ethics, Papers for the InterAction Council, Thomas S. Axworthy (ed.), 2008
Paper ISBN 978-1-55339-219-4 Cloth ISBN 978-1-55339-220-0

John Deutsch Institute for the Study of Economic Policy

Discount Rates for the Evaluation of Public Private Partnerships,
David F. Burgess and Glenn P. Jenkins (eds.), 2010
Paper ISBN 978-1-55339-163-0 Cloth ISBN 978-1-55339-164-7

Retirement Policy Issues in Canada, Michael G. Abbott, Charles M. Beach, Robin W. Boadway and James G. MacKinnon (eds.), 2009
Paper ISBN 978-1-55339-161-6 Cloth ISBN 978-1-55339-162-3

The 2006 Federal Budget: Rethinking Fiscal Priorities, Charles M. Beach, Michael Smart and Thomas A. Wilson (eds.), Policy Forum Series no. 41, 2007
Paper ISBN 978-1-55339-125-8 Cloth ISBN 978-1-55339-126-6

Health Services Restructuring in Canada: New Evidence and New Directions, Charles M. Beach, Richard P. Chaykowksi, Sam Shortt, France St-Hilaire and Arthur Sweetman (eds.), 2006 Paper ISBN 978-1-55339-076-3 Cloth ISBN 978-1-55339-075-6

A Challenge for Higher Education in Ontario, Charles M. Beach (ed.), 2005 Paper ISBN 1-55339-074-1 Cloth ISBN 1-55339-073-3

Higher Education in Canada, Charles M. Beach, Robin W. Boadway and R. Marvin McInnis (eds.), 2005 Paper ISBN 1-55339-070-9 Cloth ISBN 1-55339-069-5

Current Directions in Financial Regulation, Frank Milne and Edwin H. Neave (eds.), Policy Forum Series no. 40, 2005 Paper ISBN 1-55339-072-5 Cloth ISBN 1-55339- 071-7

Financial Services and Public Policy, Christopher Waddell (ed.), 2004 Paper ISBN 1-55339-068-7 Cloth ISBN 1-55339-067-9

Our publications may be purchased at leading bookstores, including the Queen's University Bookstore (http://www.campusbookstore.com/) or can be ordered online from: McGill-Queen's University Press, at **http://mqup.mcgill.ca/ordering.php**

For more information about new and backlist titles from Queen's Policy Studies, visit http://www.queensu.ca/sps/books or visit the McGill-Queen's University Press web site at: **http://mqup.mcgill.ca/**